KU-239-165

LEARNING RESOURCE CENTRE

This wide-ranging and compelling account surveys the exciting opportunities and difficult problems arising from the new human genetics. The availability of increasingly sophisticated information on our genetic make-up presents individuals, and society as a whole, with difficult decisions. Although it is hoped that these advances will ultimately lead the way to the effective treatment and screening for all diseases with a genetic component, at present many individuals are 'condemned' to a life sentence in the knowledge that they, or their children, will suffer from an incurable genetic disease.

This is the first book to attempt to explore and survey these issues from such a wide variety of perspectives: from personal accounts of individuals coping with the threat of genetic disease, from the viewpoint of clinicians and geneticists, and from those concerned with psychological, social, legal and ethical aspects.

SWINDON COLLEGE

LEARNING RESOURCE CENTRE

The troubled helix:
social and psychological implications of the new human genetics

The troubled helix:
social and psychological implications of the new human genetics

EDITED BY

Theresa Marteau

Professor of Health Psychology and Director of the Psychology and Genetics Research Group, United Medical and Dental Schools of Guy's & St Thomas's, London University

AND

Martin Richards

Professor of Family Research and Director of the Centre for Family Research, University of Cambridge

CAMBRIDGE
UNIVERSITY PRESS

PUBLISHED BY THE PRESS SYNDICATE OF THE UNIVERSITY OF CAMBRIDGE
The Pitt Building, Trumpington Street, Cambridge, United Kingdom

CAMBRIDGE UNIVERSITY PRESS
The Edinburgh Building, Cambridge CB2 2RU, UK   www.cup.cam.ac.uk
40 West 20th Street, New York, NY 10011–4211, USA   www.cup.org
10 Stamford Road, Oakleigh, Melbourne 3166, Australia
Ruiz de Alarcón 13, 28014 Madrid, Spain

© Cambridge University Press 1996

This book is in copyright. Subject to statutory exception
and to the provisions of relevant collective licensing agreements,
no reproduction of any part may take place without
the written permission of Cambridge University Press.

First published 1996
Reprinted 1999, 2000

Printed in United Kingdom at the University Press, Cambridge

Typeface Monotype Times 10/13.5pt   *System* QuarkXPress™   [SE]

A catalogue record for this book is available from the British Library

Library of Congress Cataloguing in Publication data

The troubled helix: social and psychological implications of the new
human genetics / edited by Theresa Marteau and Martin Richards.
    p.   cm.
Includes bibliographical references.
ISBN 0–521–46288–6 (hbk.)
1. Medical genetics – Social aspects.   2. Medical genetics – Moral
and ethical aspects.   I. Marteau, Theresa.   II. Richards, Martin.
RB155.T74   1996
616′.042 – dc20   95–23273 CIP

ISBN 0 521 46288 6 hardback
ISBN 0 521 58612 7 paperback

SWINDON COLLEGE REGENT CIRCUS	
Cypher	04.09.03
	£21.95

Contents

Part III. Social context

Contributors

Two anonymous contributors to Chapter 1, c/o Martin Richards

Elizabeth Anionwu
Unit of Clinical Genetics and Fetal Medicine, Institute of Child Health, 30 Guilford Street, London WC1N 1EH, UK

Martin Bauer
Department of Social Psychology and the Methodology Institute, London School of Economics, Houghton Street, London WC2A 2AE, UK

Angus Clarke
Institute of Medical Genetics, University Hospital of Wales, Heath Park, Cardiff CF4 4XW, UK

Hannah Bradby
MRC Medical Sociology Unit, 6 Lilybank Gardens, Glasgow G12 8QQ, UK

Charlie Davison
Department of Sociology, University of Essex, Wivenhoe Park, Colchester, Essex CO4 3SQ, UK

John Durant
The Science Museum Library, South Kensington, London SW7 5NH, UK

Frances Flinter
SE Thames Regional Genetics Centre, 8th Floor, Guy's Tower, Guy's Hospital, London SE1 9RT, UK

Merry France-Dawson
Social Science Research Unit, University of London, 18 Woburn Square, London WC1H ONS, UK

Josephine Green
Department of Midwifery Studies, University of Leeds, Leeds LS2 9LN, UK

Anders Hansen
Centre for Mass Communications Research, University of Leicester, 104 Regent Road, Leicester LE1 7LT, UK

Peter Harper
Institute of Medical Genetics, University of Wales College of Medicine, Heath Park, Cardiff CF4 4XN, UK

John Harris
Centre for Social Ethics and Policy, University of Manchester, Humanities Building, Oxford Road, Manchester M13 9PL, UK

Helen Hearnshaw
42 Higher Albert Street, Stonegravels, Chesterfield, Derby S41 7QE, UK

Mary Jennings
Department of History and Philosophy of Science, University of Cambridge, Free School Lane, Cambridge CB2 3RH UK

Anita Macaulay
11 Ash Tree Close, Wellsbourn Warwick CV35 9SA, UK

Ellen Macke
RFD Box 480-M, Edgartown, MA 02539, USA

Julia Madigan
8 Keere Street, Lewes, Sussex BN7 1TY, UK

Theresa Marteau
Psychology and Genetics Research Group, UMDS, Guy's Campus, Old Medical School Building, London SE1 9RT, UK

Susan Michie
Psychology and Genetics Research Group, UMDS, Guy's Campus, Old Medical School Building, London SE1 9RT, UK

Derek Morgan
8 Fosterville Crescent, Abergavenny, Gwent NP7 5HG, UK

Alexander Mottoh
66 Cranmer Road, Forest Gate, London E7 0JC, UK

Marcus Pembrey
Mothercare Unit of Clinical Genetics, Institute of Child Health, 30 Guilford Street, London WC1N 1EH, UK

Martin Richards
Centre for Family Research, University of Cambridge, Free School Lane, Cambridge CB2 3RF, UK

Shoshana Shiloh
Department of Psychology, Tel Aviv University, Post Office Box 39040, 69978 Ramat-Aviv, Israel

Meg Stacey
8 Lansdowne Circus, Leamington Spa, Warwickshire CV32 4SW, UK

Helen Statham
Centre for Family Research, University of Cambridge, Free School Lane, Cambridge CB2 3RF, UK

Deborah Thom
Robinson College, Grange Road, Cambridge CB3 9AN, UK

Janice Wood-Harper
Centre for Social Ethics and Policy, University of Manchester, Humanities Building, Oxford Road, Manchester M13 9PL, UK

Sue Wright
23 Alpine Road, Hove, East Sussex BN3 5HG, UK

June Zatz
Glasgow, UK

Preface

The development of new techniques for characterising the genetic material we each carry is proceeding at an accelerating pace and is beginning to affect the lives of us all. Those most conscious of this are the members of families who carry genetic disorders. For many genetic disorders, testing techniques are becoming available that indicate which individual carries gene mutations and so are at risk of developing the disorder. For those few conditions where effective interventions or treatments are available, these genetic tests may help to indicate those individuals who may benefit from these as well as the family members for whom they are unnecessary. For most disorders, however, there are no treatments but tests can be used to avoid the births of affected children through the use of prenatal diagnosis and selective abortion. Currently the main application of the new genetics is testing for the presence or absence of gene mutations, but one of the main driving forces behind the research is to develop therapies; as yet, however, very few are available. The first tentative steps are now under way to insert functional gene sequences to replace those that, through mutation, have become inoperative and so cause disease. While this research continues, for the foreseeable future, the main clinical use of the new genetics is in the prediction of disease.

More widely, the same technologies are deployed in genetic fingerprinting to trace criminals from blood and semen samples, to settle cases of disputed paternity and for a host of other purposes where the genetic identity of any animal or plant is an issue. In the worlds of agriculture, food production and the pharmaceutical industry, genetic engineering – techniques that involve changing the genetic make-up of animals and plants – is now commonplace.

This is the world of the new genetics. Its history is surprisingly brief. It is barely a century since biologists came to understand how the genetic material from two parents – the chromosomes and the genes that these carry – came together with the fertilisation of an egg by a sperm. In the early decades of this century, research work concentrated on the genetics of the fruit fly, a species that was easily bred in the laboratory and had a conveniently short generation time and rather large chromosomes. This work established the ways in which chromosomes behave at the production of eggs and sperm and showed how individual characteristics were associated with particular regions of the chromosomes: the genes.

Such an approach has allowed many human genetic conditions to be classified by their mode of inheritance – dominant where a single faulty (mutated) gene inherited from one parent is involved (e.g. Huntington's disease) or recessive, where the condition does not develop unless both copies of the gene that are inherited are faulty (e.g. in cystic fibrosis and sickle cell conditions). By drawing up a family tree of who is and is not affected in a family, this information can be used to advise family members of the chances of their passing on a particular disorder to their children. Until the 1960s this was about all that clinical genetics had to offer families who carried the burden of a genetic disorder.

Throughout this first phase of the history of genetics, it was not this application of the understanding of the mode of inheritance of specific genetic diseases that had an impact on society, but rather the notions of eugenics – the idea that, like animal and plant breeders, we should strive to improve the human stock by encouraging the reproduction of those held to be superior and discouraging from having children those believed to be inferior. Charles Darwin set out the argument in his *The Descent of Man*, which was first published in 1871.

> With savages, the weak in body or mind are soon eliminated; and those that survive commonly exhibit a vigorous state of health. We civilised men, on the other hand, do our utmost to check the process of elimination; we build asylums for the imbecile, the maimed, and the sick; we institute poor-laws; and our medical men exert their utmost skill to save the life of everyone to the last moment. There is reason to believe that vaccination has preserved thousands, who from a weak constitution would formerly have succumbed to small-pox. Thus the weak members of civilised societies propagate their kind. No-one who has attended to the breeding of domestic animals will doubt that this must be highly injurious to the race of man. It is surprising how soon a want of care, or care wrongly directed, leads to the degeneration of a domestic race; but excepting in the case of man himself, hardly anyone is so ignorant as to allow his worst animals to breed. *(pp. 133–4, 2nd edn.)*

In many industrialised societies, including Britain and the United States, eugenic ideas became a potent force in social legislation until political and scientific criticisms of these began to hold sway in the 1930s and 1940s (Kevles, 1986). The eugenic movement was very important, both for its social consequences and also because it provided the context in which clinical genetics developed. It was not until the 1950s that the two were disentangled, although some (e.g. Duster, 1990) have argued that this may be more apparent than real. This history is recounted by Deborah Thom and Mary Jennings in their chapter in this book (Chapter 10).

By the 1950s it had become clear that the chromosomes were composed of DNA. With Crick and Watson's description of the structure of its molecule which also indicated how it might replicate, the research programme of molecular genetics began, which has led to the cloning of the genes related to many human genetic disorders and has created what is now known as the new genetics. The pace of research over these last decades has been extraordinary. One of us well remembers being required to write an undergraduate essay presenting the evidence that it was DNA rather than a protein that transferred genetic information between generations. Today the letters DNA are part of popular culture (Nelkin and Lindee, 1995) and many genetic diseases are understood in molecular terms.

This book is about the social and psychological consequences of these developments in molecular genetics. As the new techniques come to be applied there are new dilemmas and issues for those whose families are afflicted by genetic disease, and new problems for those who aim at providing clinical services for them. A growing body of social scientists and psychologists has begun to analyse these problems, and this book represents some of the first fruits of their labours. The book had its origin in a group that began to meet five years ago. At that time there were a mere handful of us working in the area in Britain and we began to meet on a regular basis to exchange ideas and information about what we were doing. The field has grown rapidly and has received support from research-funding bodies, in the UK and the USA, as clinicians and others have appreciated the threats as well as the promises that the new genetics poses. In this book we have attempted to describe this new field by asking established researchers to provide interpretative and critical overviews that summarise current evidence and to point to gaps in our knowledge. The academic contributors reflect a range of disciplines: psychology, sociology, anthropology, history, philosophy and law, as well as medicine. They also vary in the relative optimism or pessimism with which they view the deployment of the new genetics.

The book opens with a series of personal accounts by people whose lives are lived in the shadow of genetic disease. It seemed right to begin with these personal experiences and give them a prominent place in the book, because these are the people who stand at the centre of current debates. But despite their situation, all too often their voices are not heard because they are drowned out by those of the professionals. We wanted to avoid this; hence their position at the start of the book. From them we turn to a geneticist, Marcus Pembrey, who describes what is meant by the new genetics and outlines the basic techniques that are involved. As a later chapter shows, public understanding of the new genetics is very limited (Chapter 11). Professor Pembrey aims to provide the basic genetic literacy we all need in order to discuss the issues. Throughout the book we have tried to avoid unnecessary jargon and technicalities, but a minimal understanding of the genetics is required and this chapter sets out the basic issues.

The next five chapters of the book describe the main areas where psychologists and social scientists are contributing to the practice of clinical genetics. These are genetic testing and its consequences, the genetic counselling that usually accompanies this, and a discussion of the ways in which the information provided by geneticists is used – or not used – by those who consult them. Although much of this research is still at a preliminary stage, these chapters show how central the psychological and social science research is to the effective and humane deployment of the new genetics.

The third section of the book brings together chapters that analyse the social context within which the new genetics is practised. Topics are varied, from Derek Morgan's discussion of legal controls, to a feminist view from Meg Stacey, by way of some history, ethics, and issues related to families, and kinship, among other things. Our intention here is to indicate not only ways in which our social worlds are changed by molecular genetics but also how the research and practice of the new genetics has been shaped by the world in which it has been developed.

Although some of the consequences of the new developments were quite predictable – like the difficulties that some of those who have been shown by testing to carry a genetic disorder have had in obtaining insurance – others have been more surprising. An example here is the very variable interest that has been shown by families for testing for different genetic conditions. For conditions that are incurable, such as Huntington's disease, it would seem that most do not wish to know their future fate. Rather than wanting to know in order to plan their lives, as many suggested would be the usual reaction, it would seem that most decide to preserve uncertainty and so the possibility of hope. But for other conditions, such as hereditary breast cancer, where there are strategies that may well be effective in reducing risk, there seems to be much more interest in testing.

The new genetics is very young; social and behavioural scientists' analyses of its consequences are even younger. This book represents a start in providing a social perspective on what molecular scientists have created.

References

Darwin, C. (1871) *The Descent of Man*, 2nd edn. London: John Murray.
Duster, T. (1990) *Backdoor to Eugenics*. New York: Routledge.
Kevles, D. J. (1986) *In the Name of Eugenics: Genetics and the Uses of Human Heredity*. Berkeley, CA: University of California Press.
Nelkin, D. and Lindee, M. S. (1995) *The DNA Mystique. The Gene as a Cultural Icon*. New York: Freeman.

Theresa Marteau
Martin Richards
January 1995

Preface to the paperback edition

It is now approaching four years since we completed work on the original edition of *The Troubled Helix*. How have things changed since? As we commented then, each week brings reports of the identification and cloning of new genes, but as we also said, from the perspective of families these have been little more than promises of future developments that might alleviate some of the burden of inherited disease that some carry. Both comments remain accurate today. This year has brought the inauguration of the American Society for Gene Therapy and it is thought that a couple of hundred gene therapy experiments may be underway, but it is unlikely that more than a handful of people have yet benefited from these techniques. This number will, of course, rise but there is every indication that it will be a much slower and rockier road that many researchers and their funders would even now want to admit publicly.

Direct genetic testing has also proved rather more complicated than some of the earlier examples, such as that for Huntington's disease, suggested. This is well illustrated by what has happened in the testing for the two genes that have dominantly inherited mutations associated with breast and ovarian cancer, BRCA1 and BRCA2. BRAC1 was cloned in late 1994 and BRCA2 soon followed; however, today most of those who on the basis of their family history would appear to be at high risk of carrying mutations of one or other of these genes still cannot be offered genetic testing because no mutation can be found in their family. The reasons for this are probably partly technical – current techniques miss a significant proportion of the mutations in these big and complex genes, but also partly biological – that there remain other genes to be discovered. In addition, it seems increasingly likely that much of the inherited risk is produced not by a single gene mutation effects but rather by a much larger number of susceptibility and protective genes each contributing small effects. While testing for large numbers of such susceptibility genes – once identified – may be technically feasible in the near, future individual predictions based on such information is likely to be much more complicated to provide. Indeed, for the foreseeable future, the most accurate risk information for most of those with a family history of breast and ovarian cancer, as well as the other so-called "complex" disorders, is likely to be the estimates based on their family histories, not on the interpretation of genetic tests.

The relatively well-understood Mendelian genes and their mutations are likely to become a much smaller part of the pictures as more alleles are discovered which confer susceptibility or protective risks for diseases (Bell, 1998). These are likely to be widely distributed genetic variants associated with particular diseases, which may be carried by significant minorities, if not majorities, of some populations. There are likely to be many of them associated with each disease – some increasing risk and others decreasing it. The risks they confer for an individual will be a complicated result of the interaction of several gene variants as well as environmental factors. The basic problem is that risks cannot be directly attributable to the genetic susceptibility variants but are the overall result of their interactions, which can at present only be estimated for a population basis, not for an individual. Until the interactions, including those with environmental factors, are unravelled and understood, risk figures derived from population estimates may well prove to be a very poor basis for the interpretation of individual genetic tests. It is for these reasons that testing for the best known of these alleles identified so far, APO4, which is associated with increased risk of Alzheimer's disease, is strongly discouraged by bodies that have examined the evidence carefully (Roses, 1997; Nuffield Council on Bioethics, 1998).

A slowly growing body of psychosocial research on genetic counselling (Hallowell and Richards, 1997; Michie *et al.*, 1997), genetic testing (Marteau & Croyle, 1998) and the ways families cope with inherited disorders (Richards, 1998) has continued to appear. Even in diseases such as breast cancer, where there are at least potentially effective ways of reducing chances of developing the disease for those with inherited risk, uptake of testing for that small group for whom it may be possible is quite modest at around 50% (Lerman *et al.*, 1996). The overwhelming demand from women which many predicted would follow the publicity surrounding the identification of BRCA1 and BRCA2 never materialised.

Hitherto, research on the psychological consequences of predictive genetic testing has focused upon the emotional consequences. This is understandable when genetic testing is used to predict conditions for which there is no treatment (such as Huntington's disease) or those for which treatment is not very successful (such as breast cancer). As genetic information is used, alongside other information, to assess risk of common conditions which may be prevented, so the emphasis of research in this area will expand to include the presentation of risk information and methods of facilitating behaviour change to avoid disease onset (Marteau, in press).

Debate about the ethical and social consequences of the new technologies and their control has grown. The debate has to varying degrees been informed by empirical research on the effects of deployment of the new techniques (Harper and Clarke, 1997 provide a good example). Many countries have established statutory as well as voluntary bodies to examine the issues. However, while many members of the public remain open to the possibility that the new genetics will lead to benefits in health care, many commentators detect a growing wariness about the hazards that may lie ahead. As yet the sharpest debates are confined to the genetic manipulation of agricultural crops rather than to the field of human genetics. The summer of 1998 has for the first time brought sabotage in the form of crop destruction to the quiet fields of East Anglia – indeed to the same fields where the machine breakers of the previous century had fought for their livelihood as agricultural workers. At the same time a referendum in Switzerland about using genetically modified plants and animals brought a majority in favour – but only a 2 to 1 majority. Some of the reactions to the cloning of the sheep,

Dolly, could suggest that genetic manipulation even in the service of preventing human disease may produce some very strong public reactions.

There are growing attempts to improve the public's "literacy" on genetic matters. Certainly we need more informed public debate about what is going on as it is hard to escape the conclusion that there is a growing divide between those deploying the new technologies and much of the public. However, as children know from fairy stories, it is hard to persuade genies to return to their bottles. There are real questions about the effectiveness of controls of techniques such as genetic testing in a global economy. As Harper and Clarke (1997) describe, there are powerful commercial pressures to recoup the cost of developing testing and to exploit patients and if some members of the public choose to use these tests, attempts at control may well be thwarted by international trade, not least through the Internet. As someone remarked recently, offshore biotechnology may well be the investment opportunity of the future.

It was our intention in *The Troubled Helix* to provide an overview of the psychosocial research which illuminates the ways in which the new human genetic technologies are beginning to impinge on individuals and families that carry genetic disorders and to give a platform for a few of those from these families to describe their experiences first hand. While debates may have become a little sharper since the first edition of the book appeared, very little of what was said there has been overtaken by subsequent research. So beyond a handful of corrections and the additions of the few remarks in this new Preface, the text is unchanged.

References

Bell, J. (1998). The new genetics in clinical practice. *British Medical Journal*, **316**, 618–20.

Hallowell, N. and Richards, M. P. M. (1997). Understanding life's lottery: an evaluation of studies of genetic risk awareness. *Journal of Health Psychology*, **2**, 31–43.

Harper, P. S. and Clarke, A. J. (1997). *Genetics, Society and Clinical Practice*. Oxford: BIOS Scientific Publishers.

Lerman, C. *et al.* (1996). BRCA1 testing in families with hereditary breast–ovarian cancer: a prospective study of patient decision-making and outcomes. *Journal of the American Medical Association*, **275**, 1885–92.

Marteau, T. M. (in press). Communicating genetic information. *British Medical Bulletin*.

Marteau, T. M. and Croyle, R. T. (1998). Psychological responses to genetic testing. British Medical Journal, **316**, 693–6.

Michie, S. Bron, F. Bobrw, M. and Marteau, T. M. M. (1997). Non-directiveness in genetic counseling: an empirical study. *American Journal of Human Genetics*, **60**, 40–7.

Nuffield Council on Bioethics. (1998). *Mental Disorders and Genetics: The Ethical Context*. London: Nuffield Council on Bioethics.

Richards, M.P.M.(1998). Annotation. Genetic research, family life and clinical practice. Journal of *Child Psychology and Psychiatry*, **39**, 291–305.

Roses, A. D. (1997). A model for susceptibility polymorphisms for complex diseases: apolipoprotein E and Alzheimer's disease. *Neurogenetics*, **1**, 101–8.

Theresa Marteau
Martin Richards
July 1998

Acknowledgements

The editors thank all the contributors for their cooperation, their compliance (more or less!) to our deadlines and their good humour in the face of our editorial demands. We are particularly grateful to the authors of the personal accounts for the candour with which they have been prepared to describe often very painful aspects of the lives of themselves and their families. We hope that they will feel that the book, as a whole, may help in a small way to improve the lives of people in their own situation. Our thanks also go to those friends and colleagues who put us in touch with potential authors for this section of the book.

Editorial work makes great demands on secretarial services. We are grateful to Hazel Showell in London and Jill Brown and Sally Roberts in Cambridge for the very efficient way in which they have coped with the extra work that the production of this book created. T. M. acknowledges financial support from The Wellcome Trust and M. R. from the Medical Research Council and the Health Promotion Research Trust.

M. R. would also like to acknowledge the support received during the long hours of editorial work from Doc Watson, Quicksilver, The Seldom Scene and many others of the same tradition. The songs of Woody Guthrie, who died in 1967 from Huntington's disease, have a particular poignancy in the context of this book: 'I've been hittin' some hard travellin', Lord'.

Part I
Personal stories

1

Daily life and the new genetics: some personal stories

1.1 Introduction   *Martin Richards*

This section presents the voices of family members whose lives have been deeply touched by genetic disease. It seems entirely appropriate that this should be the starting point of our analysis of the social and psychological consequences of the new genetics. These are people who carry the burden of genetic disorder and it is in their name that the whole clinical enterprise of the new genetics has been mounted. But we hear too little of their views: much more often it is the clinicians and researchers who tell how they think the lives of such people will be affected. In this section of the book we put those directly affected first and let a clinician have an afterword at the end.

A series of personal accounts such as these cannot be said to be representative of all those who come from families with genetic disorders. But, of course, neither can quantitative studies that set out to assess the experiences of a group of people in terms of means and median of a group. Such approaches conceal the diversity of response. There is a great deal of individuality, personal history and accident in the ways in which genetic disorder is experienced.

In planning this section we first selected a number of genetic disorders that varied in their features. Huntington's disease was chosen because it develops late in life and predictive testing is now available. Once, a family member had to wait, either for the first signs of this incurable disease, or until they had lived long enough to be reasonably certain that they had not inherited the gene mutation. Now testing is available that can reveal whether someone is carrying gene mutations leading to the development of the disease, but relatively few have chosen to know.

Hereditary breast and ovarian cancer shares some of these same features. These are dominantly inherited conditions and usually the cancers do not develop until middle age. Unlike Huntington's disease, although there is no cure, there are treatments and risk-reducing strategies available. Screening may detect a cancer at an early more treatable stage, or prophylactic surgery may remove breasts or ovaries before a cancer can develop. Whereas Huntington's disease is entirely genetic – only those with the gene mutation get the disease – only a very small minority of cases of breast or ovarian cancer are associated with a dominantly inherited gene mutation. A minority of those who carry one of these gene mutations linked to breast and/or ovarian cancer (the two identified so far are known as *BRCA1* and *BRCA2*) never develop a cancer.

3

The other conditions that we chose, sickle cell and Werdnig–Hoffman's syndrome, are recessively inherited. With these, only when two carriers (who are normal and symptomless) have a child together is there a chance of producing a child with the condition. This means that most children with these conditions are born to parents who are unaware of any family history of the condition. But these recessive conditions are very different. Werdnig–Hoffman's syndrome is invariably fatal in the first months of life. Despite its being one of the most common recessively inherited fatal conditions, it is little known to the public. Sickle cell conditions are better known and are much more variable in their consequences. Some people live to a ripe old age with little or no problem, whereas others' lives are dominated by chronic and painful disability. Sickle cell mutations are particularly common among those whose families originated in Africa, and attitudes to the disease have often been much influenced by racism.

Having decided on the conditions that we wanted represented in the book, and therefore some of the issues likely to be raised, we set about looking for family members who were prepared to write about their experiences. Colleagues and friends put us in touch with people. We sought those who had adopted different strategies in dealing with their genetic inheritance – those who had opted for predictive testing and those who had avoided it, for example – but otherwise we did not select at all. We simply sought people willing to tell their own stories.

Potential authors were given minimal instructions; we asked them to recount their experiences in their own way. The resulting accounts appear as they were written with only some light editing to clarify any points that we felt might be unclear to readers. The section concludes with a comment from a clinical geneticist. He, too, was given a free hand to comment in whatever way he felt appropriate.

Our hope, as editors, is that these accounts will remind readers not only of the pain and misery that genetic disorder can cause and the dilemmas that they may pose for family members, but also the very varied ways in which people choose to live their lives in the face of an inherited condition.

1.2 Huntington's disease

Huntington's disease is a degenerative condition of the central nervous system that generally develops in middle age. People gradually deteriorate to a point where they require total care before death finally occurs. As someone from an affected family described the disease,

> the body gradually becomes engulfed in a panoply of abnormal movements. Mood is altered, usually becoming depressed sometimes to the point of suicide, occasionally becoming manic, often irritable, explosive, hypersensitive, withdrawn and apathetic. . . . Thinking, reasoning, organizing and planning become disrupted, judgement goes awry, memory is impaired but some insight into their own and family members' conditions and even a sense of humour can be maintained until the end. Speech is lost, independent care is impossible, choking is frequent and death may be a welcome relief. (*Wexler, 1984*)

The disease is dominantly inherited and occurs in Britain and the USA with a frequency of about one per 5000 births. Genetic markers have been available from the mid-1980s, which made linkage testing possible in some families. This test, which involves taking samples from affected and non-affected family members, can be carried out during pregnancy or at any later point and indicates whether an individual has a high or low probability of carrying the gene mutation. It is thought that virtually all those who carry the gene mutation will develop the disease. The gene was finally identified in 1993, making direct testing possible. This means that a sample of DNA taken from an individual can be tested to see whether or not the gene mutation is present, so avoiding the necessity of involving family members in the testing. In most, but not quite all, cases direct testing gives a result that is highly accurate, more so than linkage testing.

Since testing began – either by linkage or directly – it is thought that only a small minority of those who have a 50:50 chance of carrying the gene mutation – i.e. those with a parent who has the condition – have chosen to undergo testing; probably around 5% of such people. Those who seek testing usually undergo extensive counselling and discussion before a geneticist will proceed with the test.

Wexler, N. (1984) Huntington's disease and other late onset genetic disorders. In A.E.H. Emery and I.M. Pullen (eds.), *Psychological Aspects of Genetic Counselling*. London: Academic Press.

It's a yo-yo- type of existence　Sue Wright

When my father was first diagnosed as having Huntington's disease (HD) it was a shock to us all, as we did not know that it was in the family. I was fourteen years old when I knew I was 50% at risk of developing HD, and that my two sisters were also 'at risk'.

I am now thirty-nine years old and have lived with being 'at risk' for many years. It's a yo-yo type of existence: one minute convinced you have HD; the next minute feeling guilty you might be worrying about nothing – you might be in the 50% who does not have the HD gene. It's always simmering in the background, becoming worse at certain times of life. It's another complication to consider when forming relationships with boyfriends. Life insurances and endowments have also been a problem. In applying for a nursing job in New Zealand I realised they would not accept me on a long-term basis. In practice, being 'at risk' meant that I was treated in the same way as if I definitely had the gene. Those who are HIV positive who don't have AIDS are in a similar predicament.

Howard and I had known each other about five years when we decided to go for genetic counselling in January 1993. We were serious about each other but unable to make any decisions regarding marriage or having a family. It was one thing to know I could get HD myself, but quite another problem knowing that any children were 'at risk' too. I was not keen to have a child 'at risk'. We needed to know all the options available before deciding whether or not to have a family.

When we first started genetic counselling it was confirmed that it was impossible to tell if I had the gene because two live relatives with HD were needed to compare the genetic material for linkage testing. No one was yet able to predict how close they were to cloning the gene. The only other test that was available was an exclusion test. This compared my genetic material with my mother's. The fetus could then be tested and if it had my mother's genetic material it was less likely to have my father's.

The counselling was on a regular basis and we really appreciated the professional way in which it was carried out. It was non-judgemental and we had complete freedom to hold whatever view or opinion we liked. We felt well supported and able to speak easily and openly.

Although Howard and I had an easy and open relationship it was often less painful to avoid speaking about important issues. During the counselling we were able to reveal our worst fears to ourselves and to each other. The counselling set deadlines which we aimed for, but we usually put off the difficult task of discussing the issues until just before the counselling session; however, it all helped us to realise what each other wanted. Did we really want to have a child and could we cope with having an abortion?

When the HD gene became cloned in March 1993 we had already had a few sessions with our genetic counsellor. Just because the gene had been cloned didn't automatically make us feel that I should be tested. Our counsellor confirmed this with us and said that the exclusion test was still available for us if we wanted it. I had been waiting for years for the gene to be identified and now it was . . . it was a strange feeling. We had to seriously consider whether we were better off living 'at risk' or living knowing I had the gene. There would be no going back. In the past I had considered this option but never imagined I would actually be offered it.

The fact that we wanted a child put a slightly different angle to the situation. At least if I knew in the first place that I had the gene I would then know if I could pass it on or not. It seemed the first most sensible step. I also felt that later if I had the gene and had any regrets in knowing it then I would know that I had done it for the sake of a possible child. The exclusion test was a possibility, but we could still have been aborting a fetus 50% in the clear if I had not got the gene.

We decided to go ahead with the test. However, it meant at least two counselling sessions with three months between them, and one clinical doctor and one genetic counsellor would be present. This was to protect us and ensured we considered things carefully. We were fairly certain at the time that we wanted to go ahead with the test so we were disappointed at the delay. As the counselling progressed our decision became clearer and more certain, which was later helpful. The delay was worth while.

Before the test result we tried to prepare ourselves and tried to imagine how it would be if I did have the gene. Before the test we both felt that if I did have the gene we would not have a child. Part of me was afraid of knowing that I might have the gene, but then there was the feeling of being fed up with being 'at risk'. It was also difficult not to think of the possibility of being clear. We tried not to raise our hopes. When we had the date of our test result we booked a simple holiday so that neither of us were working after we had the result.

When we had the result that I had the gene it was still a shock. We had feelings which were similar to a bereavement: firstly we knew I would get HD and secondly the loss of not having a child. However, we felt we had been well prepared and we were relieved to see that each of us was coping well and this reassured us despite it being a difficult time. The fact that we supported each other so well was very encouraging. The counsellors were also very supportive to us and we had further counselling sessions after the result. We both felt more at peace now we knew I had the gene. I'm sure fear can build up around something that might or might not happen and sometimes it's better to know the truth. In the months since this time neither of us has regretted the decision at all. We can now see clearly how helpful all the counselling has been in guiding us to see exactly what we wanted to go ahead with and in helping us to cope with the consequences of knowing the result. We continued with the counselling and were surprised that we still wanted to have a child. Initially I felt quite guilty about this. I felt that because I have the gene I should not have a child as well. Perhaps I saw Howard looking after me and a needy child or perhaps I just felt it wouldn't be fair to the child knowing that I could be ill for many years. However, at least we knew the child would not be 'at risk' as well. All along abortion has not been an easy option, but at least now it would be done in the certainty that the fetus had the HD gene.

When I was 'at risk' I took care in deciding which friends I could tell, as not all would understand. This choice became more important as we approached the test because we knew that we needed the support of friends, but we also knew that we needed privacy and space – particularly if we had to come to terms with a positive result. We therefore decided to confide in just two couples with whom we were close and whose views we respected. Family were too close to share with immediately, but we will tell them when the time is right.

Needing it like this   Julia Madigan

What is the point of predictive testing if there is no cure for the disease? This question I have asked myself a thousand times, but only after I had taken the plunge and was waiting for my result. Since my late teens I have wanted to know: Have I? Will I? Won't I?

My first encounter with Huntington's chorea (HC) was at about the age of 13. Mother's only sister was coming for Sunday lunch with her three young children. Before they arrived, Mother gathered together my brother, sister and me, and said that Aunt Jane was very concerned that we would start talking about HC and frighten her children. We all said 'HC, what's that?' Mother said simply that HC was in the family, and Jane liked to worry about it; she told us that her Mother had died of something else. At that age it's in one ear and out of the other, and who's for a game of football in the garden?

About eight years later, I saw by chance a five-minute appeal on TV about HC. They gave an address to write to for information or help. I wrote, without telling my family – we were not a close family – giving them a brief history. All I knew was that my

Grandmother had died at 59 and, according to Mother, had no symptoms of HC. I had a reply from a volunteer, who thought it unlikely that my Grandmother would have developed it after this age.

Needless to say, a little knowledge is a dangerous thing and, although his letter allayed my fears for a while, I did not receive the help I wanted. Every three months or so, I received a magazine (*Combat*) giving the latest information on HC, and whenever I could afford it, I would send some money in return. This was my only contact with HC, apart from looking at my Mother with a different eye.

I was about 24 when I started to notice that Mother wasn't quite the ticket, but only very slightly.

My fears were not of a personal nature, they were purely for Mother and how she would cope with life and HC. Only time would tell.

There was no one in the family to talk to about it who would not think I was over-reacting. I did not want to ring Aunt Jane, because I did not know how much she was concerned. So time went by and by.

August 1990

Aunt Jane came to visit. Before they returned home, I took the opportunity to mention my Mother's moods, and we were soon talking about Mother, HC and the family. I told Jane that I had received *Combat* magazine for several years. After a while, Jane asked Peter, her new man, 'Shall I tell her?' He said 'Yes.'

Jane told me that she had had tests at a hospital – a formal neurological assessment – and in their opinion she had HC. She went on to tell me some things about my Grandmother. Apparently, she had a drinking problem, and when she was drunk, would say to Jane, 'You'll end up in a wheelchair and a mental hospital.' Knowing you have HC is enough to drive anyone to drink! Although, according to her death certificate, she died of alcoholism and bowel problems.

Aunt Jane told me a lot more about the family history because she thought I could cope with the news, and that by helping my understanding I would be more likely to do something about it. She had told my Mother that she was going to have tests, but did not tell her the results. I remembered Mother saying to me at the time, 'Why doesn't she leave things as they are?' That just about sums my Mother up.

Jane and I decided that, at an appropriate time, I would tell the rest of the family and discuss what action to take. We thought it best to do this without Mother's knowledge, until we had all decided whether or not to tell her.

I spent the next month trying to find out more about HC from *Combat*, but with little success. I could never find the right person to speak to!

In October 1990, prompted by what Jane had said, I went to the family general practitioner (GP). He had experience of HC, and told me about the Regional Genetics Centre. He was due to see Mother soon, for a monthly prescription of HRT, and he would judge how she was then. He also wrote to the hospital for a report on Aunt Jane.

During this time, late one evening when Mother was in bed, I told my Father that

Jane had had a formal neurological assessment and it was thought she had HC. He was very sorry for Jane and sad for her, but said, 'There's nothing wrong with your Mother. She hasn't got it. I know by my Gaelic intuition.' I don't think he wanted to believe or accept it, and who could blame him?

I told Steven, my brother, who was concerned. He said he knew nothing about it, but had noticed that Mother wasn't 'quite the same'. He has two children, which must be a great worry. He said he would be happy to attend a family meeting at some point.

I went to tell my sister Karen and her husband. They had just arranged for Mother to collect their little boy after school. I didn't think it was safe for her to be driving him around, and Karen agreed. She then told me that she was pregnant, so what I had to say about Mother wasn't the best news I could give. She listened, but did not want to know more about HC and said, 'I'll just live life, and get on with it.'

The family meeting was set for the beginning of October. Gathered were: my Father, Aunt Jane and her friend Peter, Steven and his girlfriend Caroline, my friend Suzy, and myself.

At the meeting Jane explained all about the tests she had undergone. I remember saying, 'Given the family history, and now the results of Jane's tests, surely there is only one conclusion to be drawn.' Yet still I don't think that the family wanted to know about it, or wanted to believe that Mother had HC.

It was left that no one would tell Mother of the meeting or the facts about Jane. Father felt, as we all did, that she wouldn't be able to cope with the knowledge. He suggested that we should all meet in two years to see how things were. I felt completely let down. Needless to say, life went on!

27 February 1991

My first appointment at the Regional Genetics Centre was a great comfort. There was someone who understood what I was talking about and could answer my questions in a way that was clear.

The geneticist made me feel at ease as much as one could be in this situation. If there were long silences, that didn't matter. Friends had been very sympathetic but I wanted answers, and needed my fears put into some sort of context.

During my early visits, most of the conversation was about Mother and how she was. I remember being told that a full clinical examination by a neurologist is the best way of determining whether the diagnosis is correct. So, once a year, I would go up and see the geneticist and chat about life. At first, I was doing this for Mother's benefit, to try to understand what was happening inside her mind and her body, in the hope that I could have more patience with her and myself. Much later on, the conversations turned to me.

At home, just after this, I remember hearing Mother phoning. She told the doctor that HC was in her family, and asked if it was safe for her to make a blood donation.

17 March 1992

Time for a review and my second appointment with the geneticist. It was decided that I would ask Mother whether she would like to come up to the hospital. We also talked about the possibility of asking my Grandfather if he would be willing to give a sample of blood to be banked and used in the future for performing predictive tests. Granddad was 85 at the time so Jane and I went together to ask him very gently about this. There was no problem and he agreed without question.

It was also decided that I would ask my parents whether they too would give blood for testing. When I spoke to Mother about this, she said she would think about it, but I knew deep in my heart that she was only saying that to keep the status quo. Mother told me that she didn't want to know if she had HC, so I explained that if she had the test, she wouldn't have to know the result. She said that she *would* want to know having been through all that. All very understandable, but very frustrating for me.

To have a chance of a 'normal' life, marriage and children, I need to know whether or not I am going to be clear to give someone the best years of my life, and be around for them and any children that might come along.

There is now a period of some 18 months when I didn't speak to my parents, and had no contact with either of them except for cards at birthdays and Christmas, and fleeting sightings of them about the town. The reasons for this were only partly to do with HC, there were also many other reasons not associated with this.

I remember seeing Mother. I was looking around an antiques shop and I saw her standing there talking to some people. I so badly wanted to go over and hold her in my arms and never let her go. To take her away and pull all the HC from her body and throw it as far away as I possibly could. But I didn't – or I couldn't – as I was scared that she would ignore me, or push me away. Life was hard enough so I couldn't risk it because it was too painful for me. I remember how much I wanted to, but just couldn't bring myself to go to her. All within a few minutes and she was gone. I felt happy that I had seen her, but very sad and upset that I didn't make a positive move. Silly, but a real pride, and a fear of being hurt.

I do know that she kept in contact with the lady living opposite where I lived for a short while, and I remember ringing her sometimes just to hear her voice. It all seems such a silly waste of time, but perhaps I needed that time away from her and the family.

January 1993

The geneticist wrote informing that my Grandfather's blood had been safely banked, but that so far, no samples had been received from my parents.

My next appointment was on my Father's 60th birthday.

Twice during the week before this appointment, I heard on the radio that the HC gene had been discovered. The first time I was cooking at home, and thought to myself, 'oh, how wonderful', and turned the radio up so loudly in order to glean all the facts, that the sound became distorted. I can remember thinking 'well now, maybe they will be able to offer some sort of test using only my blood.'

At the appointment I think the first thing we said to one another was . . . 'Have you heard the news about the discovery of the HC gene?' There were so many questions I wanted to ask – needed to ask – like, what were the implications?

I wasn't sure how good the test was – you don't like to get too carried away on a cloud of hope! The geneticist explained that hopefully a test would be available when we met next year. I didn't have to think about it; I knew what I wanted. This was my chance. Because I look at things in black and white, there was never any doubt in my mind that I wanted the answer more than life itself.

The geneticist also described the pre-test counselling, which would be required according to the national protocol. This would involve three appointments with her and a psychiatrist.

The first would be to talk about the good and bad implications of predictive testing and to find out if you could cope with the result, good or bad. I felt that I must plan ahead for both a positive and negative result.

My positive plan was that I enrolled on an access course with the hopes of doing a degree in psychology.

My negative plan was to sell my house, spend, spend, spend, and then run off the cliffs at Beachy Head. If I was penniless I'm sure I would have run . . . not having the security of money would be enough, let alone the HC gene! Knowing my luck, I would spend, spend, spend and then meet some wonderful man who loved me for what I was, and I would have to stay!

The second meeting would be three months after the first, to give you time to think about what you were doing, and to make sure it was really what you wanted. Time, also, to change your mind. In the end, the decision was yours as they couldn't refuse you the test.

At the last meeting they would take the blood sample used for testing. They would then make another appointment for a month's time to give you the result. You have to be there in person for your result, not over the phone or via other people or letters in the post. So, if you don't come, you don't get! They also recommend that you bring someone with you, to be there if needed . . . I told them I wanted the test as soon as possible.

26 June 1993

I had cycled over to visit a friend who lives very close to my parents' house. As I cycled past my parents' house, my father was in the driveway. I said to myself, 'Do it! Just do it!' I walked into the drive and said, 'Dad'. He said, 'Hello,' and my dog Pip came rushing out to see me. It was one of the nicest sights I have ever seen. She went completely mad with that funny noise dogs make when they are pleased to see you. Mother heard this and came to the front door (our tears were already falling), and she held out her arm and said, 'Hello'. I rushed to her like I never rushed before and held her. I told her I loved her and that I missed her. She said she loved me, and she missed me. I didn't want to let go of her. My heart was aching and my body felt heavy, and yet I was happy. . . . We talked for a while, and I told her I was scared that she wouldn't want to know me any more, and that she didn't love me any more. She said that she would always love me.

One of the reasons I went to see them was because I had heard that they were going to a genetics department and I felt they were beginning to take a little interest or action. My Father told me that the hospital had sent through an appointment for the very day on which the letter had arrived. I told them of a nearer centre. Eventually an appointment was set for 8 October 1993.

The next day when I went to see her, she was on her own. She acted – and has done so to this day – as if nothing had ever happened; as if the whole situation had never occurred. And now when she talks, she says 'you remember two years ago. . . .' I find it very strange, but that's as it is.

29 July 1993

My Mother was stopped by the police. Apparently she cut somebody up in her car, and the policewoman obviously thought she was drunk and asked her to breathe in the box, which proved negative. Mother had the dog with her, who was getting upset, which didn't help her. They made her get out of the car and arrested her for possible use of drugs, and took her to the Police Station – after she had left the dog at my brother-in-law's mother's house!!

Mother did explain to the police that she may be suffering from HC, and her GP confirmed this. They still kept her there for two or three hours in total and got the police doctor to see her; in no uncertain terms he told my Mother that she had HC. The Police told Mother they were recommending that her licence be taken away, but she was free to go. . . . All this prompted her to make a private appointment to see a neurologist.

29 August 1993

My mother didn't really want to see him, and she kept saying she would cancel it. Blackmail was used. I told her, 'Either you go and he'll say "yes, you've got it, but you can still drive," or you'll have your licence taken away because you won't have seen a doctor.' This was what she was so scared of – not being able to drive her own car and being a prisoner in her own home, which is in a small village with only a few buses a day.

On the evening before the appointment, we had a very big argument about whether she would go or not. I arrived to take her and she still didn't want to go. All very understandable: to be told something you don't want to hear, to be told you must not drive any more and to have confirmed that you are going to suffer a horrible slow death. I think she already knew she had HC and was more concerned about her driving.

I don't think she would have gone in had I not been with her. We only waited a few minutes . . . I went in with Mother. The neurologist was very sweet and very gentle. He asked Mother questions and he made her walk up and down the hall. He asked about her walking and she said she had trouble with her left knee, and felt her balance wasn't right. He then examined her for 10 minutes.

Then they sat back down, and Mother asked him the question, 'Do you think I've got HC?' He replied, 'Going on your family history and my examination of you today, yes, you have HC, but I think you are only in the early stages. Your involuntary movements aren't such that you need any medication for them.'

My mother asked if she would stay like that for ever – meaning perhaps she wouldn't get any worse. The neurologist replied that he thought it very unlikely and that he had not known anyone who had, and that it was just a slow process. The neurologist then asked me if there were any questions I had, and I told him there were thousands, but I didn't want to upset my mother. He smiled, and she said, 'No, I don't want to know how I'm going to die.' Then he told her that she would be OK for driving and she cheered up no end. I asked if we should think about getting an automatic car, but he said that no, the more she had to do the better and an automatic would cut down the work. I asked about moving house, and he said that she would be better off in her own environment which she's used to, and as her bedroom is downstairs, that was fine. He told her to do as much as possible, it wasn't proved that it helped, but it would make her feel better. We thanked him and we left much happier than we arrived. Well, Mother was, and I was and I wasn't. I was pleased that she could drive but sad that she had HC, although I had known it all along.

The other day I was making some tea, and I said, 'Sugar isn't good for you . . .' My mother replied, 'It's bad enough having HC, so while I'm here I'm going to enjoy myself.' It's funny how sometimes she refers to it, but always 'one liners', never a proper conversation.

Please don't think that we all sit around thinking about HC all day long. We have our moments, I'm sure, at different times, but that's only normal, isn't it? In our individual worlds I'm sure we all think about it, but as a family we never discuss it or bring these worlds together into one big world.

7 September 1993

At my next appointment I was to meet the psychiatrist, for the first time.

Unfortunately, the train was delayed and the psychiatrist had left to go to another appointment, but the geneticist was there so we had a quick chat for about 20 minutes or so.

We talked about making another appointment; unfortunately the next one wasn't until November 16th, all of which I found very frustrating. It just meant more waiting and I thought I had waited long enough.

I asked the geneticist if it was really necessary for me to go through all these visits, knowing I had been going there for three years. Could I not just have the test and be done with it? She explained that she understood my frustration that the predictive test for HC was very nearly ready and that they hoped to offer a clinical service before the end of this year. She also pointed out that there had been a big international meeting in

America the week before in order to discuss the protocols concerning such screening. Everyone agreed that, however long they have known patients in the past, it was nevertheless appropriate to arrange at least two more visits prior to running the test.

The geneticist also explained that she and the psychiatrist would like to spend some time concentrating on me and my needs, and less on my mother and family.

So more waiting and more wondering. I can't say that it took over my life, I can't say that anything ever has. It just builds up and maybe for a few hours or a few days I feel annoyed and frustrated, but then it goes away.

16 November 1993

Now, the start of a means to an end. This is what I've dreamed of for 15–20 years. Have I/haven't I? Will I/won't I? During this visit we discussed why I was doing it, what did I hope to get from the test? What if? What if? What if? Things like insurances. All the basics first.

I wanted to know the truth, that's all. I wanted to put my mind at rest. I felt as if I was searching for a reason to stay. What is life all about? You grow up, you work, you marry, you have kids, you grow old and you die . . . so the story goes.

Some of that wasn't possible for me, except grow up, work and maybe grow old or die of something called HC in my prime of life. No thanks, I'm not brave enough. I have always been so active. I always want to be doing something. I can't just sit around, and to end up in some wheelchair without a life was not very appealing. Not to be able to communicate, to meet people, would be too horrible to bear. I think I would have to shut myself away. I like, and need, my independence.

From the age of 19 or 20 I made a conscious decision not to have a life into which men, marriage and babies came; especially the last two. Yes, there could be a cure in the years ahead, and yes, there well might be, but *might* and *could* are too grey for me. I wanted more and needed more and I wasn't prepared to inflict this possibility on the man I would love, and the children we would have together. Aren't there enough people suffering in this world without having more and more?

I must admit that at first this decision was hard. All girls dream sometimes in their life of the day they will fall in love, marry, set up home and have children, but as time has passed by, I have grown used to the idea and now I think that if it happens then it was meant to be.

What if the result was good? Well, there's a question! I think good or bad, all these questions are hypothetical. If it were this and this, I would do that and that, but we all know that perhaps it never quite works out like that. Perhaps I would do my course and pass with flying colours, go on to university and of course get a degree. Maybe I would have a marriage, kids, etc., but most of all, I'd have peace of mind that I wasn't going to have to live through a living hell. I would be there for my mother, and my life would have some reasons which it had never had before: I would start living in the real world.

But what if the result is bad? To spend, spend, spend and run off Beachy Head, or take my life one way or another. I have too much pride to depend on someone else doing the simple tasks in my life. In a funny, ironic way, the second option seems more appealing. Enjoy yourself and go with a bang!

At the end of the meeting it was left they would send me an appointment for three months' time, and perhaps I might like to write down any questions which occurred to me during the coming weeks, to bring next time. I think I always find talking helpful, and this meeting was no different.

8 February 1994

During this appointment we went over old ground, the what if's. I remember the psychiatrist wanting me to give my word that I would 'phone in' on the evening of the result if it was bad, because he feared for me . . . because I was adamant about taking my own life. This I refused to do, and told him I didn't know where I would be, how I would feel and was it any of his damn business? I do what I want, when I want.

This all became somewhat annoying, more to him I think than to either myself or the geneticist. Needless to say, I said, 'No, I can't give my word to something quite hypothetical, that might not even happen.' Also I said, 'Why not on the phone?' The geneticist said, 'We have found that it's better, good or bad, for the result to be told in person. . . .' I said that I would rather hear it on the end of a phone line, or by letter, even if it meant coming to the clinic and the geneticist phoning or writing to me in the building. I can't really explain why I felt this way, except to say I wanted to guard my result very carefully, as if I was the only one to know the contents of the letter or the call. Good or bad, this is how I wanted it. I remember saying that perhaps they could treat each person as an individual and meet them half-way with this; after all, what works with one doesn't always work with another. Thankfully, we are all different.

At the end of this appointment, the geneticist took my blood sample, and I signed the consent form for the test to finally get the ball rolling in which ever direction, good or bad!!

I must say I owe one special person many thanks. A nicer, more understanding person you couldn't wish to meet. She listened. She answered in a way that told you she cared, and just knowing she cared meant more to me than words could express. I remember telling her about what happened in the counselling and what I felt about it. She never said I was silly, or tried to stop me. The only point she didn't agree with was my determination to go up for the result on my own, and of course she offered on many occasions to come with me! Her support, out of everyone else's around me, was the only one I relied on. Everybody needs someone sometimes in their life, and if you go through life without needing someone, then your life must be very sad. Fortunately, Pauline was there for me.

During the time leading up to the third appointment, some five weeks, I must admit I didn't think that much about what I was going through. I never once asked myself, 'Am I doing the right thing?' I knew I was – and a little bit of me hoped I was! I remember it only being the last five days that I started questioning over and over again: what if?

I know earlier I said I had my plans – you must have something to cling on to – and you must have a sense of humour, otherwise you're in heaps of trouble. You always hope against hope, no matter how small. When I did silly things or felt 'busy' in the head, I would wonder if I had it. In fact, when I put orange juice in my tea one day, I was convinced! I would go into rooms and wonder why, or say words that wouldn't make sense.

For the last five years I kept thinking to myself that I had HC. I thought, 'Yes, you have', for all the reasons I have laid out before and, in a way, it was easier to cope by convincing myself I had it. It wasn't hard to do, believe me! The worst never is, I guess.

And that's how I went into it.

The last five days were the most head-spinning, and as the day approached – Saturday, Sunday, Monday – a million times I asked, 'What if? What if I haven't? What if I have? Then what? What will it be like? How will I react? How will I cope? What will I do?'

Tuesday, 15 March 1994, 12 Noon

As I drove to the station I kept trying to determine my fate by car registration numbers. Bloody silly I know, but that's what I did. I got on the train and read my book – *The Pelican Brief* by John Grisham. I felt fine going up and although at first I found my book difficult to read, I did manage to get into it.

I walked from the station to the hospital. Everyone was going about their normal business but, for me, this day was far from being normal; it could change my life forever. I reached the hospital at 11.55 and explained that I was early – my appointment was for 12.30. The ladies who work there are very sweet and always had a big smile and welcome.

I sat down and began to read my book. The minutes seemed to pass quickly and every time someone came out from the doors and past me, I hoped and prayed that it wouldn't be the geneticist. I was sitting with my back to the doors and just for a few minutes didn't want her to come through. Not before half-past anyway. As it was, she came through at 12.10. By her greeting and reaction I wasn't very hopeful. She didn't smile, she just said, 'Hello'. Whether she made conversation with me, I don't honestly remember. As we went through, I felt my whole life flash before me – they say it happens when you're about to die! We walked through the double doors, into the dark corridor and through into the familiar room where I had been coming for four years. There were four chairs and a good view of the city.

The psychiatrist was there waiting. He stood up as we entered and turned to us and smiled (but not really), and I think we shook hands. I turned and put my coat on a couch behind the chairs and put my paperback into one of the pockets trying to delay for a few seconds more. My tummy was beginning to wind up in a big knot. As I sat down I rubbed my hands in nervous anticipation, almost making a joke out of it with a silly smile, as if to say, 'Well, let's have it', or 'Have you heard this one?'

The geneticist didn't waste any time. She looked me straight in the eye and spoke. It seemed as if the noises in the room and the noises outside had disappeared, and as if her lips were talking in slow motion. She said, 'Well, we are smiling. The test shows you don't carry the HC gene'. Before she had finished her sentence, as she said, 'we are smiling', I knew it was good. I said, 'Oh really, oh God, I can't believe it, I really can't believe it!'

By this time, tears were flowing freely. I cupped my face in my hands and bent forwards to my knees almost in prayer and my alleviation from pain, distress and anxiety washed over me like a large wave which hits you when the sea is at its roughest. It pushes you one way and pulls you in another, you are scared, but afterwards you feel quite ecstatic. Such feelings you only have a few times in your life – they really are quite wonderful.

After I had gathered myself, I said, 'There's no chance it's wrong, is there?' Half joking, but half not, always the optimist! 'No, the test quite clearly shows you do not carry the HC gene.' I then said, in a rather self-congratulatory way, 'Well, there you are Julia, take the bull by the horns.'

There wasn't much conversation this visit and I wasn't in the room for more than 30 minutes. I don't think there was much point in saying too much to me, as I don't think I would have taken it all, if any, of it in. I remember saying, 'Welcome to the real world. Now I must start to live my life for real and that's a very daunting prospect. Now I have to make something of my life, whereas before I could blame it all on HC, come what may!'

We arranged as part of the testing protocol that I would return in a month's time. I left the room all smiles and hand shakes, said goodbye and headed for the stairs, where I cried like I've never cried before. I stood, for a good 10 minutes, looking out of the window in a dream. A million things passed through my head, and instead of what if, it was what now? I didn't even know if I was pleased.

I left the hospital and walked back to the station in a dream. As I approached the entrance to the station, a man stopped me and asked if I was married or single, how old I was and would I be interested in joining a dating agency. I told him, 'No, and no thank you,' but I laughed to myself as I went to check the times of the trains. It just seemed so ironic that after all these years of not thinking seriously about men, marriage and babies, the very first person who talked to me was a man about a dating agency!

During the journey I felt as if I wanted to keep bursting into tears. The emotion passing through my body was quite unbelievable. I controlled this by writing down my thoughts and my account of the morning – My Big Day.

I arrived back home and drove to Pauline's. She was out with the dogs so I knew I had a good chance of finding her on the Common. I found her car and began to look for her; it wasn't difficult with three dogs. I called the dogs, William, Rosie and Jess, who all came bounding over and jumping up. Pauline said something about which I can't remember. I don't even know if she finished, when I said, 'Well, I haven't got it, I haven't got it. The test proved negative.' She gave me a big hug, and the emotion and tears flowed; once more that large wave came over me. I kept saying, 'I can't believe it, I can't believe it.' Pauline said, 'I'm sure you can't, but this is wonderful. I'm just so delighted.'

I now talk from a privileged point of view having had my test and the result being a happy one. I worried about whether or not to tell my parents and Steven and Karen. I felt my father would be very happy – there would be no other way to take it.

But I don't feel I can tell Mother, much as I want to at times. I feel such a huge sense of guilt, happiness, relief and isolation all mixed up together. Guilt because I was a lucky one. I've got no children to pass it on to, why me? What makes me different? I'm no better than them. Also 'sod's law' says that one in three has a good chance of getting it, and I feel I've narrowed their chances, although the geneticist keeps telling me, 'They have the same 50:50 chance as you.' I just feel they may say, 'Oh trust her, she has all the luck, and where does that leave us!' Also I don't want to encourage them to have the test and theirs to be a bad result.

It's so very difficult to understand for myself at the moment, perhaps I need more time. It has been my crutch for half my life, which has had its uses, believe me! It's easy to blame HC, even if you are only saying it to yourself. I will never know now what I would have done had my result been a bad one, but does that really matter any more? At one point in my life I felt as if I was searching for a reason to stay. Now I feel as if I'm waiting for my life to begin. I never once asked myself, 'Am I doing the right thing?' Luckily for me it was, and my life now can only get better.

Don't believe a good result is all good – I lost something that day. It took away a very big part of my life which was there to fall back on. Life wasn't real in certain ways, then BANG, welcome to the real world, life begins here.

My high spirits lasted about 10 days, enjoying the result, the feeling, the relief and sharing the result with a handful of friends. And as high as these spirits were, they soon came crashing down on top of me. The same force which put me up there in my ivory tower pushed me deep down into a big black hole, so big I felt I would never get out of it. Such a deep depression that nothing means anything to you any more. Life just isn't worth living: where the smallest thing gets blown out of all proportion; when you burst into tears for no reasons (wherever you are – your tears and emotions aren't fussy); when you go to bed at night hoping you'll never wake up in the morning, and the pain in the morning when you do. I can only describe it as if someone or something is pushing you down with their hands on your head and shoulders. You just carry on with life machine-like and you wonder what keeps you going when no one and nothing gives you pleasure. It is just one big mess.

This is how I felt for four to five weeks. I always wonder why; when I see things in black and white and know what I want, why does this affect me so badly?

I remember the geneticist phoning me about ten days after my third appointment to check I was O.K. and I remember her message on the answerphone saying, '. . . and enjoying living with the good news?' At the time I thought, 'If only she knew how fed up and unhappy I feel.' It was rather funny in a way as it wasn't that long ago she had written to me to confirm my good result for the HC gene – something to frame and put on the wall.

It was some time before I had the courage to tell Aunt Jane my results. I was nervous because I felt guilty that I didn't have HC, and she did. She told me not to be silly, and was delighted and relieved for me. Jane said to tell people when I felt ready, and not before. That included my family. I keep in contact with Jane by phone, and visit her whenever possible. She's understanding and gentle, and she cares. I just feel sorry that there's not more I can do for her.

Eventually I told my father about my test result. He was pleased, I think, but he didn't say much. I told him I didn't want anyone to know, not Mother, Steven, Karen, nobody; I was very clear about this. He said, 'Your mother would be pleased', but I said I wasn't sure of this. I didn't feel happy about telling her. He told me not to be so silly. Regarding Steven and Karen, I said I didn't want them to know because of their children's chances and their own chances. I also said that's why I've never had any children or got married, and the first thing he said (and I knew he would, that's why I didn't want to tell him) was . . . 'Well, you can now, can't you, there's nothing to stop you . . .' I don't want pressure on me to do anything. I'm trying to come to terms with the news, I'm still wondering whether it's good or bad. I think, however, he understood about not telling Steven or Karen.

13 April 1994

This appointment was to see how you feel after a little time, and how life has been since the test.

When I arrived, the geneticist asked me if I minded an observer sitting in on the consultation. I said, yes. I felt a bit guilty at this, but she said that was fine.

The first question I had was to ask if the test was 100% accurate. I had read in the *HDA Newsletter* (formerly *Combat*) that a small number of tests may be ambiguous. There may not be a clear division between the HC range and the normal range. I wanted to know where I stood. The geneticist explained that my test result was far from being ambiguous, in fact I couldn't have been in a better situation. My repeats were 14–17 and the hospital 'grey' area as 30–36[1]. After hearing this I was much reassured that my result was right and that I had nothing to worry over.

I told her I felt wonderful for a little while, and then after some time I felt terrible and went back to my black hole, but was feeling just a little better than at its blackest. The geneticist tried to reassure me that the swings in mood I had been experiencing since my result were not at all unusual, in fact most people in my situation went through them. She asked me whether I was angry or felt anger that I had wasted my life, or had made a decision I didn't need to. Sometimes I look at the clock and think time has passed me by, and sometimes I think it's only the beginning. It all depends which side of the black hole I am. I can't say if I regret my decision to have the HC test because it was a good result. Whether I would be saying that in the light of a bad result, I will never know.

We also chatted about arrangements for some psychotherapy, which the psychiatrist thought might be a good idea. I told her I would be interested for two reasons: the fact

---

[1] The Huntington's disease mutation consists of a repeated sequence of DNA. Individuals with a low number of repeats are normal. There is an intermediate range where it is thought that individuals may or may not get the disease – the grey area.

that I wanted to see how things were done, especially as I was hoping to study psychology, and also I did believe it could help me in the long run – especially when I'm feeling down and in my big black hole. I felt guilty about this because it looked as if I had an ulterior motive. The geneticist said that there was nothing wrong in how I felt and we made an arrangement to meet again in three months.

It was around the beginning of May when I started to feel better in myself. Life had a meaning again and things were looking brighter. Then I used to think, 'God, I hope I'll never slip back down into that big black hole' – the thought of it left me cold.

It always overwhelmed me with wonder that, however black things become and however many times I say I don't want to live any more . . . I'm still here. I can never quite seem to give up, and I hope in the future that proves a good thing for me. Why can't I throw a cup or something against the wall and not care for the consequences, but no, good old me, there's always that little voice at the back of my head saying, 'If you throw that, you'll break it, and you don't really want to do that, do you? Remember who gave that to you, remember where you got it.' Sometimes I just wish I could let go.

I remember Pauline saying that people who go through life without any problems never really 'grow'. When I think of my short life I've had enough problems to last the whole of it. I've heard that setbacks and problems in life make you stronger to cope with things in the future. If that's true, I'm surely set up to cope with anything.

Would it be easier for me, would I have less mood swings if I had someone special in my life? I can't help thinking that this test result can't, or doesn't, mean as much because I had nothing or less to lose. It was just me, it only touched my life and no one else's. If I was married, there would be my husband and possible children, but it's only me. So I've only gained something (call it peace of mind, or call it happiness) for myself which doesn't mean anything to anyone else. Yes, my friends are pleased; yes, friends, but no one special.

I don't know if what I'm saying means anything to you reading this, or if you can understand one word of what I'm trying to get across. If you look at the other side of the coin, if the result had been bad, then because I'm all alone (that must have made my decision easier to take – I'm not wallowing in self-pity here), only I would have been affected in the true sense of the word.

So am I saying: what a waste, you've got nothing to show for your time: no husband, marriage, babies and all the social acceptances that society holds dear?

27 July 1994

I drove up to the hospital for my next appointment as there were no trains. My friend Suzy, who was over on holiday, was with me. That was the first time I had ever taken someone with me, I had always wanted to go alone. She would have been the only person whom I would have liked to be there on the day of my test result. She's my best friend; I know that sounds childish, but we are very close and it's almost as if she is part of me. We understand one another so well sometimes it's uncanny, but her living in Switzerland makes for very big phone bills, and it just wasn't possible for her to be there in person. Suzy's support, encouragement and total patience with me never ceases to amaze me. She really is a wonderful person and no better friend could you find.

I talked to the geneticist about how I was feeling, my mood swings and depression, and I explained that I couldn't help thinking that the test result didn't mean as much to me because I didn't have someone special in my life.

I remember telling the geneticist about a friend of mine, Wendy. She has a friend who suffers from HC, who is 34 and quite a long way down the road of no return. After listening to Wendy's account of her friend and reflecting on it, you think to yourself, 'Why her, why did she have to suffer, why not me? What makes me different?' What have I done or not done to be let off? God, it's so frustrating, it makes you want to scream from on high . . . like all those people who suffer from other genetic disorders; why us, why me, why her, why him, why anyone?'

I told the geneticist that I was feeling better than in April, but not 100% as yet. I also told her about when Mother and I were walking down a street where I live. We had just been to see the GP for him to sign Mother's insurance papers to say that she was fit to drive, so I guess things were on her mind, and she said, 'Did you ever go for that test for HC?' It stopped me dead in my tracks for a moment. I was lost for words, especially the right ones that sounded natural, and almost machine-like I said, 'No, I haven't got round to it yet.' She never said any more and neither did I.

I also explained that to people who I had told I was having the test, I had not yet told them the answer. I didn't want to. It was a strange feeling to have, I just said, 'Oh, it's the end of the month, if I go and get it.' Sometimes I feel I have to justify my single existence. You must be something in life, you are given labels whether you deserve them or not. I'm not sure really sure what mine is, or even what I want it to be.

We also talked about my mother. I'm always amazed how good her sense of humour is. She is always laughing, perhaps she has nothing to lose . . . perhaps she has already lost it.

The geneticist asked how the writing was going. I said that it was fine and that when Suzy got back I would be getting back into it to finish it. She then asked Suzy if there were any questions that she would like to ask. Suzy didn't have any, I don't think she thought she was going to be asked anything, although afterwards she said she wanted to know where HC came from, where did it start? I really didn't know, I could only guess and say, 'families who inter-marry.'

30 August 1994

Now here I am, I hope at a new stage in my life. Where it will lead me I'm not sure, and I'm not sure if I really care, as long as I'm doing something that keeps me busy. Perhaps sometimes I dwell on things or think about life too much, so much so that it becomes an obsession. Perhaps if I just got on with my life, maybe my life would be easier.

Was I brave or stupid in going ahead and having the test? Maybe I wasted my life, or the last 20 years, worrying, and maybe Steven and Karen are right and have just got on with their lives in the last five years. I wonder which is the easier or better way? I think to be honest, we are all different and it's 'each to his own'.

Sometimes I think life is about losing. Yes, you might win, I've won more time. Or maybe I've won a better quality of life – no HC for me – aren't I lucky? We all lose in the end, it just depends how much you make of your life while you're able. I think the most difficult thing for me now is accepting that my life has a meaning, or chance. It is worth living if I can make something of it, because now I know I'm perfect or healthy. I've always been a perfectionist, almost to a fault. With anything I do, everything has to be perfect to the point of obsession.

Over the years, I've become very independent and now I wonder if it's possible for me to have a 'normal' life – that word 'normal', I hate it! It makes things sound right, giving myself a label. As I said before, people like giving labels, and I'm sure I'm no exception to the rule, then we can file you under whatever, and forget about you. If you don't have a label, we have nowhere to place or file you. We are conditioned to have everything neat and tidy. No rough edges, dot the i's and cross the t's!

Will I be able to let somebody into my life? I really wonder about this, and have serious doubts that I will. It was funny; about six to eight weeks after the result, I was asked out at different times by four different men of various shapes and forms. I thought at the time, 'Do I have a neon sign flashing on my head?' I was flattered, but none of them really moved me; perhaps I'm waiting for the right one in my eyes. There is someone I like, but whether he likes me, I don't know, and perhaps it's just the ideal I like and not the reality.

Sometimes I just wonder where I will be in five or ten years' time. I would just like to know whether it's all been worth it. I have no crystal ball to forecast my future, I can't concentrate my gaze to see the perfect hallucination, although I try hard to do just that.

I'm glad I've been able to write this all down; it's been very useful to help clear my head, and if any of this helps someone else in the least possible way, then that's all for the better.

All this was for me: 'JULIA MADIGAN NEEDED IT LIKE THIS'. This was good for me, this was how I coped and how I was thinking it worked for me. But then I had a good result – please God, you will too. I can't bear to think of people suffering; there aren't any magic wands.

Some people who may read this, *my* account, may think, 'She's had it easy, what a breeze!' and I would agree wholeheartedly. When I read some people's stories in the HC magazine, I think, 'God, poor them, they really have suffered and been through far worse.' Really we've been let off lightly. We have been quite lucky, but not as lucky as some.

When I think that if it hadn't been for my Aunt Jane wanting to know, then would we ever have known anything about HC? I may have been well on the way by now to husband, children and 'normal' life, and given my test result it wouldn't have made any difference to my life. I don't like to think of the other side of the coin. They say fate lends a hand in most things, so perhaps it has here! All I know is that I don't want to ache or hurt or to be sad any more; I just want to be happy.

Please let this be the start of something new – the first day of the rest of my life.

Living with the threat of Huntington's disease   *Anon.*

I first heard about Huntington's disease shortly after my grandmother had been diagnosed as suffering from it, when she was in her sixties and I was 15 years old. My father
was able to give me, together with my brother and sister, clear details of the genetic
basis of the disease, including our own chances of having the faulty gene, as well as an
idea of the major symptoms. From the diary which I kept at the time, it would seem the
initial impact this information had on me was relatively slight, with other aspects of my
life continuing to dominate my thoughts. Since then, although I have always thought
about and worried about Huntington's disease from time to time, on the whole it has
seemed distant from my current concerns. For the most part life has been far too full,
and indeed far too full of more immediate problems and dilemmas, for something
which I imagined might happen to me at around the age of 60 to worry me unduly.
Although my sister's experience of the disease is objectively much the same as my own,
it's effect on her has been quite different – she recently told me she feels the knowledge
of having Huntington's disease in our family has blighted her whole life.

Within my family of origin the subject of Huntington's disease has never been taboo
in any way, but, on the other hand, neither have we spent a great deal of time talking
about it: during the last few months I have been able to discuss previously uncovered
ground with most family members. When my father was young, however, the subject
was never mentioned within his family, and he was told nothing about the disease until
well after he had married and had children. My mother thus married into a
Huntington's family without being warned about it, something which she fiercely
resents.

I told my partner about the disease at an early stage in our relationship, well before
any real commitment had been made, but he accepted the risk of the disease without
any great difficulty, viewing it as 'part of the package'. For some years our lives were
too busy with other things for us to think about children, but wanting to have children
together did gradually become part of our relationship. We both felt this was too
important, too fundamental an aspect of life, to be given up because of the risk of
Huntington's disease – a risk which seemed fairly remote, both in terms of probability
and in terms of time. My sister recently told me she spent most of her twenties assuming she would never have children because of the risk, but I do not remember ever
thinking in these terms myself. As it turned out, my sister did have a first child when in
her late twenties, and my brother also had two children by this time, which probably
made it easier for us to choose to have children as well. Our first child was born when I
was 30.

Having become increasingly worried that my father was showing symptoms of the
disease over a period of some three years, my mother finally suggested he should
consult a geneticist when he was in his early sixties; he was duly referred to a neurologist
and the diagnosis was made. By this time I was already pregnant for a second time, and
I was not told about my father's diagnosis until some time after our second son was
born. Any question of my partner and I having more children had certainly not arisen

when, little more than a year later, contraception failed and I became pregnant for a third time. At first I was not especially pleased about this for a number of reasons, of which the risk of Huntington's disease was only one. My partner and I did discuss the possibility of an abortion, though not particularly seriously – for me, to abort a child because of an increase in the perceived risk of Huntington's disease, when the child's actual risk would be no greater than that of its brothers, made little sense. And anyway, it was not long before I began to feel that, rather than being unwelcome, this third child was in some way special – while we had chosen to have the first and second, he was a gift which had been bestowed upon us.

I am aware many people consider having a family of four children a little excessive, perhaps somewhat inappropriate in today's world, maybe even irresponsible. I am therefore doubly aware that many people would judge having four children when at risk of having a genetic disease to be the height of irresponsibility. I certainly cannot justify now having four children myself, nor do I have any great expectation of others' understanding this choice. All I can say is that despite having three children I still longed for more, and particularly, having three sons, I longed for a daughter. In these circumstances, it somehow seemed acceptable for us to impetuously 'try' for a baby on just one or two occasions – and somewhat surprisingly, the result was our daughter.

Thinking about my family of four now, I frequently wonder how on earth we managed to get ourselves into this situation – yet I cannot have any regrets. Indeed, when I contemplate my daughter, in whom I take an intense delight, I feel enormously privileged to have her as well as my three beautiful boys. I thank goodness we did take that wild chance, rather than being sensible, that we let ourselves focus on life and hope, rather than on illness and despair.

On the whole my life continues to be too fully occupied – with children and work, not to mention worries about a failing business and new job – for worries about Huntington's disease to hold sway. Nevertheless, since the birth of my daughter I have noticed a change in my attitude to being at risk of the disease, with a considerable increase both in the extent to which it preoccupies and concerns me, and in the extent to which I look out for, and find, possible symptoms in myself. One reason for this may be that, although my father's symptoms are still far from overwhelming, they have become more marked recently. In addition, some shift has taken place in my mind with the certainty that my child-bearing days are now over, which to me implies no longer being 'young'. I have reached a point where I must acknowledge that Huntington's disease commonly first manifests itself in people of my age.

An important element in my experience of coming from a family affected by Huntington's disease is the relatively late age at which both my grandmother and father first developed symptoms of the disease. My understanding has always been that although the commonest age of onset is between about 35 and 45, and while there are no guarantees that the pattern of later onset in my family will continue, 'the chances are' if I do have the Huntington's gene I can nevertheless expect to live a normally healthy life until around the age of 60. In my late 30s I am suddenly acutely aware I *could* develop the disease at any time. This feeling was recently reinforced when I learnt that, although her own father also developed the disease later in life, my great-grand-mother's symptoms began when she was about 45. I am all too aware that if this happened to my great-grandmother, there really is no certainly it will not happen to me, and to have based my life on the assumption that I would not develop the disease until retirement age was extraordinarily rash. The possibility of succumbing to the disease while my children are still young is one which fills me with the utmost dread.

Looking back, it seems quite clear I have spent over 20 years in state of 'denial', of taking a 'head in the sand' approach to the whole question of Huntington's disease – avoiding thinking about it, avoiding focusing on the potential significance it has for me, indeed refusing to accept it really has anything to do with me. In retrospect I realise I was aware of my father's symptoms for some time before he was diagnosed as having the disease, and indeed before I had my first child, but succeeded in thinking of these as slightly irritating habits or isolated incidents, rather than really considering the possibility of them being symptoms of the disease. Although my mother did mention some of her worries about my father to me, before he was diagnosed, this had no real impact on me either; I certainly did not take on board her growing conviction, not that he might get the disease, but that he had already got it.

Until recently I have actively sought relatively little information about Huntington's disease myself, and have relied on my father to keep me up to date about advances in research. While he has been a long-standing member of the Huntington's Disease Association and, since being diagnosed, has also visited a geneticist from time to time, I have always avoided any situation of being identified as 'at risk'. Thus, only very recently have I considered becoming a member of the HD Association in my own right, even now I would strenuously resist putting myself in a position where my 'at risk' status would be entered in my medical records, and have never considered seeing a genetic counsellor. Once having achieved a clear understanding of the genetic basis of the disease, and therefore knowing my own risk of having the Huntington's gene, and that of passing it on to my children, I never felt there was anything to be gained by seeing a genetic counsellor. Perhaps I have also assumed that most people, including genetic counsellors, would feel some criticism of the choice to have children when at risk of having a genetic disease, and have had no wish to open myself to being judged in this way. Indeed, it is one of the reasons why, since telling two or three good friends when I was in my teens, I have even avoided telling any friends, or indeed my partner's family, about the disease.

My 'head in the sand' approach to dealing with the subject of Huntington's disease probably was facilitated by the absence of a test for the gene during my childbearing years. Although I have known about the availability of linkage testing, via my father, since the HD Association informed its members in the mid-1980s, I always knew this was not a possibility for my family: my father is the only member of the family currently manifesting the disease. I do wonder whether my partner and I would have made different decisions had testing been available in my case. I suspect I might have found the prospect of aborting an affected fetus unacceptable, and especially if this would have involved a relatively late abortion following amniocentesis (rather than an earlier one, following chorionic villus sampling). Yet, I suspect I would also have felt it unacceptable to bring a potentially affected baby into the world when this could have been avoided with certainty. In addition, I am not sure I would have wanted to risk finding out that I definitely had the gene myself. I feel fairly sure I would not have taken the option of exclusion testing, finding the prospect of aborting a fetus with only a 50% chance of having the gene altogether too cavalier. (In contrast, my sister recently told me that had this test been available to her, she might well have used it – her fear of finding out her own genetic status, combined with a profound desire not to pass the Huntington's gene on to another generation, making the possibility of aborting an unaffected fetus worth the risk). It seems just possible, had testing been available, that the only option I would have been able to contemplate with any equanimity would have been the one to avoid having children at all.

When I first learnt – from a newspaper, shortly after the birth of my daughter – that the Huntington's gene itself had been located, I had a number of immediate thoughts. The first, perhaps shockingly, was to be thankful we already had our daughter, since I would have found the choice to have another child so much more complicated, perhaps impossible, with testing soon to be an option. Another was the thought that this might expedite the development of treatments for the disease – I have always hoped (perhaps even expected) that if I, and certainly if my children, do develop the disease, treatment will have become available. In addition, I had the clear thought that, with a test for the gene soon to be offered, I would have to face the question of whether to be tested myself.

Since then I have thought about whether I want to be tested, and although the test is now available (as the geneticist recently told my parents), I have still not come to any clear conclusion. Sometimes I think the obvious course is to have the test, to 'grasp the nettle', to *know* one way or the other. I feel strongly that I do not want to spend the next 20 years worrying about something unnecessarily; and should the test prove positive, this would enable me, and my partner, to plan our lives accordingly. At other times, I think the knowledge that I definitely have the gene, and hence the knowledge that in all probability I have passed it on to at least one of my children, would be too much to risk imposing on myself, and on my relationship with my partner. In addition, by the time it actually struck, treatment for the disease might have become available, which could render any plans we had made obsolete. There might even be disadvantages to finding out I was free of the gene. At present my sister and I can share the worry and uncertainty of being at risk – if testing showed I was free of the gene, but she was not, it would be enormously divisive, and could seriously damage our relationship.

Since the birth of my daughter, first the Huntington's gene has been found, now testing for it is available, and at last I am full of questions about the disease. Questions about the reproductive choices other people who are at risk have made, about variability in the age of onset within families, and, given the nature of the gene, about the extent to which it may in future be possible to predict the age of onset for the individual. The most important question, of course, is whether *I* have the Huntington's gene; but I am still not sure I want to know the answer to that one yet.

1.3 Hereditary breast and ovarian cancer

About 5% of cases of breast and ovarian cancer appear to be linked to dominantly inherited gene mutations. The commonest of such genes, which probably accounts for about half the hereditary cases of these cancers, is located on chromosome 17 and has been called *BRCA1*. Since 1993, linkage testing for *BRCA1* has been offered to families where it seems likely to provide useful information and, now that the gene has been cloned and some of the mutations identified, direct testing should soon become available. Women who carry *BRCA1* mutations probably have about an 85% chance of developing breast and/or ovarian cancer. The inherited form of these cancers may develop at younger ages than the common non-inherited disease. Recently a second gene *BRCA2*, has been located on chromosome 13. This seems to be less associated with ovarian cancer than *BRCA1* but may be linked to male breast cancer.

Women who are likely to be carriers of these gene mutations, either on the basis of their family history or through linkage testing, are offered a number of possibilities that may help to reduce their chances of developing a cancer. These include various forms of screening, mammography for breast cancer, and ultrasound for ovarian cancer. Some women have chosen to have their ovaries or breasts removed surgically to reduce their risk.

I am definitely having it done   June Zatz

I remember in 1973, when I was 23 years old, sitting in the hairdresser's, reading some obscure American magazine. I came across an article about a 19-year-old girl from New York where every female on her mother's side had contracted breast cancer going back two or three generations. A decision was taken that she have both breasts removed as this was the only way to ensure that she would not contract the disease herself. I was absolutely aghast when I read this article. This was the first time I learnt that breast cancer was hereditary. My mother, her two sisters and her first cousin had had breast cancer but I never really thought of it as being a threat to me. And the thought of a 19-year-old having such a drastic operation! I never discussed the article with anyone. It was after reading the article that the implications were clearer to me of contracting the disease. I dismissed as ridiculous the part in the article about the operation, but I know that I stored it away in the back of my mind.

My mother had a radical mastectomy when she was 42 years old. As a result of the operation she was left with fluid in her right arm, which caused severe infections where she was bedridden for days. She had bouts of ill health for years but in between she was a determined lady who never complained. She had a second mastectomy 13 years later followed by various operations and bouts of ill health. After a benign tumour was removed from her liver when she was 58, her consultant warned my brother and me that 'cancer will get her sooner rather than later'. Her ultimate illness was when she was 66 years old, short and sharp – cancer of the spine – paralysis – painful death – even now, seven years later, I cannot think about this period without immense pain.

My mother's younger sister developed breast cancer when she was 46, and after numerous operations and great pain died two years later when 48 years old. This was prior to the days of the hospice (where my mother died) and so my mother looked after her, but the pain was not well controlled.

My other aunt died when I was only seven years old. My mother told me she had breast cancer in her early forties, but also had a heart condition from which she died in her mid-forties.

After finding out that breast cancer can be hereditary I read everything I could get my hands on. Only once when in my twenties I attempted to bring up the subject with my mother, as I felt perhaps she could get advice for me from her specialist. But she just dismissed it and refused to discuss it. I never attempted to mention it again.

My husband, however, got to know of my fears because I often imagined I felt a lump in my breast. Then I became depressed, lay awake at night imagining all sorts of things and, having two young children, I naturally worried about them – what would happen if I weren't around for them? Throughout the years I only ever went to the doctor twice when the 'lump' didn't seem to disappear and I was literally going round the bend. Even when at the doctor's I managed to conceal my true feelings, was very matter of fact, and he was never aware of the family history as I didn't tell him.

My mother died seven years ago, when I was 37 years old. Within a few weeks of her death I realised I had to do something about my fears and get them out in the open. It was almost as though with her death this now released me to discuss my fears out in the open. I wrote to mother's oncologist detailing the family history and asking whether there was anything else I should be doing apart from self-examination.

This resulted in my being referred to an oncology clinic for regular monitoring, initially six-monthly, then annually. These visits to the clinic I can only describe as horrendous. First of all, there was the waiting area where females of all ages (many surprisingly young) were waiting to find out whether they had breast cancer or because they already had it. Then, in the cubicles while waiting to be examined, you could hear the registrar in the next cubicle talking to the patient and I knew from the conversation that all was not well. By the time I was due to be examined I would be a wreck – then the mammogram; then the waiting a week for the result. After the result I would be elated, my first thought would be thank God, I will be around for another year. I would immediately calculate my children's ages. This was what was of prime importance to me – I wanted to be around to see them into adulthood.

As I neared the age when my mother developed her first breast cancer I became more fearful. Over the years I had read many articles and I did not feel that treatments were any more effective today than years ago, even if it were detected at an early stage. I felt that I would get breast cancer as my body was similar to my mother's in many ways – she had fibroids and a hysterectomy – she had gall stones and had her gall bladder removed. I have had both these operations by my late thirties.

My mind kept casting back to the article I'd read all those years ago about the 19-year-old who had both breasts removed as a preventative measure. I had never read of this again but decided it was time to pursue it further.

In the summer of 1991 I wrote to my oncologist asking if such an operation were possible. His answer took a fairly long time coming, mainly due to administrative procedures. While I waited for the reply I was on tenterhooks: my feelings were that I would be told I was crazy, that this operation could only be done in America. Eventually when his letter arrived stating that yes, I could have this operation – I literally couldn't believe it – I read the letter over and over for days. Then I wrote another long letter with lots of questions. There followed several letters to and fro: a referral to a plastic surgeon to discuss reconstruction and a referral to a specialist cancer geneticist to discuss the genetic side of things.

It took me months to make up my mind definitely. After the second letter from the oncologist I told my husband what I was investigating and he was horror-struck. He just could not understand why I should have both breasts removed when they were perfectly healthy. To him the logical thing was to wait and, if cancer appeared, then it will have been caught early, then I should have a mastectomy. He didn't understand that, the way I saw it, it was too late by then. At this stage, however, I hadn't made up my mind and in fact because of my husband's attitude I put things on hold for several weeks. One day, however, I woke up early feeling such anger and resentment towards him – he and I had always until now agreed on major issues.

I discussed it with him again. This time I had a different attitude: he had made me feel selfish contemplating such an operation but what I said to him was: What about me, my fears; this is my life and who's going to look after me if I get cancer – will he nurse me? At the same time I said I hadn't yet made up my mind but I was going to continue my investigations whether he liked it or not. He listened and said he'd like to come with me when I next had a consultation with either the oncologist or the plastic surgeon. I refused as I knew he was still against it but said I'd tell him everything they said.

Some days I would wake up and say, 'I'm definitely having it done.' Other days I'd think I was crazy – I tried to imagine what it would be like without breasts, although there was the chance of reconstruction.

Gradually my husband came round to my way of things. The more information I gathered the more he understood what I could be facing. One day I said, 'I don't think I'll have it done', and then to my surprise he said, 'You should go ahead – I'd rather have you alive and well with no breasts.'

In April 1992 I made the decision to have the operation. Throughout the six months it took me to come to this decision I had several consultations with the three specialists. The clinical geneticist put me in touch with another lady who had had a similar operation. She had an almost identical family history to mine. I had a long discussion with her on the phone. It was invaluable to talk to someone who had been in my situation. She and her sister had had the operation and were greatly relieved that the chance of breast cancer had been removed.

The only people who knew about it were my husband, my two children and two friends, one of whom had a medical background and the other was a cousin to whom I'm very close and with whom I'd often discussed the 'family history'. I instinctively knew from the beginning that I didn't want anyone's advice on the matter, I wanted facts on which I could base my decision. The professionals were wonderful. Everyone was patient, answered my questions, was quietly supportive and never offered advice on what I should do. There's no way the outcome would have been as successful without their help and support. I should also point out they were all males.

At the time my daughter was 18 years old and my son 16. They knew what was happening but my daughter could not understand my contemplating such an operation, so I didn't discuss it with her, and my son couldn't deal with it at all – perhaps due to his age and sexuality.

The plastic surgeon had discussed the breast reconstruction he would be performing, which involved a lengthy operation. At the same time as performing the mastectomies, he would remove fat and muscle from my stomach and use this to form new breasts. It was a complicated procedure and also there was a chance it wouldn't work. I asked myself, could I cope with having no breasts? I'd decided not to have silicon implants because of recent bad publicity. I concluded that disfigurement was of secondary importance to the certainty of not getting breast cancer. Since 12 years of age I'd been surrounded by this disease and here was a way to get rid of it.

From the time of making my decision in April 1992 I never looked back once. Now I had to decide when to tell people because I'd be in hospital for two weeks and off work for a couple of months. I had lunch with a close friend and started to tell her. She was so aghast that I didn't finish telling her about the reconstruction. I couldn't wait to get away from her. She just couldn't hide her feelings. After this I told a few more people but got more or less the same reaction. I was furious and upset. The way I saw it I had made a very difficult decision and these supposed close friends were completely unsupportive. The made comments like 'Do you know what you are doing?' or 'Why remove your breasts when you don't have cancer?'

I told my doctor friend who had known from the beginning and she said, 'They're so busy trying to deal with their own feelings that they can't focus on you and provide support for you.' I realised just what a threat this seemed to be to other women – to have one's breasts removed. I told my husband I couldn't tell anyone else so he said to make a list and he would sit down and phone them. I gave him half a dozen names and he did it immediately as I didn't want these people to find out from others. Needless to say I received a few phone calls from people we'd not told, to say they'd heard about it; again, they were very unsupportive. Despite the lack of help I never faltered.

The operation was performed in September 1992. I was really quite ill, which was to be expected. The plastic surgeon was so attentive and kind that somehow, even though I felt very ill, so long as he was looking after me I knew everything would be OK. I was fortunate in that the reconstruction was successful. While lying in hospital I became obsessed with the imminent pathology result of both breasts. What if I'm too late and the cancer was already there? Prior to the operation I remember saying to the surgeon, 'Please cut everything out.' I didn't want any chances taken. It took about ten days for the pathology results and when I was told everything was O.K. I just couldn't take it in.

After the operation people's attitude was how brave I'd been to have such an operation, but I didn't see it like that at all. I'd had the operation done through fear – fear of getting breast cancer.

I've never looked back since the operation. It's the best decision I made in my life or am likely to ever make.

This may sound odd, but it took about 6–12 months after the operation for the full psychological effects to be felt. I now read articles about breast cancer and watch programmes on TV without dread and despair. Something else that also became apparent was that prior to the operation I had never made any serious plans for the future – subconsciously I had thought I would die at a relatively early age from breast cancer. Now this inevitability had been removed along with my breasts. I now felt a person given their freedom, with a whole new quality of life.

Once the decision to proceed with the operation was made, my husband's support was invaluable. He is delighted that I will never get breast cancer, the bonus being that the reconstruction was successful.

With the advances of genetic knowledge more women are going to be identified as having a high risk of contracting breast cancer. The dilemma that I faced will be much more common. I can only say that the radical surgery I decided on was for me by far the lesser of two evils.

A family history of breast and ovarian cancer   Ellen Macke[1]

When I first telephoned Boston's Dana–Farber Cancer Institute in 1991, I didn't have a cancer diagnosis. I didn't even have a referral by another doctor. I was a 41-year-old female in a family where four other women first had breast cancer in their mid-thirties to mid-forties: my grandmother, my mother, one of my mother's younger sisters, and one of her three daughters. The Dana–Farber was the only 'cancer institute' in my area – I telephoned the switchboard asking whether any program existed there for women with a family history of breast cancer like mine.

I was looking for something more than screening. Through a 'breast clinic' at the facility that provided my employer's Executive Health Program, I was already receiving regular mammograms and manual examinations by breast surgeons. Although I'm not particularly shy, I didn't feel it was fair to ask a surgeon running behind schedule from one examination room to another to spend time delivering a tutorial on current trends in breast cancer research to a patient without a suspicious mammogram or a lump.

The recent diagnosis of my younger cousin, the first diagnosis in the family among my generation, made me feel more vulnerable to the disease than I had before. Through the Dana–Farber, I hoped to find a program that would offer an ongoing series of lectures or seminars to keep me up to date with the latest research about the role of family history, minimizing risk and treating breast cancer if it did occur. I wanted to learn more about the disease in a general way, outside of a clinical setting where the focus inevitably narrowed to whether I was free of symptoms at the time of my appointment. I thought a lecture or seminar program might also put me in contact with other women who shared similar family histories so that I could learn from their experiences. A

[1] © Ellen Macke 1996.

better understanding of the mechanism and manifestations of the disease, and the choices other women had made, would help me prepare for my own encounter with it.

As a result of my first telephone call to the Dana–Farber, however, I learned only that a doctor there was working on establishing a 'high-risk clinic' for breast cancer. The person to whom I spoke did not know whether it would encompass the type of program I was seeking. Before anyone got back to me, I left to live in London.

When I returned to the United States for the summer nearly a year later, two events prompted me to follow up to see what had become of the plans for a high-risk clinic at the Dana–Farber. First, I received a letter announcing the retirement of the surgeon who had been following me at the clinic where I received my mammograms, so I needed a new doctor. More importantly, my younger sister telephoned me to say that she had discovered a lump in her right breast and was scheduled for a biopsy.

The biopsy indicated that my sister had invasive ductal carcinoma. The panel of doctors that she consulted first surprised both of us by suggesting that she have both breasts removed and her ovaries as well (with a hysterectomy). This was the first time that doctors had recommended prophylactic mastectomy directly to either of us and the first suggestion that either of us consider prophylactic oophorectomy and hysterectomy. For the panel, my sister's diagnosis in the context of our family history set off alarm bells. I wondered whether I also needed to pay attention to them.

The second time I telephoned the Dana–Farber, the high-risk clinic had started to function. My description of my family history quickly led to an appointment with the oncologist directing it. She listened to me reciting my family history again. Solemnly, she sketched my family tree with her own notations and began an explanation of the suspected role of a gene named *BRCA1* in family histories like mine.

Although I knew that a family history like mine was a risk factor for breast cancer, the source of risk was a nebulous 'genetic predisposition' in my mind. When I was young, my mother attributed her own breast cancer diagnosis to birth order. She talked about being the affected first-born daughter of an affected first-born daughter of an affected first-born daughter. She told me that as a first-born daughter in this line, I should expect to encounter the disease as well. With the diagnosis of one of my mother's younger sisters when I was 25, my mother stopped talking about the disease as a problem for first-born daughters. Instead, she dwelt on the personality traits that her affected sister shared with their mother – a certain intensity and vulnerability to stress looming large among them. Her focus implied that if family history increased risk, it operated through some common temperament, either environmentally or genetically shaped.

Very little information about the mechanism of inheritance appeared at that time. I still had hope that the disease in my grandmother, my mother and my aunt resulted from a common environmental exposure or a configuration of independent risk factors that my generation did not share. The diagnoses of my cousin and then my sister severely diminished this hope. I began to pay more attention to reports of a search for a gene in familial breast cancer, but they contained few specifics about how it might operate.

Now, as the oncologist at the Dana–Farber spoke to me, a vague 'genetic predisposition' became a dominant pattern mutation of a gene with a name, even though scientists had not yet identified it. I didn't find a series of seminars at the Dana–Farber. Instead, as the conversation continued, I learned about genetic linkage analysis.

Understanding the suspected genetic pattern of inheritance and the potential of linkage analysis was exciting and frightening at the same time. I was excited to learn that inheritance of this 'genetic predisposition' was not inevitable, but a 50:50 chance for any family member. Even more exciting was the possibility that my unaffected cousins and I might learn through linkage analysis whether we were likely to carry *BRCA1* before we made decisions about prophylactic surgery – and perhaps contribute to progress in identifying the elusive gene as well.

The risks to which *BRCA1* might expose us, however, were greater than I had appreciated. I had not linked my grandmother's and my aunt's ovarian cancers to any risk that I, my sister, my cousins and future generations might share. Nor had I appreciated that the family risk might pass through unaffected males or that males in the family might themselves have an increased risk of particular cancers.

As soon as I left my meeting with the oncologist at the Dana–Farber, I telephoned my father, my sister and my mother's sisters and brother (all of whom lived 3000 miles away). I told them what I had learned and what participation in the linkage analysis study would require from the family. As the bearer of news with potentially calamitous ramifications, I was lucky to be dealing with a family that, led by my aunt, did not panic. Although cancer had killed both her mother and her older sister (my mother) at the age of 61, she had not reacted to her own diagnosis of ovarian cancer at the age of 59 as a death sentence. She took on this diagnosis as a challenge just as she had taken on successive encounters with breast cancer in her mid-forties.

My aunt's attitude greatly helped to sustain all of the family through the diagnoses and treatments of her daughter and my sister. She taught us to face cancer without self-pity as a challenge in our lives. She and other members of the family gave wholehearted support to participation in the linkage analysis study – not just for what family members might learn about their own risks, but even more as a contribution to knowledge that might some day lead to more effective treatment, if not elimination, of the disease.

This support did not mean that we had no apprehension about what the results of the study might mean for us. If the study showed that we were a linked family, we would know that my sister's three daughters were vulnerable to being carriers as well as my cousin's two sons. My mother's youngest sister might still turn out to be at risk even though she had survived to age 55 without a diagnosis. Four unaffected females remained in my generation, all in their mid-thirties to early forties. If my mother's brother turned out to be an unaffected male carrier, even his daughter could be at risk. The concerns spread to the four unaffected males in my generation as well – for their children as much as themselves.

Data collection for the linkage study took several months to complete. We had great hopes that we would not have to wait too many more months for some results.

Unfortunately, my aunt's ovarian cancer progressed in the meantime and she died in April 1993 while her daughter was undergoing a program of extremely high-dose chemotherapy for her metastatic breast cancer. My cousin was unable to leave the program for her mother's funeral. My husband and I decided to organize a family reunion for May which would celebrate not only our recent marriage, but also my aunt's life, my cousin's completion of her high-dose chemotherapy program and my sister's completion of her chemotherapy at the end of the previous December. The reunion gave us an opportunity to discuss in person our participation in the study and to share our concerns.

My sister scheduled a prophylactic mastectomy of her left breast immediately following the reunion so I could be there to help with her daughters. I still felt comfortable postponing serious consideration of prophylactic mastectomies until I learned the results of the linkage analysis. My next annual mammogram was scheduled for August.

We were still awaiting results from the linkage analysis in August when microcalcifications appeared on the mammograms of both my right and left breasts. A surgeon performed a biopsy to remove a cluster from my right breast. She called with the pathologists' report a few days later. Associated with the microcalcifications, they found ductal carcinoma in situ ('DCIS').

I had not expected a diagnosis before the results of the linkage analysis. My diagnosis did not necessarily mean I was a *BRCA1* carrier even if the linkage analysis indicated that *BRCA1* was responsible for the breast cancer in my family. Breast cancer sometimes occurs sporadically in linked families. Whether or not I was a carrier, however, the DCIS diagnosis began my personal encounter with the disease.

Because my biopsy was, in effect, a lumpectomy, my surgeon suggested three options for further treatment of the right breast: (1) close observation alone, (2) radiation therapy, or (3) mastectomy. She asked me to schedule an appointment with a radiologist to see whether I was a candidate for radiation therapy.

Scattered microcalcifications remained in my left breast. They did not cluster enough for a needle-localization biopsy. No one could say for certain what would happen to those remaining microcalcifications; so much about the process of the disease remains a mystery. My doctors were suggesting mammograms of the left breast every six months and an exam every three months. How often was often enough to detect any malignancy at its inception?

I thought back to the lump that had developed in my sister's right breast in less than six months. Even after it was palpable, it failed to appear as anything suspicious on a mammogram. I thought about how massive chemotherapy and extensive radiation had disrupted my cousin's life, but failed to arrest her cancer. She had only a one in four chance of surviving the next five years. Then I couldn't help but remember images from the slow, painful deaths that cancer had brought to my grandmother, my mother and my aunt.

The more I thought, the more that bilateral mastectomies began to seem like an opportunity. If I chose mastectomies now, my doctors did not consider chemotherapy or radiation necessary as follow-up. My lymph nodes would be left intact unless the surgeon saw something suspicious, which she considered a remote possibility given my diagnosis. Because no surgeon can be certain of removing every single breast cell, some risk of breast cancer always remains after mastectomy. My surgeon estimated the risk as 1 or 2%.

Without surgery, I seemed to face much uncertainty. How often would an exam or mammogram lead to a biopsy, forcing my husband and me to rearrange our schedules and put our future plans on hold? How likely was a future biopsy to result in a diagnosis that required not only six weeks of recovery from surgery, but also months of chemotherapy, weeks of radiation, or both? If I escaped chemotherapy and radiation, would I lose lymph nodes and experience swelling (lymphodema) after vigorous exercise or gardening?

I looked for information to assess my risks, but found that little existed. I learned that the estimated risk of further breast cancer was 5–8% for women diagnosed with DCIS who receive radiation treatment. For women who had neither radiation nor mastectomy, the estimated risk was 20% or so. These estimates, however, all seemed to come from one study that followed women with DCIS for only five years. No published study had followed them longer or tracked DCIS for any length of time in women believed to carry *BRCA1*.

In the absence of extensive data, statistical decision analysis helped me to evaluate my options. I drew a decision tree that began with three branches, one for each of the options among which I had to choose: simple mastectomy, radiation therapy or close observation.

At the end of the branch for each option, I drew two further branches to represent two possible outcomes that could follow each choice: 'recurrence' or 'no recurrence'. Except where the original choice was simple mastectomy, I drew another two branches at the end of each 'recurrence' branch. One of these represented 'a recurrence that required treatment involving a modified radical mastectomy with removal of lymph nodes followed by chemotherapy, radiation, or both'. The other represented 'a recurrence that required only simple mastectomy'. Before I could test different assumptions about the risk of recurrence after each choice, I had to rate the relative impact on my life of traveling along the branches to the end of each path.

Having assigned values to the end of each path, I returned to the uncertainties about where each choice would lead. Before I began to fill in probabilities on each of the branches that followed a choice, I made several copies of the decision tree so that I could compare the impact of making different assumptions about these probabilities. Decision analysis could not tell me which assumptions about the probabilities of recurrence were more likely to be correct, but it showed me that simple mastectomies were the best choice for me over a considerable range of those probabilities – given the values I had assigned to the end of each path.

Before going ahead, however, I looked for reassurance that I had realistically assessed how simple mastectomies would affect me. Over the following week, I telephoned my sister, who lives 3000 miles away, almost daily.

My sister patiently answered each day's questions. She compared the results of her two surgeries for me: swelling occurred sporadically on her right side where she had the modified radical but not on her left where she had the simple mastectomy. She told me that her self-supporting prosthesis was still comfortable; she had only minor adverse reactions to the adhesive strips that it required. We talked about how she felt without prostheses since the second surgery and how other people reacted, including her husband. We discussed the pros and cons of reconstruction.

When my sister thought that I had missed an issue during these conversations, she raised it for me. When I telephoned my cousin for her perspective on the surgery and life without breasts, she did the same. We talked frankly about what choice she would make if she could go back in time far enough to consider prophylactic surgery. Talking to the two of them reinforced my inclination to take advantage of my opportunity. I scheduled the surgery.

As it turned out, however, I will never know for certain whether I would have followed through with the mastectomies on the basis of my DCIS diagnosis alone. About a week before the scheduled surgery, my oncologist gave me some preliminary results from the linkage analysis. Markers believed to be associated with *BRCA1* appeared on chromosome 17q in DNA samples from all of the members of my family diagnosed with breast cancer, including me.

I will also never know for certain whether I would have had prophylactic mastectomies if results from the linkage analysis had been available before my diagnosis. Given the limitations of linkage analysis, the probability that I will carry *BRCA1* when scientists finally discover the gene itself is very high, but not 100%. A breast cancer diagnosis, even DCIS, made me risk averse. I would probably have felt equally risk averse in the absence of a diagnosis knowing only that linkage analysis indicated that I had lost a bet with 50:50 odds.

With or without a diagnosis, my evaluation of the choice between 'close observation' and simple mastectomies could well have turned out differently if I had feared that the surgery would threaten a career or important relationships, now or in the future. I was extremely fortunate to have a very supportive husband, family, friends, and physicians. Choosing mastectomies would also have been more difficult had I been contemplating bearing children or had I lacked adequate medical insurance.

This account has tried to convey something of the process involved in making a decision about mastectomies. For me, it seemed to have three stages. The first stage was very emotional as I identified with all of the family members whom I had seen struggling with the disease and suffering for so many years and then contemplated the impact of similar struggles on my life. The second stage was more analytic: I focused on collecting whatever information I could and tested different assumptions about the uncertainties using decision analysis. The third stage combined elements of the first two as I drew on the experiences of others who had already faced such decisions (or lacked the opportunity I had) and were living with the consequences.

These stages might occur in a different sequence or with different features for other women. Statistical decision analysis helped me to put my emotions together with the limited data available in a framework. With this framework, I could test the impact of various assumptions about risks on the outcomes of the various choices available to me. Women without training or coaching in decision analysis would probably evaluate the impact of different assumptions about their risks and outcomes less explicitly.

When *BRCA1* is located, screening may identify carriers in families where the gene has not previously expressed itself. Lacking a family history of *BRCA1* cancers, only the experiences of friends, reports in the media, and the attitudes of professionals will shape their emotional reactions. If I had been the first among my family or friends to have confronted a diagnosis or knowledge that I was likely to carry *BRCA1*, I hope that I would have found a network to connect me to other women's experiences. Although my doctors were women who caringly shared their medical knowledge and observations of other patients' experiences, they could not give me the same insights and support that I received from my sister and cousin.

The decision-making process itself is part of a journey. Every woman's journey to learning whether she carries *BRCA1* will be unique. I now understand that this journey began for me in the hushed voices of my mother and her sister discussing what happened during my grandmother's stay at the hospital when I was 9 – although I didn't realize it at the time. The journey continued through every diagnosis in my family and all that I read or heard about high-risk families over the years.

The journey does not end with learning the results of the linkage analysis or even with surgery. I still live with uncertainty about the extent to which I have protected myself from the consequences of *BRCA1*. (Less than four months after my mastectomies, I scheduled prophylactic oophorectomies and a hysterectomy.) I still wait eagerly for the discovery of *BRCA1* and an understanding of how to mitigate its adverse effects. When I am undressed, I still miss my breasts – but unlike so many others in my family, I may live to an average life-expectancy free from debilitating episodes battling *BRCA1* cancers. This is the destination that I hope my choices will help me reach.

An ordinary experience Anon.

Mine is an ordinary experience, nothing too unusual or dramatic but looking back I feel extremely relieved that I contacted a cancer genetics specialist.

My mother had had an ovarian tumour removed and was undergoing treatment. Although she wasn't certain and in fact didn't talk much of her own mother's cause of death, she did mention to me that she thought that her mother and possibly at least one of her mother's sisters had died as a result of ovarian cancer. I decided to contact a cancer genetics specialist, after reading about his research in 1986, to see if his project would be interested in our family history. At this stage I really didn't have much concern for my own health in this respect; I just felt that if the family could help the research work being done then we would be more than willing to do so. A genetic research nurse visited my mother before she died and I think played an extremely important role in my mother's acceptance of her death. Mother found the nurse perhaps one of the most understanding professionals she had spoken to during the course of her illness and at the same time felt that she was contributing information which might help the research project and therefore others in the future.

My own personal involvement with the research was, I suppose, highlighted when I developed breast cancer and had a lumpectomy followed by radiotherapy which finished just prior to my mother's death in 1987. My own vulnerability to cancer made me more aware of the importance of the work of the research group. Although I don't feel my life is dominated by thoughts of my risk of ovarian cancer (the project team tell me this is a one in two risk), I do feel relieved that the family history is known, that my own ovaries are screened regularly, and therefore any problems will be detected earlier than they would have been if we had been unaware of the possible familial disease. My 'ovaries' and whether to keep them until after natural menopause or have them removed has faced me with a decision that was not easy to make. I received conflicting advice from professionals. The oncologist who treated my breast cancer, and whom I made aware of the ovarian cancer history, advised removal of my ovaries to reduce the risk of disease. I was referred to a gynaecologist; I explained my dilemma and he advised that I keep the ovaries. He explained that I would not be prescribed hormone replacement therapy because of the breast cancer and therefore would be plunged into the menopause at 43 years of age, this would bring the possibility of all the discomforts that menopause might bring plus the risk of osteoporosis and greater risk of heart disease. I was not offered help in the form of a neutral counsellor but only the chance to talk to two health professionals, each with a different first priority for me as a patient. My decision has been to rely on screening and to keep the ovaries until after a natural menopause, when I will consider the situation again. The screening process is quite acceptable to me; it has never held any fears for me. If it was possible to test me and see if I had the gene responsible for ovarian cancer I would be more than happy to have the test. It would help me to make a decision – I feel I can cope better with things I know and am certain about rather than uncertainties.

I have a daughter of 24. We can talk quite openly about the family history and my risks and hers too. Initially, she has said that she wouldn't want to know if she carried the gene but is now able to consider the advantages of knowing. We both have confidence that the research being done will benefit not only us but many others. However, there are questions for her. She hopes to become a mother herself, if she could be told that she carried the gene how would that affect her decision to become a mother of daughters?

My husband has never found it easy to talk about the issue. I think he would rather not think about the possibilities and he has never initiated discussion on the topic with me. I have no other blood relatives that I keep in regular touch with to share the situation, and my parents-in-law do not feel comfortable talking about cancer at all.

I still feel that many of the doctors and consultants I have met as a result of my cancer history do not appreciate the inner feelings I have: they have all been very matter of fact and give little opening for discussion of 'feelings'. They are much more willing to talk of physical health. I feel that if I had not been reasonably articulate and assertive I would still have many unanswered questions and even now I find I don't question decisions made about me until quite a while after they are made. Six months ago I was told that the six-monthly screening of my ovaries would be extended to one year. It is only now that I feel quite worried about this; I feel safer being seen twice a year. I will have to ask about this and try and explain how I feel when I next visit the hospital. At no time has anyone suggested talking with my family – I think this would be helpful.

I am still dealing with the unknown. Have I or have I not inherited the gene? Have I or have I not passed it to my daughter? I accept this situation as one that is pure chance: I do not think I would have feelings of guilt if my daughter had inherited the gene via me. If I had never taken an interest in this work and had lived in ignorance of my own and my daughter's risk I suppose we would have been spared some worrying thoughts but I think my overwhelming feeling is one of relief that I am aware so that I can do all that is possible to prevent my premature death due to ovarian cancer.

1.4 Werdnig–Hoffman's syndrome

Werdnig–Hoffman's syndrome (spinal muscular atrophy, type 1; severe infantile spinal muscular atrophy) is a degenerative disease of the horn cells of the spinal cord that leads to a progressive loss of control of muscle movement. Symptoms of muscular weakness are usually present at birth or soon afterwards and death is inevitable within the first two years. The disease is recessively inherited. Prenatal diagnosis using linkage methods became available in 1991.

Jennifer's story   Anita Macaulay

Our story began back in 1984, when Ken and I were thrilled to learn that we were expecting our first baby. I had a reasonably uneventful pregnancy apart from the odd uneasy feeling that something was not 'quite right'. Having wanted a baby for a very long time and having been told that the chances of me getting pregnant were very slim following major abdominal surgery when I was 18, I put it down to being my usual pessimistic self. I noticed that the baby didn't move very much but this was brushed aside when I mentioned it to the GP at the local surgery. Seventeen days overdue, I was induced. After a 24-hour labour the baby went into distress and was delivered by forceps. Evidently the baby's head was trying to come out 'ear first'. We later discovered that this was due to the weakness in the neck muscles. My dream had come true – a baby girl, whom we named Jennifer Lorraine. I remember laughing and saying that I could always recognise my baby's cry as she seemed to 'whinny' like a horse at the end of a cry: the nurses and friends said it was 'Mum's intuition'. We later learnt that this was due to the already present muscle wastage of the respiratory muscles.

On the tenth day the health visitor visited for the first time. My sister had had a baby seven months previously and already I had noticed that Jennifer didn't move or kick her legs as much as my sister's baby had. I remarked on this to the health visitor who reassured me that babies were different and that she was 'probably lazy'. Over the next few weeks she did not begin to move or develop any head control. At her six-week check by the local paediatrician she was pronounced fit even though I pointed out that she seemed 'very floppy' and that she had no leg movement. I was told that I should be 'grateful that I had a placid baby'.

So my nightmare began. Every week I went to the health visitor and said that I was worried about her lack of movement and every week the health visitor told me she was all right – the paediatrician had said I had nothing to worry about. This went on until she was four months old. My mother had told me that I should take Jennifer to my GP as there seemed to be something wrong and that she should be holding her head by now. Although I knew deep in my heart that something was wrong, I didn't really want to face up to what it might be, so when my mother confronted me, I lashed out to somehow alleviate my fears. However, I did take her advice and went along to see our GP. By this time Jennifer's respiratory muscles were wasting badly and she was diaphragm breathing. I told the GP about her floppiness and lack of movement in her legs and as an aside mentioned that she was 'breathing peculiarly'. He reassured me about the breathing and said that lots of babies breathed like that. However he said he would refer her to the paediatrician for a check up. She was not marked as a priority case. Our appointment did not arrive for several weeks. I contacted the hospital who told me it would probably be in a couple of months! In the meantime we had come to realise that there was something seriously wrong. I remember when she was 11 weeks old looking at her one night in her cot and thinking, 'I know you will never be able to hold your head up'. I combed the medical books at the local library and came up with my own diagnosis: cerebral palsy. Ken and I had accepted that she would never be able to walk and had started to think about life with a child who needed to use a wheelchair – we believed that we would cope with whatever the paediatrician told us when we finally got to see him.

However, events then took a turn for the worse. We had been invited out and for the first time in Jennifer's life we asked Ken's Mum to babysit. When we got back Ken's Mum was distressed. She said that Jennifer had suddenly 'gone into a short of choke and turned blue'. She had snatched her up and Jennifer's colour had returned and she seemed as if nothing had happened. I half thought that it was probably wind and that she was over-reacting. The following morning I went to get Jennifer out of her cot and as I did she choked and stopped breathing. I snatched her up, thumped her on the back and blew into her face. She revived quickly. I phoned my Mum and said that Jennifer would be dead before anyone took me seriously at this rate. I asked my Mum to come with me to the local casualty unit. I said I was going to sit there until someone believed what I was saying. My Mum was with me within minutes and we rushed off to the hospital. Of course by this time Jennifer looked a picture of health, or so I thought. I explained what had happened and was told by a nurse that she probably had too many clothes on and had got hot and had a bit of a fit. I thought, 'Still no one is going to take me seriously.' Then the doctor came along, took one look at her and said he would call the paediatrician. The paediatrician was wonderful. He examined her and in a matter of seconds asked why he had not seen her a long time before. I explained what had been happening at the local surgery and all the 'reassurances' that there was nothing wrong and I was a 'neurotic first-time Mum'. He was quite concerned by this. The paediatrician explained that he would like to keep Jennifer in for tests. We stayed in for four days during which time she had X-rays taken, a lumbar puncture, blood tests and goodness knows what else. On the Friday evening he came along and said that he would like to talk to Ken and me the following morning. I knew immediately that it was bad news.

At the appointed time he came with the Ward Sister and told us that she had Werdnig–Hoffman's syndrome. He told us gently the facts such as they were, that she wasn't in any pain, that it was a muscle-wasting disease and that she probably wouldn't live beyond 18 months. I remember thinking throughout the conversation, 'I can cope with all of this as long as you don't tell me it's hereditary – I couldn't bear the thought of having to go through all of this again.' Finally he told us that WHS is a recessive condition and once you are known to be at risk you have a one in four chance with each pregnancy that the baby will be affected. My world just fell apart, my beautiful daughter whom I had brought into hospital thinking she had a problem with her legs was going to die and there was nothing we could do to save her. We were told to take her home and make the most of her. He asked us to go to another hospital the following Monday for an electromyogram test to confirm his diagnosis. This we did and as we walked through the doors into the unit I saw the doctor through a window look at Jennifer, his face drop and he said, clearly not realising I was looking at him, 'Oh my God . . .'

The paediatrician also wanted us to see the paediatric neurologist for his opinion. We waited four weeks and then, having heard nothing, rang the hospital and were told that they had no record of referral. At this point I lost my temper completely and screamed at his secretary, 'My baby will die before anyone does anything and I am not having it.' She was very rude to me. We complained formally to the hospital about her attitude. Months later we received a letter saying that no action had been taken: the secretary denied having any such conversation with me. In the meantime I contacted the paediatrician and told him what had happened and said that in the circumstances I wasn't going to take Jennifer there so could he refer us elsewhere. He asked if we would go to a London hospital. The following week Jennifer was admitted for three days while they carried out a whole range of tests including the final confirmation test: muscle biopsy. On the last day, one of the specialist team came and spoke with us at length. He told us that Jennifer did have Werdnig–Hoffman's syndrome and that there wasn't anything we could do. I remember asking what was happening in the research field for this condition and he told me that it was so rare no one was researching into it. That was my final hope destroyed – while Jennifer was alive and there was research under way I had hope, but to be told that there was no such work made me realise that even my hope was gone. We were told that the results of the biopsy would take six weeks and we would get an appointment to return for this result but they were 99% certain of the diagnosis.

We came home devastated. Every time I looked at Jennifer I just cried. These babies though seem to have an uncanny knack of looking into your eyes and reading your very soul; Jennifer would look at me when I was crying as if to say, 'Stop it – don't cry for me now.' After a few days I decided to save my crying for after she had gone and to make every day a pleasure for her and to cram as much into her short life as possible. This we did. However, just four weeks after our visit to the London hospital Jennifer developed a bad chest infection and was admitted into hospital. She fought hard but it was all too much for her and she died three days later in my arms. She was just seven months old. My first reaction was that I felt cheated – I had been told she would live perhaps 18 months.

Not only were we left with the devastation of losing our precious little girl, we also had to deal with the legacy she had left us – the genetic implications. We were not offered genetic counselling, although we did inquire about it at the London hospital. The doctor who explained about the condition and how it would affect Jennifer told us about the genetic implications and said it was probably not worthwhile having genetic counselling as they would not be able to tell us any more than he had.

Our GP told us in no uncertain terms that we shouldn't ever have any more children – he even thumped his fist on the desk as he said so. We asked about adoption and artificial insemination. He said he would refer us for artificial insemination by donor (AID). The appointment came through and we went for the counselling and then I began treatment. After three years we called it a day. We knew that I had fertility problems and it seemed that we were throwing good money after bad. While I had been going for AID we had also begun to go through the adoption application process. We were interviewed on all sorts of issues and finally after about eighteen months our application was turned down. We were told 'any child would grow up in Jennifer's shadow'! So Ken and I decided to fight this rejection. Our fight went on for two years before finally we threatened to go to the press and our Member of Parliament to investigate adoption policies, and all of a sudden we were permitted to go through the application process with a different social worker. After six months of interviews we were approved and went on the waiting list. Nine months later our social worker called unexpectedly and told us that we had a baby boy – nine months old and actually born the day after we had been approved! Stuart is the apple of our eye and very, very precious. We are so lucky to have him.

Ken and I believe that our experiences seemed unduly bad. It is hard enough to cope with the diagnosis and the inevitable outcome that severe WHS has, to give birth to a child and then to have to watch them slowly die is no experience a parent would wish to go through. But to cope with the professional inadequacies that seemed to follow our every step along the way is awful. Of course if we had known at the beginning, it could be said that we should have fought harder with the health visitor and local doctor, but what parent really wants to admit that their child has some dreadful condition? Who wouldn't accept being told they were neurotic and that their child was lazy and placid – you would want to believe it, wouldn't you?

We were, however, very lucky with our support from the local hospital. The nursing staff on the children's ward became our friends and advisers; although they had not had any experience of nursing a baby with this disease, they helped us with the practicalities and lent us equipment to do as much as we could for Jennifer at home. We had an 'open door' agreement with them so that we could by-pass the local surgery and take Jennifer straight to the ward if we were worried. This friendship continued after Jennifer's death and even today nearly ten years on I still see the ward sister regularly.

Our family and friends were as distraught as we were but they were always there for us. I remember my sister telling my Mum that she wished it could have happened to her instead of me as she thought she would have been able to cope better than I could. She changed her mind when after Jennifer's death I went on to establish the Jennifer Trust: a support group for other families whose babies had had this dreadful disease. She said she realised then why it happened to me and not her – because she couldn't have done that. After Jennifer's death I came to realise that other parents would perhaps welcome the opportunity to talk with someone else who had had a baby with the same condition. Thus my way of coping with Jennifer's loss was to start up a group, which today supports many hundreds of families affected by WHS.

At the time we were making our decision regarding future children, prenatal diagnosis was not an option, as the test had not been developed. However, shortly after it had been introduced, Professor Kay Davies, who led the research to identify the gene for WHS and then developed the test, advised us that she had located Jennifer's Guthrie test card[1] and that DNA may be obtained from this card should we wish to pursue prenatal diagnosis. Whilst we were thrilled that such a test had been developed, having talked it through we decided not to pursue this option as we were already under way with our second adoption application and because of my fertility problems.

Our family have supported us all the way with our decisions to pursue AID and adoption. I think they were all thankful that we decided not to try again, and risk another affected baby. Our decision not to try again was based partly on their feelings – they had all been so devastated by Jennifer's death we felt that it was unfair to put our parents through it all again as well as ourselves. The day we made them grandparents again by adopting Stuart was probably one of the happiest of their days as well as our own. All four grandparents had suffered, as we had, and had suffered watching us so sad, wanting a child and seemingly unable to have one by any method.

A mother's account   Helen Hearnshaw

When our little boy Daniel was diagnosed as having Werdnig–Hoffman's syndrome (WHS) at the age of seven months, the genetic implications of it all were the last things on our minds. Having just been informed that our baby's muscles were wasting away and that it was most likely that he would die from respiratory failure before reaching his first birthday, the future seemed totally irrelevant. Some days later, when we had recovered from the initial shock of the diagnosis and the realisation that there was no miracle cure anywhere in the world to help Daniel, we allowed our thoughts to return to the future.

The implications of both my husband and I being carriers of the same recessive gene were given to us, albeit somewhat briefly, by the consultant paediatrician taking care of

---

[1] Used as the basis of neonatal screening for diseases such as phenylketonuria.

Daniel. She explained about the problem of the one-in-four risk but put it to us in such a way that it appeared to be almost insignificant. She said that if we looked upon it that 25 out of every 100 of our pregnancies had a risk of being affected by WHS then the odds were really not so bad. Who were we to argue? She told us to go home and try again. Three weeks later Daniel lost his brave fight for life and in our inconsolable grief, and with the blessing of my general practitioner (GP), we rushed headlong into another pregnancy. We were fully aware that at that time there were no tests available but we truly believed that we had had it on good authority that lightning would not strike twice.

Perhaps it was the fragility of my condition, a grieving mother having recently lost one baby, heavily pregnant with twins; I can think of no other reason that fully explains why the various doctors and a genetic counsellor all allowed us to go on believing that what had happened to Daniel was unlikely to happen again. Not at any time was it impressed upon us just how serious the one in four risk really was and I breezed through my pregnancy in blissful ignorance.

When the twins Danielle and Stefanie were born they appeared to be perfectly healthy. By the time they were six weeks old they had begun to develop an all too familiar weakness in their legs. Two weeks later, and before any conclusive tests had been completed, they were totally floppy. To say that we were shocked would be an understatement but we had precious little time to waste on thoughts of blame and recriminations and we got on with the business of lavishing our babies with the happiest and healthiest of lives.

At about this time, just when I needed him most, my GP, who up until then had been supportive, began to avoid all contact with me. I was told later by my health visitor that when faced with two such beautiful and bright little girls and being a father himself, he found it difficult to deal with the undeniable fact that they were dying. This was the first but by no means the last time that we were to encounter this most unprofessional and inexcusable reaction. For a while we coped mainly on our own, having already learned the basics of nasal gastric feeds, physiotherapy and suction with Daniel. We were prepared for and able to cope with the effects that the disease was to have on the children. When eventually we did need help (e.g. oxygen), we turned to our local hospital where we discovered to our dismay that doctors and nurses who had heard of WHS, let alone cared for an affected baby, were few and far between. Their role became a supportive one and they were guided by us when ministering to the children's day-to-day special needs whenever it became necessary for them to be admitted.

For most of their short lives, both Danielle and Stefanie, though immobile, were happy and healthy babies, but inevitably as time went by and the illness progressed, their chest muscles began to fail them and pneumonia set in. For the last few days of their lives, when there was nothing more that could be done, we could only watch helplessly as strength and spirit deserted them just a little bit more each day. Every breath became a struggle, every night seemed like it would be their last. Their constant distress was heartbreaking and it went on and on and on. The end finally came as a welcome release from their suffering and the twins died within a few days of each other at the age of ten months. Never before had it occurred to me that life could be so cruel to such sweet innocents but it made me realise that never again could I take the risk of con-demning another child to that same dreadful fate.

One of the more positive and useful points to come from our previous visit to the genetic counsellor was a telephone number for a group which offered advice and support to families affected by WHS. We contacted the group, the Jennifer Trust for Spinal Muscular Atrophy, not long after the twins were diagnosed and we were amazed and comforted to learn that there were other people not too far away who were going through exactly what we were going through. We were also surprised to discover that research was already under way to perfect prenatal diagnosis with the help of blood samples taken from affected families. Then, as now, the gene which causes WHS was unidentified and DNA markers were being sought to give a near accurate result. Without any hesitation we offered our blood for research, and fortunately for us, because the samples were in the right place at the right time, we were screened for the markers as the research progressed. With help from the Jennifer Trust we were privi-leged to be among the first to be offered the benefit of the new prenatal diagnosis just four months after the deaths of the twins.

An early appointment was made with the genetic counsellor, ensuring that this time we were seen before I became pregnant. It was explained to us that there was a 5% risk of an inaccurate result. We agreed that this was acceptable. At ten weeks of pregnancy a CVS (chorionic villus sampling) would be performed, this procedure itself having a 3% risk of causing miscarriage. This also was acceptable. In the event of an unfavourable result a termination would be carried out. This was a difficult decision but one to which we had to agree. Nothing was left to chance, no words were left unsaid, this time the genetic counsellor made sure of it.

Because of the previous difficulties with my GP I decided to switch to a more senior member of the same practice who was already familiar with my history. He made encouraging noises at the confirmation of my third pregnancy and booked me in for a CVS but without first preparing me for the pain and indignity I was about to go through. All thoughts of pain and discomfort were soon forgotten when the news of a negative test result arrived less than two weeks later. It was confirmed that our baby, although a carrier, was a healthy boy. Ten weeks later and we were just getting used to the idea of becoming a family again when for no obvious reason I began to haemor-rhage. It continued ceaselessly for almost two weeks then without warning I went into labour. Inevitably at 22 weeks the baby was born dead. No one said at the time that it had anything to do with the CVS but then no one could say that it hadn't.

We had to have faith in the CVS, we desperately wanted a baby. A post-mortem proved conclusively that the test had been accurate. Encouraged by my obstetrician we tried again. My fourth pregnancy was tested as before but this time the result was posi-tive. In theory, what was to happen next should have been relatively straightforward so as to cause as little unnecessary distress as possible. But theory hadn't allowed for the reluctance of yet another GP to become involved. He was unsympathetic and uncoop-erative in making any arrangements, using a pathetic excuse about it being too close to Christmas. I took the matter in my own hands and contacted the hospital myself. They were very kind and understanding and the termination went ahead on Christmas Eve of that year. It was the strangest, most unnatural feeling in the world to have to agree to the destruction of something so very, very precious. Memories of how the twins had suffered dispelled any doubts that I might have had but didn't ease the pain of grief for yet another lost child.

In the aftermath of a disastrous Christmas both family and friends were quick to condemn the introduction of such tests that could inflict so much pain and heartache on an already emotionally battered couple. But the need to recapture that special feeling of love born only of parenthood somehow gave us the will and strength to go on.

Our spirits by now were at an all time low and it was becoming difficult to be optimistic about the future. I changed my GP yet again; I needed all the support I could get and was gratefully relieved to find it with my third choice. My husband and I had agreed that should another attempt fail then we would have to seriously consider an alternative method to have a baby. Before the prenatal tests had been perfected, my impatience had got the better of me and I persuaded my husband that we should find out more about AID and egg donation. On closer inspection, however, it became obvious that neither of these options were ideal because, given the choice, we both preferred to have children that were biologically our own. But had the prenatal tests not been made available when they were, I'm sure that we would have taken advantage of one of these options as a desperate last resort.

Although family and friends had supported us through everything that had happened, the seemingly never-ending grief was beginning to take its toll on those closest to us, so when I became pregnant for the fifth time, for their sake we decided to keep it a secret. We managed to keep it to ourselves right up until the day that it was confirmed that I was carrying a healthy baby girl and when we proudly announced our news it seemed that the whole world was happy for us and wanted to wish us well.

It was a nervous pregnancy, mainly because of the previous miscarriage, but my antenatal care by the hospital was second to none. I was scanned at regular intervals and every precaution was taken to ensure that this baby would arrive safely. And so she did at 38 weeks by elective caesarian, Alyss Daisy came into the world kicking and screaming like no other baby I have ever had. Perhaps understandably, for her first eight weeks I was a little paranoid and convinced myself on occasions that she was showing signs of weakness. But with help and support from my new GP and health visitor we finally made it to the milestone that not so long ago seemed so elusive: our baby's first birthday.

Because together we have both strength and determination, my husband and I were able to survive through more tragedy in four years than fortunately most people have to face in a lifetime. Some of that tragedy could well have been avoided if only we had received proper genetic counselling at the right time. The paediatrician who advised us so inappropriately at the beginning was no less than negligent. But the final decision to try again was ours alone, and we accept the consequence of our actions. Through our experiences we have come to realise that luck, be it good or bad, plays a major part in genetic inheritance. Each and every pregnancy, like a game of Russian roulette, is a gamble ultimately between life and death. For parents such as ourselves whose lives have been torn apart by a genetically inherited terminal illness, the risk factor so carefully calculated may be 1 in 4 or 1 in 24, but as we hold our breath with fingers crossed and wait for the results of yet another prenatal test, the odds will always be the same. An equal 50:50 chance. Why? Because at the end of the day the only results that really matter are either 'it is' or 'it isn't'.

1.5 Sickle cell conditions

Sickle cell conditions are recessively inherited mutations that cause changes in the haemoglobin molecule. Haemoglobin, which makes up the content of the red cells in the blood, is involved in the transport of oxygen to the body tissues. In sickle cell conditions, in certain circumstances many red cells may become sickle-shaped, blood becomes viscous and the capacity to transport oxygen is reduced. The severity of the condition is very variable and those with the double-recessive condition and certain others may suffer periodic sickling crises. Heterozygotes (those with only a single copy of the mutated gene) are essentially normal and have partial protection against malaria. This protection is responsible for the high frequency of the gene in some populations – heterozygote frequency approaches one in four in some parts of Africa, for example.

Prenatal diagnosis (originally using fetal blood but now usually based on DNA) has been available for many years but because of the wide variability of the disease was seldom used. In the early 1970s a screening programme for sickle carriers was set up by the Federal government in the USA (see Chapter 14). At first it received popular support. However, it rapidly attracted wide criticism, in part because clear information about the meaning of carrier status was not provided and also as many saw the programme to be eugenic in nature. Some states required children at risk to be tested before they could be enrolled in school and some insurance companies denied cover to black carriers. Carriers were rejected by the US Air Force and for employment by some airlines in the false belief that such people might pass out at high altitudes. A number of prominent scientists publicly stated that black people who carried the gene should not have children. Eventually, the widespread criticism of the screening programme led to its withdrawal. Memories of this disaster have played an important part in forming attitudes in the United States to subsequent genetic screening programmes.

Some observations about my life with a sickle cell condition   Merry France-Dawson, Ph.D.

Twenty-nine years ago, I was diagnosed as having sickle cell anaemia (HbSS, the double recessive condition), after having a series of crises that left me close to death and hospitalised for five months. As I lay unconscious in a South American hospital, I heard the doctor tell my father that I needed an urgent blood transfusion – our blood was compatible. The doctor added that there was every chance that I would not survive the night.

I listened with horror to the pronouncement of my impending death, and quickly made the decision that I had no intention of dying. Five months later, I was given a year to live. Living by instalments became commonplace, until after I reached my 21st birthday, by which time one mischievous doctor (who used me as a guinea-pig for flummoxing medical students in a London hospital) proclaimed that I would probably live till I was 80. Most likely I will prove him wrong too, although I am now more than half-way there. It is possible – I know of one man who, on his sudden death, at the age of 77 years, was on post-mortem found to have suffered his first known sickle cell crisis. He had HbSS. It was at this time that the use of new sophisticated diagnostic techniques showed that I didn't have HbSS but rather the 'less serious' HbSC haemoglobinopathy (C being another recessively inherited abnormality of the haemoglobin molecule). My only thought on hearing this was that it was serious enough the night I nearly died. Since those days, I had always been curious about my 'illness'. For a long time I thought it was unique to me, because I had grown up not knowing anyone else with 'it'. I was curious because doctors often told me things I could or could not do, and I delighted in doing precisely what I was advised not to, largely ignoring what they suggested I should. This included doing nurse training at the age of 20. Predictions of how long I would live, and other dubious advice, served only to make me determined to prove 'them' wrong. However, listening to 'learned' people talk about my dying, having friends and family treat me like I was no longer valid (though still loved), affected me more than I realised. This was manifested by my loss of interest in academic work. After all, why bother if you are going to die? It was only after I reached my 29th birthday that I returned to my studies.

I have coped reasonably well with my 'illness', and have not had a major crisis since I was 15. However, most of my coping came from developing an appreciation from what I really could or could not do, quickly learnt from experiences of feeling unwell if I overdid things. I was fortunate in that I had these experiences. Some people are not so lucky, and crises can develop without warning. I was also fortunate in that I was born to a middle-class family, able to keep me supplied with the latest American developments in medical care for haemoglobinopathies.

It was in my 30th year that I started an undergraduate course in biological anthropology. Ironically, my nurse training had provided me with no information on the conditions. Even today, nursing textbooks used in Britain are singularly lacking on information about haemoglobinopathies. The B.Sc. course introduced me to genetics and variability. It provided me with the explanations I needed at the molecular level about genetically determined illnesses, and in particular about the evolution of abnormal haemoglobins. My new-found knowledge helped me to understand why, for example, I was the only one in my family to be affected, why my condition was considered less serious than sickle cell anaemia, and so on. More importantly, higher education opened to me access to academic journals, which were full of research work involving haemoglobinopathies and other genetic conditions. As a result, my knowledge base moved from simply being experiential to being informed in terms of current thinking about the conditions. This combined knowledge reinforced my inherent belief that with a recognition of just what I am able to do physically, and with the ability to control my environment appropriately, I would be as capable as most people.

This, then, was how I came to learn about sickle and other haemoglobinopathies. I had received no genetic counselling at any time. Indeed, I soon discovered that my knowledge base was greater than that of most physicians. On the few occasions that I have needed their care (generally for minor illnesses), many would simply ask what I felt was the best way to proceed. Even the specialists left me largely to my own devices, expressing delight at annual check-ups that my hips, eyes, etc., are in pretty good shape! I have not escaped all of the complications unscathed, and have bilateral hearing loss. I wear one hearing aid, and may need another in the other ear, but I am still functioning!

In the early years I have had eugenicist advice given to me in relation to family planning. I had been told that I could not successfully give birth to a live baby, and that the risks to me, should I try, would be life-threatening. This advice was always given as proven fact. It began when I was an uninformed adolescent, and I accepted it without demur. This was an area where I was not sure that I wished to defy doctors! By the time I realised I could have had a child, I was determined to focus my own academic work on the needs of people with sickle cell conditions, and how these needs were being met by the health service. By then I had discovered that I was not alone, and that others with the condition were not invisible. I have been, and continue to be, immensely interested in the empowerment of individuals and families, so that they might optimise their abilities to cope with the conditions. This meant that I had neither the time nor the inclination to have babies.

Eugenicist arguments may be acceptable to many of us if we are found to be carrying affected fetuses. However, because sickle cell conditions are so variable in their manifestation and prognosis, it seems somewhat immoral to advise us not to have children without full consultation with us. It is true, however, that some individuals with the conditions do become very seriously ill, and may have an extremely short life-span with little quality of life. A woman who has already had experience of such a child might not wish to risk having another. Indeed, I would have grave concerns should I become pregnant and tests confirmed that I was going to have an affected baby. I do not think I would wish to terminate, but then this is hypothetical, as I would not choose to have a carrier father my child.

I have had no *evidence* which has led me to believe that my doctors were being negatively discriminating in advising me against pregnancy and childbirth. I believe they simply did not have sufficient information themselves, and would have given similar advice to anyone with a potentially serious genetic condition. Certainly, even today, the risks of pregnancy and sickle cell conditions are more likely to be highlighted in academic papers rather than in the management of affected pregnancies. Undoubtedly many of us will die as a result of our pregnancies but many more will become mothers. While the risks have to be clearly spelt out, so too have success rates. We must be allowed to make our own choices.

I have also never experienced negative discrimination on the few occasions I have had to be a 'day-patient' for minor surgery. However, I believe this to be because I was seen as a member of the 'sorority/fraternity' of health workers and knowledgeable. Furthermore, on those occasions I had been working as a researcher (on a sickle cell study) for the Royal College of Nursing of Great Britain and I was often treated as a minor celebrity.

Negative discrimination has probably had a greater effect on me in the workplace. As a researcher, I am subject (along with researchers of varying racial and cultural backgrounds), to working on short-term contracts. While this is difficult for all of us, it sometimes appears that I have been treated unfairly: whether this is because of the disclosure of my diagnosis, or because I am black, or for some other reason, it is hard to say.

Good genetic and other counselling and advice can help people with sickle cell conditions to lead relatively fulfilled lives. However, I received advice only from my father. It is advice that has stood me in good stead. On giving me some American leaflets during my sixteenth year (poor enough productions, because at the time so little was known about the condition), he said:

> This is your problem, not mine. You have to live with this thing. Read these and anything else you can find. Try to keep up to date with what's being written. It is up to you, because I cannot live for you. Read these and tell me what they say. I'm not going to, but I'm interested.

At the time, still depressed and ill, I thought him rather cruel, until I realised that he *had* read the material and was just ensuring that I not only read them, but that I also understood what they meant. It was the beginning of my interest in health care issues, and an adult appreciation of my father's worth.

My experiences are atypical of the average person with a sickle cell condition. I have been able to take some degree of control over my health and the care I receive. My only regret is that I have not had any children. I am still capable of trying. However, I am now of an age where the possibility of producing a baby with other genetic 'disabilities' is high. There is also the residual fear that I might be wanting to have a child for the 'wrong' reasons.

Today, health care professionals in Britain are more aware of the complexities of living with sickle cell conditions, and how the care they provide can affect the lives of people like me and our families. The problems of racism and differences in cultural approaches to health and illness are also being addressed by some carers. However, the emphasis on care provision for us is still largely on 'crises' interventions rather than holistic care.

We can survive in 'difficult' environments with the use of the appropriate health promoting strategies, as I have. We can learn and work, provided we do so in suitable environments, as I have. We can live reasonably active and fulfilling lives, as I have. We *can* have babies, and yes, we can live to see 80. There's still time for me. I have not done badly. Not badly at all!

Sickle cell disease: a father's view. An interview with Alexander Mottoh
Martin Richards

M.R. : Can I begin by just asking about your background? Family?

A.M. : I grew up in Nigeria until 1975 when I came to this country. I was 23 at the time. I came over here principally to study and that was just what I did in the first 11 years of my being here. I first studied for my A-levels at South London College and then proceeded to Brunel University where I spent the next four years studying for a batchelor's degree in medicinal, agricultural and environmental chemistry. I then went on to study for a doctorate in organic chemistry at King's College London. After I got my doctorate I spent a year at Oxford University as a research fellow before deciding to become a secondary school teacher.

M.R. : At what point did you marry? Was that after you got to Britain?

A.M. : Yes, I got married here in Britain just after my first degree, and I now have three children.

M.R. : When did you first learn anything about sickle?

A.M. : Um . . . I knew about sickle cell, but not the details, before I got married.

M.R. : Were there other people in your family who had had children with sickle?

A.M. : No.

M.R. : Did your wife know about it when you married?

A.M. : No. And there is nobody with a sickle cell problem in her family.

M.R. : So when you had children you had no thought that it would be an issue at all?

A.M. : That's right. We didn't have any hint at all. If we had any hint we would have gone for a test immediately after conception.

A.M. : Our first child was born in February 1982. No problem was apparent until exactly 12 months later on her first birthday when she developed pain on her left foot. We did not think much of it because the pain went away after a few days. We ignorantly thought the pain was a result of some kind of bug she had caught from her minder. At the time my wife and I were studying for our Ph.D.s. Our doctor had to refer her for a blood test after she had another painful crisis.

It was after she was found positive that we were tested and found to be carriers.

M.R. : Did you see a geneticist at the time?

A.M. : No, we saw the consultant haematologist who attends to our daughter and was aware of the consequences of the condition. The only geneticists we have come across are those my wife meets in the course of her work: the Genetics Interest Group (a voluntary charity group which brings together organisations concerned with different genetic conditions) and the Sickle Cell Society.

M.R. : Tell me about your daughter. What about her health?

A.M. : She is growing well. But it is worrying. She has crises about every three months, you know. Sometimes she would go for a longer period without any crises. But a minor infection may trigger a serious crisis. That is worrying.

M.R. : Which presumably has had quite an effect on her schooling?

A.M. : It does have an effect. It has not helped her education. Since she got into secondary school about nine months ago, she spent the first three months in and out of hospital until she was placed on a regular blood transfusion. Since then she hasn't had any crises.

M.R. : How does she feel about it?

A.M. : She understands the situation, and she is not happy about it, you know, she keeps asking, 'Why me?'

M.R. : How old was she when you first talked to her about it?

A.M. : Three or four.

M.R. : Does she talk about the future and what the condition may mean?

A.M. : Not much, no, not in those terms. But we worry about her, about her future, but she doesn't really talk much about it yet. She does understand the situation. It bothers me sometimes – she is resentful about not being able to play with other kids when the weather is cold. Even though she knows why, she still feels unhappy about it.

M.R. : It must be very hard for her. How about school? Are the teachers sympathetic?

A.M. : Yes. A sickle cell counsellor went to explain to them. But before that we had already supplied the school with some pamphlets and material about sickle cell.

M.R. : Does your daughter know other children with sickle?

A.M. : Yes, she does, especially in the hospital. She knows a lot of them because they meet in hospital on social occasions.

M.R. : Has the Sickle Cell Society been helpful to you?

A.M. : Yes, but that is not my only source of support. I try to read every literature and publication on the subject. I know that it is something one has to live with until some sort of drugs are developed that can stop sickling. I believe that is possible if the search for the drugs is well resourced. You see, the multinational drug manufacturers are not particularly interested because they are afraid of not being able to recover money spent on research. The simple reason for this shortsightedness is just that the illness is only suffered by ethnic minorities in both the USA and the UK. And most developing countries with a high rate of sufferers may not be in a position to afford such products if developed.

M.R. : What about the treatment your daughter gets in hospital? Do you feel satisfied with that?

A.M. : I think the hospital we go to has been very good to sufferers. The standard of care there is getting better and better all the time and their staff are wonderful. Sicklers are often given top priority whenever they arrive in casualty.

M.R. : Do you think that it's significant in terms of the treatment you get or the research that's done on sickle that it is a disease that occurs primarily within the black population of Britain?

A.M. : Yes, there is no doubt in my mind that had the illness been universal a more concerted effort would have been made to find anti-sickling drugs. More progress has been made in the area of social welfare after a prolonged and sometimes frustrating battle, but the key area, which is research into the cure or sickling control, has no defined organised programme.

M.R. : Could I just ask you a little about your other children?

A.M. : My first son has the trait. We opted for a test at the early stages of the pregnancy. So we knew beforehand that he has the trait.

M.R. : And that's presumably something you will tell him at some point?

A.M. : Yes. He knows about it already. He does understand that his sister has got sickle cell disease and that she needs a lot of attention, especially when she is ill.

M.R. : Have you tried to explain what it may mean for him in the future if he has children?

A.M. : I have to do that when the time is right. About age 11, I think. He should be able to understand then.

M.R. : And your third child?

A.M. : He is normal. No trait.

M.R. : What about your own family and your wife's family? They presumably know that you must both have the trait. Has that been something you've told other members of the family about?

A.M. : I have two brothers and a sister, they all have children of their own but none has sickle cell illness. Likewise, there is no such situation in my wife's family.

M.R. : Are there other things you feel are important that we've not discussed or touched on?

A.M. : I don't think so, but I can only add that we are trying to cope with the situation, especially as we have no control over it. We've tried to rearrange our lives to suit our situation. My wife and I were both research scientists in higher institutions but we had to chuck it in to become school teachers because we are able to stay close to our family. We are also able to teach her at home especially during a long period of crisis, plus the additional bonus of being on holidays whenever our kids are on holidays.

M.R. : In that sense it really has had a very major effect on your life.

A.M. : Oh yes. I can't tell you that I'm enjoying my present job because I'm not. I spent years training to be a scientist but now all that effort seems a waste of time.

M.R. : I'm interested in the Sickle Cell Society and the kind of role that that may play and in what ways you might see it developing to help other families. Do you see it basically as a pressure group on the research front, and the service front, or as a sort of support group for families? Or is it all of those things?

A.M. : It's supposed to be all of those things, but I would like it to focus more on the research front because that is where the future lies. The social service front is getting better attention now than before.

M.R. : Is there a local group here in this part of London?

A.M. : Yes, we have a local group of Sickle Cell Society in East London. My wife was once their chairperson and she is the treasurer now. Our aim was to focus attention on the research front. As far as I know there is no serious research going on in the area of anti-sickling drugs. I'm sure if there were, they would have found something by now.

M.R. : Is this largely a black area?

A.M. : There are more Asians than blacks. The black and Asian communities are encouraged to undergo screening. Unfortunately some people shy away from such screening. I think it is very important to screen potential couples. I know that it is compulsory for couples to be screened in the USA before they are issued a marriage licence.

M.R. : But there is a big backlash against that now.

A.M. : That's right. I know it is a little bit too extreme. However, I still think screening needs to be emphasised more. This I hope would help achieve two things – warning people about such problems and reducing the enormous social problems which ignorance could cause.

M.R. : What age would you recommend people to be screened?

A.M. : I think as early as possible.

M.R. : Teenagers?

A.M. : Yes, so as to be aware of the score before getting involved. In fact newborn babies should be screened and the result explained to them when they are old enough to understand what it means.

1.6 Personal experiences of genetic diseases: a clinical geneticist's reaction
Peter S. Harper

If anyone were to doubt the profound impact that genetic diseases can have on individuals and families, the personal accounts collected in this book will have removed those doubts. They are vivid and compelling, at times deeply moving, at others profoundly disturbing. No matter whether one is a professional involved in studying genetic disease and providing services, as I am, or a social scientist, or a family member faced by comparable genetic disorders, these accounts form an essential counterpart to what one can learn from systematic studies.

In trying to put down my thoughts on paper, I find it difficult to do so in a systematic or logical way; these are personal reactions to personal accounts. I am reminded of the many comparable families that I have seen for genetic counselling, with whom I have interacted for a brief period; we see in these descriptions a series of lives touched by genetic problems over long stretches of time, from birth to death and affecting most important decisions in between.

Personal accounts of this nature tell us about the people who have written them, about the particular genetic disorder and its problems, and about the professionals with whom these people have come into contact. I find it easiest to look at these three aspects in turn.

The people who have told their stories have given us, without intending to, pictures of their personalities and characters which compel not only sympathy but immense respect. They are almost all women, and this, I am sure is no coincidence, but reflects the fact that for the most part it is women who form the central focus of the family-based problems that genetic diseases create. They are all clearly very different types of person, but the common features are courage, tenacity, resilience; they are fighters, for themselves, for their family and for others. They may be deeply wounded by the attitudes of professionals or relatives, and by the diseases in their family, but they are not going to let these submerge or defeat them.

How typical are these people and their families? Can we really draw general conclusions from their own reactions and experiences? I think we can. Clearly they are educated and articulate, but I doubt if they are unusual except perhaps in being able to express their experiences so clearly. They may perhaps be 'survivors', whose strength has carried them through tragedies that others might have been shattered by, but again, I do not see them as being atypical in this respect, though they have been better able than others to find professional help that proved to be a real support. I see them as flag bearers for that great majority of individuals whose experiences and feelings we shall never fully learn about, but for whom the feelings are likely to be every bit as strong as those written here.

Turning to some of the general points illustrated by the accounts, one that is strikingly illustrated is how much trauma and suffering families are prepared to go through to achieve their goals. The two spinal muscular atrophy families show this vividly in terms of the aim of having a healthy child. Both have achieved their aim, by prenatal diagnosis in one case, by adoption in the other; but only at the cost of great anguish along the way. Both these accounts show how determined families often are, how prepared they may be to make use of new genetic developments, and how vital it is that they are allowed and encouraged to follow the course that seems best for themselves.

The accounts of women at risk for late onset genetic diseases – breast cancer and Huntington's disease – show a similar determination to know one's true risk status and to cope with the consequences. The two situations differ somewhat in these consequences, because for breast cancer there is the possibility of preventive surgery, whereas for Huntington's disease, one can as yet do little to modify the disorder. However, for both diseases the individuals writing their story clearly express their need to know, so as to plan their lives and decisions, and they are prepared to go through many difficulties to achieve this knowledge. We do not yet know how representative they are in this wish, and again, one can see differences between the conditions reflected in the family attitudes described; concerned and positive for the breast cancer families, often reluctant and even evasive in the case of Huntington's disease.

One aspect of the descriptions that makes for fascinating reading is the interactions between the writer and their wider family. Some of these are perhaps unexpected, at least in their intensity. It comes over clearly that husbands and close woman friends may see the option of preventive mastectomy as somewhat unnatural or a threat, making it impossible for them to give the support that was hoped for. For Huntington's disease, the frustration of trying to persuade family members to cooperate in providing blood samples or in facing up to their own likely diagnosis is something that may result in complete breakdown of communications within a family – as seems to have been the case for a while in Julia Madigan's graphic description. The advantage of newer specific genetic testing approaches that no longer make it essential to sample the whole family is strikingly illustrated here; the different family attitudes to information and testing may remain frustrating, but at least they no longer need to prevent the member who does wish to know her risk from attaining this.

It is easy to think that genetic diseases only begin to have a serious effect on a person's life when a particular stage is reached, such as decisions on pregnancy or approaching potential age at onset. Some of the examples given here show how far from true this is; apparently trivial events in adolescence or youth, such as family visits or reading articles in a popular magazine, are shown as raising concerns long before any question of genetic risk had been formally reached; these events are recalled with clarity decades later.

What do individual descriptions of a person's experiences tell us about the diseases themselves? All the examples chosen here are of serious conditions, but the need for both accuracy and caution in categorising a disorder is clearly shown by the accounts of sickle cell disorders. Merry France-Dawson's initial misdiagnosis of HbSS rather than the milder HbSC sickle cell disease obviously had a major effect on her own ambitions and expectations; without a determined personality and supportive family, the effects might have been irreversible.

The spinal muscular atrophy descriptions carry a poignancy that I have encountered personally in a number of families with this group of disorders. The sentence 'these babies . . . seem to have an uncanny knack of looking into your eyes and reading your very soul' will strike a chord not only with families, but with many professionals. I have often wondered why this is particularly so for this condition, and have no answer. Perhaps this is one reason why it may be difficult to remain objective and dispassionate when involved with such a family; possibly also this may account for the disastrous reactions of some of the professionals encountered by these families.

After working with Huntington's disease families for over 20 years, I find that the nuances portrayed in Julia Madigan's account ring true, and will certainly echo the experiences of many other families with the disorder. The variability and unpredictability of the condition creates so many problems for the family; one affected member approaches life in a calm and responsible way, concerned for and prepared to help others, while another is unable to confront the diagnosis until quite a late stage. In other families, including the first description given here, these problems seem to have been less evident – or perhaps they were too painful to record. In any case, these accounts show how it is not only the nature of the disorder, its age at onset, severity, its physical and mental symptoms, that influence genetic decisions of a family, but also how much variation there can be within a single disorder, owing to the differing personalities and attitudes of the individual concerned.

Of the three people at risk for Huntington's disease who have written their stories, one received an adverse result from testing, one proved not to have inherited the disorder, and the third has decided (at least so far) against testing. It is satisfying, and speaks well for the way in which the testing service has been delivered, that all three seem to be relatively content with their outcome. One can sense the grief behind the high-risk outcome, contrasting with the intensity of relief following the normal result – I am sure that no one given a high-risk result would have been able to write in such a way – but at the same time, the unexpected disadvantages of a normal result are well portrayed. No longer can all the problems and injustices of life be blamed on one's inheritance; they must be faced up to in the same way that others have to face them. It is interesting that the most dispassionate account of Huntington's disease testing comes from the person who has chosen not to be tested and who has had children knowing of the genetic risks involved. It is good to be reminded that a significant proportion of people, having weighed up the various factors and considered their own situation, make a deliberate decision not to use the genetics services. It is vital that people in this position do not feel pressurised in any way; for some, it is important simply to know that there is an option available that they may chose to use at some point in the future.

The greatest value that these and other personal descriptions may turn out to have is the insight they give into how professionals have (or have not) tried to provide services of different kinds. They make uncomfortable reading in many respects, and so they should, for in some instances, families have been on the receiving end of attitudes and advice that have been deeply harmful. Specialist geneticists come out of this quite well, but here I suspect one is seeing a biased sample and should not be complacent. At all events it is worth looking closely at some of the points where professionals have had a profound effect, for better or for worse, on the lives of the families.

For some of the families, especially those with recessively inherited childhood disorders such as spinal muscular atrophy or sickle cell disease, the initial contact with medical professionals came when their child was diagnosed as having a serious or fatal disease. The overwhelming impact of this is clear from these accounts, and the impression given is that this difficult situation was handled sensitively and with support. The same seems largely to have been true for the later onset disorders; it was encouraging to read of the considerate way in which the diagnosis of Huntington's disease was approached by the neurologist in Julia Madigan's story.

These positive experiences make the contrast all the greater when we read how these clinicians – in some instances the same individuals – handled the genetic aspects in such an inadequate, sometimes disastrous, manner. Importantly, it does not seem that these doctors got their facts wrong, but rather the way in which they gave the information. The spinal muscular atrophy families again provide striking examples. One paediatrician was perceived as having 'explained about the problem of the one-in-four risk but put it to us in such a way that it appeared almost insignificant . . . she told us to go home and try again.' Was the timing of the information wrong, coming along with a devastating diagnosis? Was there insufficient time given to allow the parents to voice their own concerns and fears? Was the paediatrician trying to inject a ray of hope into an interview dominated by bad news? We cannot tell, but the consequence of affected twins coming soon afterwards would have shattered the lives of most couples. The second spinal muscular atrophy family seems to have had equal misfortune in this respect. Their sympathetic and supportive paediatrician told them that 'it was probably not worthwhile having genetic counselling as they would not be able to tell us any more than he had'; their family doctor told them not to have more children, and 'even thumped his fist on the table as he said so'! They had no success with AID and were turned down for adoption – perhaps they would have had an advocate to help them in this series of encounters had they been seen for genetic counselling?

It is quite clear from these accounts that many clinicians find it very difficult to relay important information on genetic risks and their consequences in a meaningful way, even when the situation is factually simple. For some, this is a general problem of communication, as shown by the ultimate dereliction of duty shown by the GP who avoided the family because he could not bear to see the children dying from spinal muscular atrophy. More often it seems that they are just not experienced in handling situations of genetic risk, or that this aspect is regarded as subsidiary to care of the affected child.

As a clinical geneticist, one's first reaction is that these families should have been seen in a specialist genetics clinic, but in reality, the situation will only improve when medical genetics and genetic counselling form an important part of medical education for all medical professionals. I very much doubt whether any of the clinicians involved in these examples had the slightest idea of the distress and harm that they helped to create.

The families with adult onset diseases described here seem to have generally had happier encounters with those providing genetic information – once they were able to get it. For Huntington's disease, this perhaps reflects the fact that it is generally known to be genetic, and that genetics specialists have always seen it as part of their regular practice long before predictive tests were possible. Most neurologists and other clinicians are only too happy for a clinical geneticist to handle the difficult problems of seeing members of the extended family and weighing up decisions on testing; indeed one is often expected to make the definitive diagnosis and undertake long-term management as well!

I am sure that families could have been found where geneticists had been unsympathetic and unsupportive – or at the least incomprehensible – but having encountered many unsolicited reports of families' experience in lay society newsletters – often alarmingly frank and not at all anonymous – I think that, on the whole, genetics professionals are providing a valuable and valued service, albeit only for a fraction of those who need it. It is worth considering why the experiences of these and other families with a specialist genetics service may be more helpful than when a primary clinician handles the situation unaided; these personal accounts contain some clues.

The most important factor may be availability of time; the experiences of the Huntington's disease and breast cancer families who were seen by clinical geneticists or genetics nurse specialists were clearly very time-consuming and were often repeated over a period of years, whereas for the primary clinicians the imparting of genetic information seems to have been an isolated and brief event. It is difficult to see how the difficult and complex issues facing these people could have been handled satisfactorily without this adequate time period; at present, medical genetics services are structured so as to permit this, but will this continue in the face of pressures of services to become more 'efficient' and 'competitive'? Many primary clinicians at present rightly refer families for specialist genetic counselling because they know they cannot themselves provide this vital time element.

A second factor that is clear from these histories is how important it is perceived to be for the person giving information to be objective and non-judgemental, and to give the range of options unencumbered by the personal views of the professional concerned. This is never easy, but geneticists do at least attempt to fulfil these aims – certainly a geneticist would be most unlikely to 'thump . . . his fist on the desk' while saying 'we shouldn't ever have any more children'! I suspect that many clinicians do not realise just how much people value a stance that is impartial yet supportive, and correspondingly how much they resent being told what to do or not to do.

Even apparently trivial factors can be important. The setting of the consultation often makes a huge difference to how people perceive information given to them, as seen from the description given of the gynaecology waiting area that lacked both privacy and dignity. Again, most genetic counselling clinics try to look welcoming and non-clinical.

Finally, it is worth returning to the family who were not referred because the paediatrician considered that 'they would not be able to tell us any more than he had'. The clinician had clearly not appreciated that what this and many families needed most was not to be told anything new, but rather to have an opportunity to ask their own questions in their own way, to go over difficult, often insoluble, problems in an unhurried manner, and to feel supported even if at present nothing new could be offered. My own experience has been that it is often those families for whom one has actually 'done' very little who are the most grateful, while paradoxically (but understandably) those for whom much is done in terms of tests and procedures, can feel overwhelmed by what is going on. I think that this has many implications for the use of genetic tests, where there is an increasing danger of people being given results that have serious consequences but without adequate opportunity to decide whether they wish for this.

Many non-specialist readers will understandably question why the positive features that I have here attributed to medical genetics services should be considered specific to genetics. Of course, they are not, and should not be; they should be universal in all fields of medicine and are indeed the foundation of all medical care and interactions. To a large extent they seem to have become lost with increasing technology, but people faced with difficult medical and genetic problems do not value them any the less today than formerly. As a medical geneticist myself, I like to feel that my specialty has retained, or perhaps relearned, some of these values that used to, and hopefully in future may again, permeate the whole of medicine. Listening to what people say about their own problems, and about us, is one of the most important ways by which we can ensure that we really do try to help them in the way they wish; the histories set out here, vivid, uncomfortable and often tragic, are a part of our learning how best to help families with genetic disorders. All the people who have written them, and their families, deserve our thanks for the courage and frankness with which they have laid open their lives to us. I hope that they may take encouragement from the knowledge that their decision to do this should lead to others being helped more than, in some cases, they have been themselves.

Part II
Clinical context

2

The new genetics: a user's guide

MARCUS PEMBREY

2.1 The impact of discoveries in medical genetics

There are some 4000 known, simply-inherited genetic disorders and, in aggregate, they are the cause of much suffering in 1–2% of the population. The common disorders include cystic fibrosis, sickle cell disease, the thalassaemias, fragile-X syndrome, Duchenne muscular dystrophy, haemophilia A, Huntington's disease, neurofibromatosis and adult polycystic kidney disease.

We 'fight' germs, but such language seems inappropriate for genetic disease because our genes, whether faulty or not, are an integral part of our makeup. Genetic testing can forewarn potential parents, but also pose moral dilemmas. Genetic knowledge can impose a burden of choice: a choice of whether to forgo children, trust to luck or seek prenatal diagnosis with the option of abortion if the baby is affected.

Although it is widely recognised that improvements in the treatment of genetic disease are desperately needed, some fear that 'tinkering' with our genes and the use of modified viruses as vehicles, or vectors, to deliver new genes to the body might be dangerous in some way to the population at large. Informed public debate needs an informed public, and that includes health professionals. Families facing these issues need help: genetic services, counselling and support in coming to a decision that is right for them.

What follows is intended as a simple guide to how genes work and sometimes fail. It explains how genetic tests are done and touches on some of the issues raised by advances in genetic testing.

2.2 Cells, proteins and genes

During embryological development, various tissues of the body come into being as the cells proliferate, differentiate and specialise, their structure and behaviour being adapted to their particular function. But what accounts for the different properties of specialised cells? Basically these differences stem from the different combination of proteins with which the cell is made and with which it carries out its work. **Cells are the basic building blocks of the body, and proteins the basic building blocks of the cell.**

But what determines which proteins are produced in what cells? This is determined

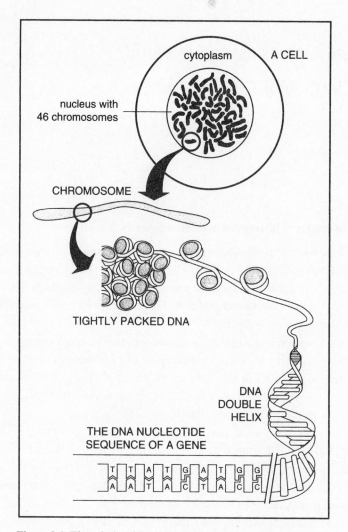

cytoplasm   A CELL

nucleus with
46 chromosomes

CHROMOSOME

TIGHTLY PACKED DNA

DNA
DOUBLE
HELIX

THE DNA NUCLEOTIDE
SEQUENCE OF A GENE

T T A T G A T G G
A A T A C T A C C

Figure 2.1. The relationship of DNA to chromosomes.

by the action of the genes. **Genes provide the instructions for the manufacture of proteins
by the cell**. With very few exceptions, each cell of the body has a full set of genes packaged into 46 tiny structures called chromosomes, the only genetic material that can be
seen directly down the microscope (Figure 2.1). Not all the genes are active in any one
cell, just those appropriate to that cell type and the functions it must perform. Our
genes not only guide our development from fertilised egg to fully grown adult, but go
on providing the information that is needed for everyday maintenance and functioning
of our bodies. For this reason an inherited genetic fault can cause errors of development that are manifest at birth or soon after, or cause a specific malfunction of the
body. A fault in a gene may disrupt the normal tissue repair mechanisms or the body's
system for coping with nutritional or other environmental stresses, leading to susceptibility to a specific disease later in life.

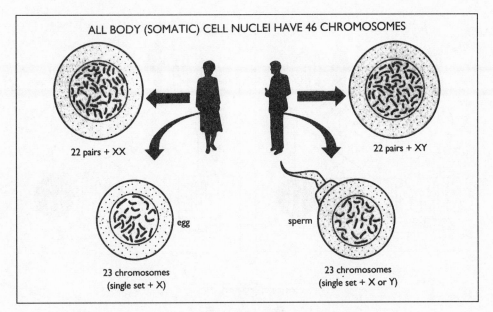

Figure 2.2. Human chromosomes in the body cells, eggs and sperm.

2.3 Inheritance

Chromosomes, and therefore the genes they carry, come in pairs; one set of 23 from mother and one from father, making 46 in total (Figure 2.2). It is the chromosomes that carry the genetic information from one generation to the next. In the formation of sperm, only one chromosome from each pair is placed into the developing sperm, and likewise for the egg, so each has a single set of 23 chromosomes. This means that when the egg and sperm come together at fertilisation, the proper number of 46 is restored, ready for the baby's development. This does not always happen correctly, and where the egg or sperm provides both copies of chromosome 21 the fertilized egg will end up with three copies (trisomy 21) and the resulting child will have Down syndrome. About 95% of trisomy 21 is due to an error in egg formation, and the probability of Down syndrome increases with the mother's age.

Twenty-two of the chromosome pairs are the same in males and females, but the twenty-third pair is different. The chromosome pairs that are the same in both sexes, numbered 1 to 22, are called autosomes to distinguish them from the sex chromosome pair, X and Y. Females have two X chromosomes, whereas males have one X and one (smaller) Y chromosome. This last fact means that sperm are of two types with respect to their chromosomes: half will have an X chromosome and half a Y chromosome. If an X-carrying sperm fertilises the egg, the baby will be a girl, whereas a Y-carrying sperm makes a boy.

Understanding how the chromosomes behave during egg and sperm formation has allowed us to explain the simplest pattern of inheritance of a character, or disease. These patterns were originally observed by Gregor Mendel in his study of peas over a

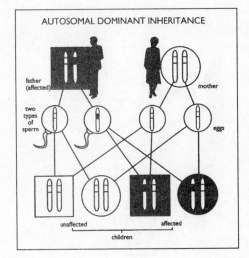

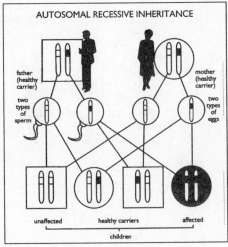

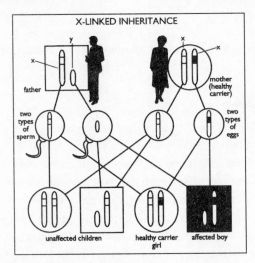

Figure 2.3. Mendelian patterns of inheritance.

century ago. There are three simple patterns of inheritance. The first two involve one of the pairs of genes on chromosomes 1 to 22 and are referred to as autosomal dominant and autosomal recessive inheritance. The third, sex-linked inheritance, involves one of the genes on the X chromosome and for this reason is more correctly called X-linked inheritance.

These three 'Mendelian' patterns of inheritance are illustrated in Figure 2.3 by way of simple chromosome diagrams depicting a single chromosome pair and just one gene upon it. The top left diagram shows that in autosomal dominant inheritance, as occurs in Huntington's disease or neurofibromatosis, there is inheritance from just one parent. An affected individual has a 50:50 chance of passing it on at each conception, no matter whom he or she marries. Autosomal recessive inheritance, as occurs in cystic fibrosis or thalassaemia, is inheritance from both parents (top right). Only if both pass on a faulty gene can the child be affected. Most often both parents are healthy carriers and therefore unlikely to be aware that they face a one in four, or 25%, chance of an affected child with *each and every* pregnancy.

Recognition that many genes, important for a variety of tissue functions, are carried on the X chromosome, whereas the Y chromosome carries only the gene for maleness and a few concerned with growth and sperm formation, has allowed us to explain X-linked inheritance as occurs in Duchenne muscular dystrophy or haemophilia (bottom diagram). In this type of inheritance only boys are affected but the disease can be passed on by unsuspecting female carriers. The female carrier is usually unaffected (or just mildly so) because, unlike the affected male, she has a second X chromosome which carries a working copy of the gene that can compensate for the malfunction of the faulty gene.

Statistical probabilities are all very well when being reassured by comfortingly low odds, but it is not much help to those facing a high chance of an affected child. Here people often want to know whether they are or they are not a carrier, whether the developing baby is affected or not. All else is just agonising uncertainty. It has only been because of the revolutionary advances of molecular genetics, and in particular the mapping of genes to precise locations on the chromosomes, that widespread carrier testing and early prenatal diagnosis on the baby have become possible.

2.4 Genes: what they are and how they work

Chromosomes and genes

Genes are sections of the enormously long double-stranded molecule, DNA (deoxy-ribonucleic acid), that is the major component of each chromosome. It may be helpful to use the analogy depicted in Figures 2.4 and 2.5. If one imagines that the chromosome is an audio cassette, then the DNA molecule is the tape inside. DNA is indeed the important part containing the information that is tightly coiled and packaged in an organised way within the overall chromosome structure. A gene is not separate from the DNA molecule but part of it, in the same way that a recorded song is an integral part of the tape. A recording is not the song proper, just the electronic information, and likewise the gene is not the protein product but the information necessary for its assembly from the pool of amino acids in the cell. An audio recording encodes a song and a gene encodes a protein. As Figure 2.1 illustrates, the DNA double helix, and therefore any gene, is made up of two strings of building blocks called **nucleotides** that are of just four varieties: adenine (A), thymine (T), cytosine (C) and guanine (G). It is the linear sequence of nucleotides of the gene, . . . AAGTGGCTTT . . . etc., that contain the 'instructions' for the manufacture of the particular protein that the gene encodes. The genetic code is read three nucleotides at a time, each triplet or codon encoding one of the 20 amino acids that are the building blocks of protein. For example, CCT codes for the amino acid proline, and GAG for the amino acid glutamic acid.

The genes, several thousands of them for most chromosomes, are spaced along the length of the DNA molecule, rather like recordings are spaced along a tape. There is a

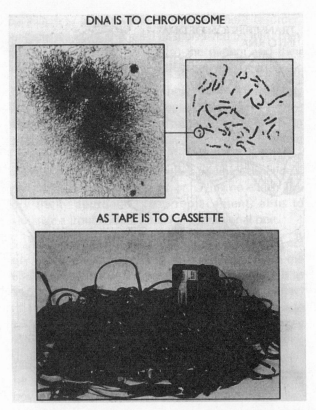

Figure 2.4. A very long DNA molecule is tightly coiled into the chromosome.

difference, however. Genes are separated by long tracts of DNA that just consist of repetitive nucleotide sequences, sometimes called 'junk' DNA because it does not encode any proteins or has, as yet, no known function. Furthermore, the coding parts of genes proper (exons) are also split into bits by non-coding intervening sequences or introns, rather as if the verses of a recorded song were separated by recordings of random noise. Genes have to be read, of course, and the molecule, whose job it is to read and transcribe the message that is written in the gene's DNA sequence, has to know when and where to start reading. For this reason the DNA a little way upstream of each gene has some special control sequences that are recognised by master controller molecules (transcription factors) that can sit on the DNA and control operations, switching the reading or transcription of the gene on and off.

The pairing rule

An important feature of DNA is that the nucleotides of the opposing strands follow a pairing rule. A always pairs with T, and C with G. This means that when the two strands of the DNA double helix separate during chromosome replication, a second (complementary) strand can be made alongside each single strand by following the pairing rule.

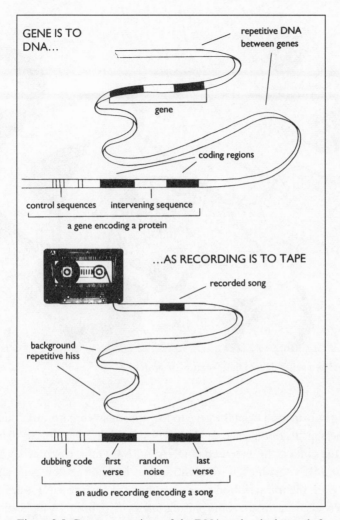

Figure 2.5. Genes are sections of the DNA molecule that code for protein.

This DNA replication process is liable to errors, and despite the body's elaborate proof-reading system designed to spot these mistakes, errors do occur. These and errors produced in other ways, such as faulty exchange of DNA between the chromosomes of a pair during egg or sperm formation, are called mutations. Mutations are the root cause of genetic disease.

How the genetic message dictates the structure of proteins

Genes, as we have seen, are an integral part of the 23 pairs of chromosomes, and chromosomes spend their life inside the nucleus of the cell. The proteins, however, are manufactured at cellular assembly plants (ribosomes) outside the nucleus in the cytoplasm of the cell. To deal with this lack of direct contact, the gene uses a messenger molecule

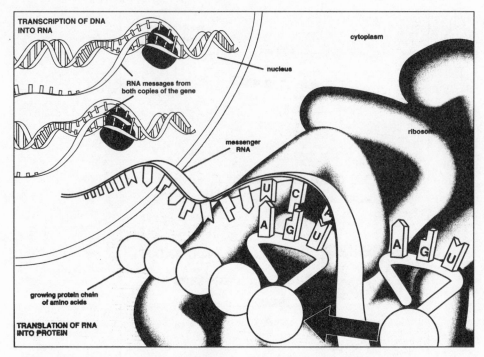

TRANSCRIPTION OF DNA INTO RNA

cytoplasm

nucleus

RNA messages from both copies of the gene

messenger RNA

ribosome

growing protein chain of amino acids

TRANSLATION OF RNA INTO PROTEIN

Figure 2.6. How genes dictate the structure of protein.

to carry the genetic information from the gene to the ribosomes (Figure 2.6). This molecule is called messenger ribonucleic acid (mRNA) and is made of a string of nucleotides much the same as the ones used in DNA: A, C, G and U (for uracil, which is similar to T). The mRNA, which is single-stranded, is made as the gene's DNA sequence is being read, the transcription process adding the correct nucleotides to the growing RNA molecule by using the same pairing rule as used for DNA replication. In this way there is a faithful transfer from gene to mRNA of the series of codes (codons) for amino acids that represent the information necessary to build the protein encoded by the gene. The ribosomes in the cytoplasm 'lock on' to the long ribbon-like mRNA molecules as they pass out of the nucleus and move along them, translating the nucleotide code into the appropriate chain of amino acids.

It is the *order* in which different amino acids are assembled into a chain that determines the primary nature of the resultant protein. This order is reflected in the sequence of nucleotide triplets of the mRNA. The ribosome 'presents' each nucleotide triplet or codon of the mRNA in turn, so that the amino acid corresponding to that codon will be added to the growing protein chain. As illustrated in Figure 2.6, the individual amino acids in the cytoplasm are guided to the ribosome by small transfer RNA molecules (tRNA) that have the key property of binding the correct amino acid for the codon they happen to have as part of their structure. Exploiting the same old pairing rule, each tRNA molecule presents its amino acid to the growing protein chain in accord with the successive codons of the mRNA. When translation is complete, the newly formed chain of amino acids breaks away from the ribosome and folds up into the mature protein. Thus, in summary, **DNA is transcribed into an RNA message and the RNA is then translated into the protein**.

The reader may well have already realised that there is an additional complication to this transfer of genetic information from gene (DNA) to mRNA to protein. As Figure 2.5 illustrates, the coding regions of genes or exons are split up by intervening sequences or introns that do not contain DNA sequences coding for amino acids. How are the coding bits of the gene brought together in a continuous, coherent message? First, the whole DNA sequence of the gene is transcribed from beginning to end into a primary RNA molecule, which includes the introns. Next, still within the cell's nucleus, the introns are spliced out and the coding regions joined end to end. This is done by a multi-molecular complex called a spliceosome, which recognises specific sequences in the RNA as signals to get splicing. Finally, the definitive messenger RNA leaves the nucleus. One way in which a mutation in a gene's DNA sequence can lead to trouble is by disrupting the signals needed to trigger the proper splicing and processing of RNA.

The gene, with its master copy of DNA-based information, sits in the nucleus firing off messages as and when more of its encoded protein is needed by the cell. As we go about our lives, eating, moving, thinking, growing, reproducing or fighting off infections, so our cells are responding to outside signals by activating some genes and silencing others, the combined activity of the genes determining the cells' metabolism and behaviour.

2.5   Mutations

Mutation is the name given to changes, or errors, in the DNA sequence of a gene such that the amount or structure of the protein gene product is altered. Returning to the audio tape analogy, a mutation is a flaw in the recording of a particular song. The recording of a song could be missing altogether, the bit of tape having been accidentally deleted during some editing process. This loss may not be too much of a problem, because one still has the recording, or gene, from the other parent. The resulting song is only half as loud as it should be, but at least it sounds correct. This situation is analogous to being a healthy carrier for an autosomal recessive condition such as beta-thalassaemia or cystic fibrosis. However, if the tapes inherited from both mother and father had the recording missing, there would be no song at all.

Although mutations often produce the effect just described, they can sometimes result in, not a missing protein product, but a troublesome one. Even this may not cause too much trouble in a single dose, as in the case of sickle haemoglobin, but often it overrides the effect of the normal gene. Imagine that one of the two recordings has the equivalent of several bars of music missing from the middle, or some wrong notes. When the tapes are played together the mutant song 'spoils' the normal one, and an awful, discordant sound is heard. Mutant genes like this mess things up even in a single dose and therefore cause autosomal dominant diseases, such as Huntington's disease or adult polycystic kidney disease.

Detection of the change in the DNA sequence that constitutes the mutation may not be a trivial matter even when the research has already identified which gene it is that goes wrong in that particular disease. Genes can be very small or small, e.g. the beta-globin gene (involved in the blood diseases, sickle cell disease and beta-thalassaemia) which is about 1600 nucleotide bases long. However, some genes can be enormous, e.g. the dystrophin gene (defective in Duchenne muscular dystrophy), which is over 2 000 000 nucleotides long. The ease with which a mutation is detected will depend on several things. First, how large a gene has to be scanned for any nucleotide change? Secondly, how obvious is that change? Is it a deletion removing thousands of nucleotide bases of coding sequence, or just a single altered base, for example changing the signal for RNA-splicing such that the definite mRNA is completely wrong? Thirdly, is the disease in question always, or nearly always, caused by exactly the same mutation (e.g. sickle cell), so one knows where to look and what DNA analysis to use to demonstrate the change? Table 2.1 summarises the gene(s) and usual type(s) of mutation that occur within them to cause some of the commoner genetic disorders.

2.6   Human genetic research and improved genetic diagnosis

There are 50 000 to 100 000 genes spaced along the 23 chromosomes that make up a single set as in an egg, and this is why it has taken a multi-million pound international research effort, called the Human Genome Project, to start mapping individual genes to particular locations on the chromosomes and defining their chemical structures. Knowing the location of a disease gene is the first step in developing a useful diagnostic test. The transmission of the section of chromosome that contains, among others, the disease gene of interest can be tracked through the family pedigree and the genetic status of family members predicted. This allows carrier testing or diagnosis of the disease state in the developing baby. Gene tracking has proved very helpful to families trying to maintain family life and reproductive confidence in the face of a known genetic risk. However, if the actual flaw in the gene that is causing all the trouble is to be defined and its effect on the body understood; if the reason why some flaws are dominant and override the influence of the healthy copy is to be determined; if overcoming the genetic malfunction by the gene therapy is to be contemplated in the future, then the disease gene has to be discovered, and its structure – the DNA sequence – worked out. To define the structure of a gene is to open the door to a new level of understanding of how it works and influences development, health and disease.

Genetic testing

There are several different ways in which the genetic status of a person can be established: DNA analysis, RNA analysis, biochemical/protein analysis or some form of presymptomatic detection of emerging pathology based on anatomical or functional changes.

Table 2.1a. *Diseases where the mutation is at the same site within the gene and therefore laboratory detection straightforward*

Disease	Inheritance	Chromosomal location	Gene name	Mutation	Comment
Fragile-X	X-linked	Xq27	FMR-1	Expansion of a CGG (non-coding) repeat	10–50 repeats normal 52–200 premutation 200–2000 full mutation Expansion causes FMR-1 gene to be silenced by methylation
Huntington's disease	Autosomal dominant	4p16	Huntington	Expansion of a (coding) CAG repeat	9–34 repeats normal 43–81 mutation
Myotonic dystrophy	Autosomal dominant	19q13	Myotonin protein kinase	Expansion of a (non-coding) CTG repeat	5–35 repeats.normal 50–80 'premutation' 80–200 full mutation
Sickle cell disease	Autosomal recessive	11p15	Beta-globin	Codon 6: GAG $\rightarrow$ T	Replacement of glumatic acid by valine enhances the aggregation of deoxygenated adult haemoglobin
Achondroplasia	Autosomal dominant (most new mutations)	4p16	Fibroblast growth factor receptor-3	Codon 380: CGG $\rightarrow$ A or C	Replacement of glycine by arginine

Table 2.1b. *Diseases where the mutations can be at one of a few or many locations within the gene, making laboratory detection complicated*

Disease	Inheritance	Chromosomal location	Gene name	Mutations	Comment
Cystic fibrosis (CF)	Autosomal recessive	7q31-2	Cystic fibrosis transmembrane conductance regulator (*CFTR*)	Various. Deletions of codon 508 common in North Europeans	Ethnic variation in proportion of CF chromosomes that are deletion 508
Tay–Sachs disease	Autosomal recessive	15q23-4	α-subunit of β-N-acetylhexoaminidase A	Various. A 4-nucleotide insertion in exon 11 is commonest of 3 mutations in Ashkenazi Jews.	Ethnic variation, e.g. 7.6 kilobase deletion commonest mutation in French-Canadians
Beta-Thalassaemia	Autosomal recessive	11p15	Beta-globin	Various	Ethnic variation. In Sardinia about 95% of beta-thal. chromosomes have codon 39 CAG to TAG mutation, but this degree of homogeneity is unusual
Duchenne muscular dystrophy (DMD)	X-linked	Xp21	Dystrophin	About 65% are deletions of one or more exons	Mutations resulting in no dystrophin cause DMD. Mutations causing an altered dystrophin usually cause milder Becker muscular dystrophy
Familial polyposis coli	Autosomal dominant	5q22-3	*APC*	Various	
Neurofibromatosis	Autosomal dominant	17q11	NF – GAP-related protein	Various	
Some familial breast cancers	'Autosomal dominant' (reduced penetrance)	17q12-21	*BRCA1*	Various	

Mutation detection

Mutation detection is the most direct way to analyse the DNA sequence of the gene in question and determine whether there is a harmful mutation present. This is fine when the expected mutation is discovered. However, a negative result can be difficult to interpret, often leaving disturbing uncertainty. Does it mean that the clinical diagnosis of the disorder is wrong – it isn't cystic fibrosis, or familial polyposis coli, after all? Or is the clinical diagnosis correct but, in this particular family, the disease is caused by an unusual mutation in the gene not covered by the standard mutation detection test?

Gene tracking

This is the term given to a DNA-analysis based test that predicts the genetic status of a member of an affected family by asking the question, 'Did they inherit the same (relevant) chromosome region as a previously affected family member?' There are variations of this question, but they all need a family linkage study using DNA markers that are known to co-inherit with the disease gene in question (in other words, are closely linked on the same chromosome). Gene tracking can be used once the disease gene has been mapped, but before the gene itself has been cloned and mutations detected. Despite the elegant, definitive nature of mutation detection, where a disease can be due to one of many different mutations in a gene (allelic heterogeneity), gene tracking is often the method of genetic prediction still used in practice. A comparison of the advantages and disadvantages of mutation detection and gene tracking are given in Table 2.2.

Types of genetic test

Recently some mutation detection tests have been introduced, for Duchenne muscular dystrophy for example, where the starting point for the analysis is the RNA produced by the gene. Although the dystrophin gene, as one might expect, is not really active in white blood cells, a tiny amount of RNA is produced (by so-called 'illegitimate transcription'), which is sufficient for analysis. The advantage over starting with RNA is that subtle DNA mutations that cause major abnormalities of RNA splicing show up more easily, and with all the coding regions of the gene end to end in the RNA it is easier to scan for mutations.

The great advantage of DNA analysis is that the DNA is present in all tissues, from the early preimplantation embryo or placenta to cells scraped from the inside of the mouth (buccal cells). However, a protein (or the RNA corresponding to it – illegitimate transcription notwithstanding) may only be present in certain differentiated cells that are not always accessible. However, sometimes a genetic test based on a specifically altered or absent protein is the most practical approach to genetic screening. Detection of abnormal haemoglobins in a blood sample, e.g. haemoglobin S in sickle cell disease or trait, is an example.

Finally, the genetic status of an individual can be inferred from a more indirect test.

Table 2.2. *Two approaches to genetic prediction by DNA analysis*

Mutation detection	Gene tracking
Advantages	
Definitive DNA diagnosis	Mutation independent
Population screening (autosomal recessive, X-linked) possible	Can be used as soon as disease locus mapped
Presymptomatic screening (autosomal dominant) possible	Exclusion of risk often achieved
Few, if any, family studies	
Disadvantages	
Allelic heterogeneity	Affected families only
Exclusion of risk may be difficult	Family studies needed
Causative mutation or not?	Uninformative markers
	Recombination errors
	Paternity?

Raised blood phenylalanine indicates a form of phenylketonuria (PKU) and is the basis of neonatal screening for PKU using the Guthrie card. Within an affected family, renal cysts detected by ultrasound scan can indicate the presence of the mutant gene for adult polycystic kidney disease long before any symptoms arise.

2.7   To know or not to know?

The advances in gene mapping and molecular genetics are leading to more and more situations where people can be offered a reliable genetic test, either on themselves or on their unborn baby. The technical aspects of the testing process are getting simpler year by year, but this does not mean the decisions that people have to make are any easier. *Testing must be linked to adequate counselling, exploring what they would feel and how they would cope if they tested positive.* The test may be to discover whether or not they are a carrier for cystic fibrosis. They would need to consider what their partner's reaction would be if they were shown to be a carrier. Would they tell their brothers and sisters about their carrier status, knowing that this gives them a 50% chance of being a carrier too?

Family ties can take on a new meaning in genetics and challenge our usual view of confidentiality. Where genetic matters are concerned, the information in some way belongs to the family as well as the individual. What right does one family member have to genetic results of another family member? What obligations are there on people who, through their own test result, discover that other family members are at risk of affected children? Where can they get advice on how to broach the subject, if they think the approach is going to be unwelcome?

Sometimes genetic testing means facing up to both risks for their own health in later life, as well as for any future children. They may themselves be at risk of a late onset degenerative brain disorder, such as Huntington's disease, because of an affected parent. Can they go on living with the uncertainty? Can they cope with knowing their own future?

Another area that is developing fast is our understanding of the genetic influences in some types of cancer. In particular it has become possible to test for mutations in one of the susceptibility genes (*BRCA1*) for breast and/or ovarian cancer. Already many women with a sister or mother, who had early onset breast cancer (say before the age of 45) are wondering if they should be having breast cancer screening mammography, even though they are younger than the cut-off age of 50 years used for the routine screening programme. Will they want to be tested for the susceptibility gene? What counselling support will be needed?

Testing the family

Family history is an important 'screening test' for many genetic diseases, especially those inherited in a dominant or X-linked fashion. Although it has to be remembered that many affected individuals are the first to be so in the family, with the genetic condition arriving 'out of the blue' as it were, there are also many people who secretly worry about a particular family illness or birth defect. There may be a family history of progressive muscle weakness, of mental retardation, or of bony abnormalities and so on. A couple may have previously given birth to a child with malformations of some kind. Their child may have died. In these situations, referral to a Clinical Genetics Centre can do a lot to sort things out and provide genetic counselling. Many families with a worrying family history or reproductive history are in urgent need of genetic counselling and although many of the disorders are individually rare, in aggregate they represent a very large number. Helping such families is the major activity of the specialised Clinical Genetics Centres.

Carrier screening in the community

When a condition is inherited in an autosomal recessive fashion, the majority of affected children are the first and only sufferer in the family. If couples unknowingly at risk are to be forewarned, carrier testing will need to be offered on a selected population basis. This approach is being developed for the commonest of the autosomal recessive diseases, some of which are highly specific to various ethnic communities. Carrier screening programmes, such as those for the thalassaemias and sickle cell disease, have been established in certain areas for some time. Others, such as those for cystic fibrosis carriers, are just being introduced. For organisational reasons the usual setting for the offer of carrier screening has been the antenatal clinic, although the ongoing pregnancy poses additional difficulties in achieving a careful, unhurried decision that is right for the couple. There is a move to try to offer more preconceptional screening at the primary healthcare level, e.g. through family doctor practices and family planning clinics.

Prenatal diagnosis

Diagnostic genetic tests on the fetus were developed in response to requests for help from couples facing a high risk of having a child with a severe genetic disorder. To be able to offer such tests is an integral part of clinical genetic services. The objectives of such tests are as follows:

(1)   to allow the widest possible range of informed choice to women and their partners at risk of having a child with a genetic disorder;
(2)   to provide reassurance and reduce the level of anxiety associated with reproduction;
(3)   to allow couples at risk to embark upon having a family, knowing that, if they wish, they may avoid the birth of seriously affected children through termination of an affected pregnancy;
(4)   to prepare a couple that wish to know in advance that their child is affected in order that they can continue the pregnancy prepared, and to ensure early treatment for the child.

As with all genetic services, counselling is of paramount importance in helping couples cope with the situation and come to a decision that is right for them. In making the decision of whether or not to terminate a pregnancy when their baby has been shown to be destined to suffer a genetic disease, the couple will be influenced by many factors. It is often assumed that if effective treatment becomes possible, then this will reduce the demand for prenatal diagnosis and selective abortion. Gene therapy and other specific treatments offer great hope for the future, but they are by no means the only consideration. A couple may accept that advances in treatment will occur, but fear that the health services they will need for their child will be denied them, on the grounds of cost, for example.

2.8   Gene therapy

Gene therapy is a new way of treating genetic disease that aims to tackle the root cause. If a gene is missing it aims to provide one. Gene therapy has got to the stage of clinical trials, because of decades of research into understanding how genes work and how the key bits of DNA can be 'cut out' of the total DNA extracted from blood or other samples and rearranged to order in the laboratory. Without the development of these so-called genetic engineering techniques, and without the use of experimental animals at certain stages, this form of help for patients would still be science fiction. It would be wrong to imply that there are no standard treatments, such as drugs, blood transfusions or organ transplants, that can help people suffering from a genetic disorder, but these are often of limited or short-lived benefit. They may represent a real burden to the patient and their family with repeated visits to hospital.

There are two types of gene therapy: tricky and very tricky! Which approach is adopted depends to a large extent on the type of mutation and whether it results in just an absence of a protein or a troublesome one (Figures 2.7 and 2.8). The least difficult approach, gene insertion or augmentation, involves adding a functional gene when that

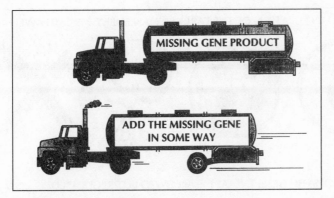

Figure 2.7. Gene insertion.

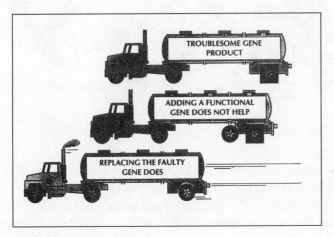

Figure 2.8. Gene replacement.

gene is missing or having no effect. The very tricky approach, gene replacement, aims at exchanging a troublesome gene for a functional one.

Ideally, gene therapy requires that the new DNA not only gets into the right cells, but replicates with the cellular genes so that it is passed onto progeny cells as the tissue grows or renews itself. Otherwise the effect will 'grow out' like peroxide blonde hair. The aim is to reach the self-renewing tissue stem cells while avoiding the germline cells – those leading to eggs or sperm (Figure 2.9). The simplest way to add new DNA to the cells is to copy what the sperm does and 'get in at the beginning'. Genetic modification of laboratory animals (often to create animal models of human genetic diseases in order to devise new treatments) is usually done by adding the DNA at the beginning of development. *However, this approach is, rightly, outlawed in humans because it would lead to genetic modification of the germline cells.* The fact that genetic manipulation of the early human embryo is forbidden does not 'close the door' as far as families facing genetic risks are concerned. It is now becoming possible, in conjunction with IVF

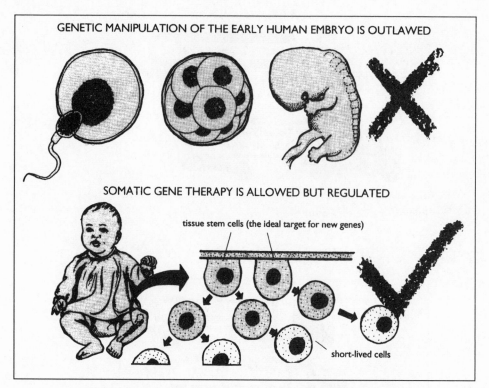

Figure 2.9. New genes must be put into the right cells at the right time.

('test-tube baby' techniques) to test which of several embryos have the genetic defect and only put the unaffected ones back in the mother's womb. Pre-implantation diagnosis is still at the trial stage and in its infancy, but it means the needs of families could be met without contemplating gene therapy in the embryo (sometimes referred to as germline therapy). In clinical practice the aim is to use *somatic gene therapy*, targeting only those tissues of the body where the genetic disease is disturbing function.

The method for gene delivery to cells may be a direct physical approach or exploit a virus that naturally enters cells. It depends, in part, on the tissue that needs treating. For blood diseases there is the opportunity to remove the bone marrow cells from the patient, use a disabled non-infectious virus to carry the new gene into the appropriate cells and then transfuse the corrected cells back into the patient. In treating the lung in cystic fibrosis, an inhaler might be used in conjunction with a 'membrane merging' method of gene delivery. The new DNA is placed in artificial or natural lipid vesicles that will fuse with the membranes of the cells lining the airways. It is always difficult to predict how quickly this type of research will be translated into effective routine therapy. It could take many years and therefore is difficult for families to take into account when making decisions here and now.

As with all experimental treatments, possible dangers have to be considered. It might not work, the correcting gene might go to the wrong type of cell or be expressed inappropriately, or the new gene might disrupt a normal gene when it inserts, causing a cancerous change in that cell. The main safeguard is adequate supervision and monitoring. In the UK, following the Clothier Report on the Ethics of Gene Therapy in 1992, the Government established a Gene Therapy Advisory Committee in November 1993.

2.9   Genetic services and education

There is a lot to be done to translate what is being discovered as a result of advances in molecular genetic research into services for families. In countries such as the UK, the existing network of Regional Clinical Genetics Centres provides a framework on which to build, but major developments in the genetic counselling services will be needed to meet future demand for help from families who through no fault of their own face a high genetic risk. Education, both of the health professionals and the public at large, is a prerequisite for such developments. Genetics raises a number of ethical issues that, although not different in kind from those found in other areas of medicine, are highlighted because of the special qualities of genetic information. Debate on how that information is used should be in the public arena.

2.10   Further reading

Weatherall, D. J. (1991) *The New Genetics and Clinical Practice*, 3rd edn. Oxford: Oxford University Press.

3

Decision-making in the context of genetic risk

SHOSHANA SHILOH

The significance of genetics would have remained purely in the scientific realm had it not begun to affect decisions made by and about individuals. The inseparability of clinical genetics from decision-making is illustrated in reviews of the literature on genetic counselling, which are devoted in large part to studies about the process and outcomes of counsellees' decisions (Evers-Kiebooms and Van den Berghe, 1979; Reif and Baitsch, 1985; Kessler, 1989). Decision-making was recognised as one of the key elements in genetic counselling (Fraser, 1974), and was further emphasised by Bringle and Antley's (1980) elaboration of genetic counselling into a model for counsellee decision-making. Kessler (1980) considered the increasing importance of decision-making in genetic counselling as part of its general shift away from eugenic values towards a psychological paradigm. The underlying logic behind these views is that individuals should be appropriately informed about their genetic risks and behavioural options to deal with those risks, and that they would use that information vigilantly to choose among the alternatives made available by newly developed genetic technologies. Before genetic counselling, most people do not even know about these options: they do not perceive the situation as a decision until explicitly told it is by the genetic counsellor (Beeson and Golbus, 1985).

Putting individuals in a situation of choice about genetic risks fits the broader trend of the evolution of behavioural mechanisms that serve to regulate chance events – a trend expressed most vividly in the growing importance of individual choice over reproduction (Miller, 1983). Increased human freedom to choose, which has a strong positive value in Western culture, also entails psychological costs such as elevated anxiety and guilt (Marteau, 1991; Lippman, 1992). The dilemma between increasing people's mastery over their reproductive lives and the incumbent psychological burden received expression in a recent statement on genetic counselling (Thompson and Rothenberg, 1992) recommending provision of another option: informed *refusal* of genetic counselling. Absurdly, this new option further augments individuals' responsibilities, and traps them in yet another decision: do I or do I not accept genetic counselling?

The rapid progress in clinical use of recombinant DNA technologies further heightens the significance of decision-making by presenting many more options: more tests for more diseases and more populations. These developments are not just quantitative.

The public has become accustomed to considering genetic risks when making reproductive decisions, like utilising prenatal diagnosis for severe genetic conditions. Now,
totally unprepared, they may be flooded with new and qualitatively different options
for which no social guidelines exist: the possibility of prenatal diagnosis of a wide
range of less severe conditions, presymptomatic testing for late onset genetic diseases,
and carrier screening for populations at risk for chronic diseases (Rowley *et al.*, 1984;
Roberts, 1990; Decruyenaere *et al.*, 1993).

This chapter will begin by describing and characterising the types of decision involving genetic risks. Then, questions about who decides and how decisions are reached will
be discussed. Next, genetic-related decisions will be examined from the perspective of
decision-making theory, and implications for genetic counselling and future research
will be considered. Throughout the chapter, links will be explored between genetic-
related decisions and the literature about decision-making in general, and risky decisions involving health consequences in particular.

3.1   What is decided?

Decisions involving genetic risks revolve mainly around family planning and reproduction. The key issues are: whether to take a chance and marry and/or bear a child in view
of increased genetic risks; whether to use prenatal diagnosis (if so, which procedure –
chorionic villus sampling? amniocentesis?); whether to continue a pregnancy when test
results are positive or inconclusive. The introduction of DNA analysis that enables
carrier screening and presymptomatic diagnosis has brought other questions: Should I
do the tests? At what age? What protective or preparative measures should I take in the
event of an increased risk?

Although most of our knowledge about genetic-related decisions comes from
studies about reproductive decisions, choices among the new non-reproductive options
have become the focus of a few recent studies, which will also be reviewed. We shall find
that despite diversity in content, genetic-related decisions share a number of features
that lend themselves to generalisation: severity; relevance to consequences significant
to oneself, to family members and even to future generations; irreversibility; and relationship to deep-seated values about the most basic and important matters of life.

Genetic-related decisions also share certain structural and contextual features. Most
entail a great deal of uncertainty, being based on probabilistic outcomes. Options are
risky in the sense that any choice involves negative outcomes and results in an avoidance–avoidance conflict. Some decisions are of the 'risk versus risk' type in which one
decides between two risks, such as between two prenatal diagnosis procedures. Other
decisions belong to the even more agonising category (Lave, 1987) of 'risk versus cost',
like choosing between a risky pregnancy and not having children. Outcomes are frequently remote and multi-dimensional, involving consequences on different levels (e.g.
health, emotional well-being, financial losses) for different individuals. They are often
linked sequentially, like deciding to diagnose and then deciding what to do with the
results. Decisions usually have to be made by at least two individuals, and sometimes

within a limited time, like the post-amniocentesis decision. Regardless of the content, these structural attributes bring high decisional conflict (Janis and Mann, 1977).

Numerous surveys have addressed the question of *what* people decide when faced with genetic risks. Such studies typically report the percentages of subjects deciding one way or another in a given situation. Though interesting, they have limited general-isability and often yield inconsistent results. They often bear methodological weak-nesses such as inadequate sampling procedures, over-generalisation from self-selected groups, and the use of measures of questionable reliability and validity. More impor-tant, however, are the fundamental issues revealed by comparing contradictory find-ings between studies, or among different samples within a single study. What appear at first to be methodological weaknesses in fact involve some of the major findings in this field – findings uncovering factors that influence decisions.

Time differences between studies can be meaningful. Gamberini *et al.* (1991) found that most (68%) of the families segregating for Cooley anaemia stopped reproducing after the birth of an affected child, in contrast to findings obtained a decade ago (46%). The authors attribute this change to therapeutic improvements that extend life expectancy of patients, and reduce parents' need for reproductive compensation. Changes over time were also reported in a longitudinal study on families with a cystic fibrosis child (Evers-Kiebooms *et al.*, 1990): one third of the parents held a different opinion about reproduction and use of prenatal diagnosis in 1984 compared with 1987, when DNA analysis made the option accessible. The authors attributed this discrep-ancy to the difference between the cognitive level, which fairly well accepts pregnancy interruption for genetic reasons when it is still a hypothetical question, and the per-sonal–emotional level when it comes to carrying out the decision. Changes observed over time in people's decisions about prenatal diagnosis and abortions may also reflect changes in social norms regarding what is an acceptable choice.

Discrepancies in the findings reported by different studies may also stem from how the decision is defined, and from how the choices are presented to the subjects. Decisions may be implied by intentions, or by actual choices and behaviour. In some cases choices are hypothetical, in others actual. The phrasing of the questions can vary considerably and yield different results: 'What do you plan to do . . . ?' 'What in your opinion is the best choice . . . ?' 'What would you choose if . . . ?' 'Would you change your decision if . . . ?' For example, in a study on prenatal diagnosis for adult polycystic disease (Hodgkinson *et al.*, 1990), 75% of 190 subjects from affected families felt that a prenatal test should be available; only 23% at high risk of passing on the disease and contemplating children felt they would be interested in such a test; and only one requested prenatal diagnosis. Many centres providing predictive testing for Huntington's disease report that the proportion of people asking to enter the pro-gramme is much smaller than expected on the basis of previously expressed intentions to undergo the test (Quaid *et al.*, 1989). When Evers-Kiebooms (1990) asked at-risk people if they would take the predictive DNA test when it became available, 66% said definitely or probably yes, but one third of these respondents did not plan to make use of it immediately when it became available. Adam *et al.* (1993) reported that among pregnant individuals at risk for Huntington's disease, only 30% requested prenatal testing, some of them withdrew before performing the test, and only 18% actually performed it; this was much lower than the 32–65% expected based on early intention surveys (Markel *et al.*, 1987; Meissen and Berchek, 1987; Kessler *et al.*, 1987; Mastromauro *et al.*, 1987).

Such discrepant findings cannot be regarded as inconsistencies, but rather as expressions of both the links and the true differences between attitudes, intentions, decisions and behaviour (Fishbein and Ajzen, 1975; Kendzierski, 1990). Kessler (1980) maintained that decision-making after genetic counselling occurs on multiple, simultaneous and sometimes unconscious levels, and results in what appears to be a paradox: deciding one thing and doing the opposite, such as the couple who state they have decided to have another child, but continue to use contraception successfully. The author suggested that such situations can be understood as attempts to deal with the highly conflicted needs and wishes of differing levels of cognitive–affective and interpersonal functioning, which could be satisfied by the 'paradoxical' strategy adopted.

Another source of contradictory findings is the misinterpretation of research findings by inferring subjects' decisions or motives from their choices in related, but non-identical, decisions. This is especially likely to happen with dynamic decisions that follow one another and are incorrectly perceived as one decision. For example, Schiliro *et al.* (1988) reported that 90% of patients with thalassaemia expressed interest in prenatal testing, and the authors erroneously interpreted this as indicating their intention to abort affected fetuses. Numerous studies have found large differences between intentions to test and to abort an affected fetus. Among mothers of children with haemophilia (Kraus and Brettler, 1988), 43% intended to use prenatal diagnosis for the disease, but only 17% would consider abortion of an affected fetus. Among subjects at risk for adrenoleukodystrophy (Costakos *et al.*, 1991), 86% would like to use prenatal diagnosis, 57% would abort an affected male, and 14% would abort a carrier female.

Decisions also vary according to who is asked. Notable differences were found in decisions about genetic risks reported by patients, their parents or spouses, individuals at risk, and different cultural groups. For example, 75% of partners of Huntington patients were in favour of prenatal diagnosis of the disease (Evers-Kiebooms *et al.*, 1991), compared with 29% of at-risk individuals (Bloch *et al.*, 1989). Differences of opinion were found (Beeson and Golbus, 1985) between husband and wife in 38% of couples at risk for children with X-linked conditions regarding prenatal diagnosis and abortion of affected fetuses, with husbands more opposed than the wives to risking the birth of an affected child. Differences in couples' attitudes and decisions about genetic risks were also reported in other studies (Sorenson and Wertz, 1986; d'Ydewalle and Evers-Kiebooms, 1987).

Intentions to use prenatal diagnosis by parents of affected children also depended on the kind of defect studied. Only 18% of parents of children with PKU expressed willingness to have prenatal diagnosis in subsequent pregnancies (Barwell and Pollitt, 1987), compared with 81% of women at increased risk for fragile-X (Meryash and Abuelo, 1988), and 82% of carriers of haemophilia (Lajos and Czeizel, 1987).

Choosing to bear another child by couples after the birth of a genetically handicapped child varied according to whether the affected child was first (46%) or not first (17%) in the family (Steele *et al.*, 1986). Among parents of children with cystic fibrosis, a study conducted in Wales found that 52% would abort an affected fetus (Al-Jader *et al.*, 1990), whereas only 28% of American parents would abort (Wertz *et al.*, 1992).

To summarise, there is no general answer to the question: 'What do people decide?' The diversity in the literature is not arbitrary. It indicates the existence of different perspectives about decisions that are held by the different parties involved. These perspectives are associated with a wide range of psychosocial factors related to the process of decision-making with genetic risks. Differences of perspective can cause communication problems among clients and professionals, and decisional conflicts among family members. Recognition of their magnitude and logic is basic to a meaningful counselling intervention.

3.2 Who decides?

One of the first steps in decision analysis is to identify the decider. In the context of genetic risks, this is not straightforward. Traditionally, genetic counselling has been non-directive, with the counsellor refraining from recommending which option should be chosen, and passing the responsibility for the decision to the counsellees. This approach, based on values of freedom and individual mastery, is not always welcomed. Lippman-Hand and Fraser (1979*a*) described frequent requests for guidance by counsellees. Similarly, Karp (1983) described the frequently asked 'terrible question': 'What would you do in my place?' Denying counsellees' request for advice may impede the counselling relationship, be interpreted as lack of care, and even become a bitter struggle (see a case report presented in Chapter 8 of Applebaum and Firestein (1983)). It can also be interpreted by counsellees as indicating the counsellor's judgements about the severity of the information provided (Shiloh and Saxe, 1989).

Antley (1979) differentiated between non-directive counselling, a psychological strategy to enhance counsellees' self-esteem, and terms more relevant to genetic counselling, like 'prescriptive' and 'non-prescriptive' approaches, which focus on the impact of the counsellor's goals on decisions. He claimed that genetic counselling may take the form of overt-prescriptive counselling, with the counsellor convincing the client to accept his or her evaluations and come to the same conclusions. It can also take the form of 'covert-prescriptive' approaches, where the counsellor consciously or unconsciously attempts to bring the counsellee to a particular decision by techniques intended to be unperceived, such as selective emphasis of facts, and enhancing and/or allaying certain fears and anxieties. Marteau *et al.* (1993) described such an approach in recordings of genetic counselling given by obstetricians.

It is questionable whether non-prescriptive counselling is really possible. Lippman and Wilfond (1992) wrote that 'no single story, however balanced, can ever be neutral or value free.' Some studies showed that different ways of presenting genetic risks (e.g. percentages–odds, words–numbers, positive–negative presentations) result in differing perceptions and choice of options by subjects (Kessler and Levine, 1987; Marteau, 1989; Shiloh and Sagi, 1989; Huys *et al.*, 1990). Others studies showed that counsellees' decisions depended on who provided genetic counselling. Robinson *et al.* (1989) found that 39% of couples who received information about intrauterine diagnosis of sex chromosome aneuploidy from a genetic counsellor, and 67% of those who received the information from their family doctor, decided to terminate the pregnancy.

The meaning of decisional control in genetic counselling and in counsellor's versus counsellee's responsibilities may need to be re-evaluated. Deci and Ryan (1985) argued that people may not always prefer to control what happens to them. Rather, they are motivated to maintain a sense of choice over what happens to them, and sometimes that self-determination includes the choice to relinquish control to another person in a specific situation. Miller (1980) put forth the 'minimax' hypothesis whereby individuals are motivated by a desire to minimise the maximum danger to themselves. Control, she states, is often preferred when an upper limit can be put on how bad the situation can become. But, when faced with a situation like a medical decision, a lay person may prefer to relinquish control over the decisional process to an identified expert whose decision is perceived to be a more reliable guarantee of minimising aversiveness than one's own. Such a preference cannot be considered loss of control, but rather self-determination (Fisher, 1986). Recent research findings support these notions. Individuals who prefer to have detailed information about their medical condition ('high monitors') do not seek this information for its instrumental value, but as a means to reduce uncertainty and gain access to experts who are in a position to give advice and solve the problem. They are more likely to opt for a passive role in their medical care, which cannot be regarded as loss of personal control (Miller *et al.*, 1988; Carver *et al.*, 1989).

The practical implications of these notions for genetic counselling are far-reaching. The passive role that many genetic counsellors prefer to play may have to change. Genetic counsellors may not be able to continue regarding themselves as purely 'information providers' (Hsia, 1979), first because neutral information provision is not possible, and secondly because clients need and desire more help. This is not to say that genetic counsellors should start advising which decision is best for their clients. Attempts to develop general guidelines for appropriate decisions about genetic risk has proved impossible (Fost, 1989). It is more reasonable to expect genetic counsellors to take the role of 'facilitator of counsellees' decision-making', a role that was explicitly detailed by Antley (1979). This requires that a genetic counsellor acquire expertise in decision-making theories and counselling techniques aimed at helping clients reach a decision wisely, rather than reach a wise decision.

3.3 Decisional process in the context of genetic risk

In addition to *what* is decided with regard to genetic risks, another area of research deals with *how* decisions are reached. Analyses of information processing leading to decisions involving genetic risks are still scarce. The only comprehensive study of this kind was published by Lippman-Hand and Fraser (1979*a,b,c*) more than a decade ago. These authors analysed transcripts of 30 tape-recorded genetic counselling sessions, and open-ended, semistructured, post-counselling interviews with 60 genetic counsellees. They suggested a model for post-counselling decision-making about reproduction that can be seen as parents' response to uncertainty. Uncertainty is experienced by parents as a global sense of 'being at risk'. In trying to solve the problem, parents focus on second-order judgements of probabilities and utilities that are not viewed independently. They tend to translate factual recurrence rate information into 'binary' views – it either will or will not happen. This process helps them focus on what the outcome will involve for them, and imagine the most important consequences. Most of the parents develop a scenario of 'trying out the worst', in which they explore ways of neutralising the perceived consequences and limiting their uncertainties. They use factual information to try to determine how they would manage with all the issues perceived as problematic, were chances against them. In doing so, they search for a least-loss alternative, one whose maximum loss would be acceptable. To the degree that the search is successful, a decision may be reached. In the absence of a least-loss option, a decision is less likely. Their decision cannot be described as a well-structured choice situation. The whole process is described by the authors as dynamic, not necessarily sequential or orderly, and not based solely on the facts of the situation as they might be defined by an outsider. Parents simplify the information available to them and use a heuristic information-processing model to deal with the complex cognitive task of resolving the problems created by being at risk. Most often an acceptable rather than an optimal solution is sought.

In contrast to this study, most research on how genetic-related decisions are made attempts to disclose factors influencing such decisions by examining the correlation between specific choices and relevant variables. Researchers noted that personal variables such as reproductive drive have a strong effect on post-genetic counselling reproductive decisions, regardless of the information on genetic risks provided in counselling (Côté, 1983). Sophisticated multivariate studies were designed to explain a larger part of decisional variance by covering a wider range of predictors, both personal and medical–genetic. Variables found to associate with the decision to use prenatal diagnosis included willingness to abort, siblings' approval of abortion, no perceived accomplishments for the affected child (Wertz *et al.*, 1992), perceived benefit of the test in reducing uncertainty, and perceived severity of the defect (Sagi *et al.*, 1992). Other factors most related to reproduction involving genetic risks were: having an affected child or any personal experience with the disorder, woman's age, counsellees' precounselling reproductive plans, desire to have children, and level of genetic risk (Sissine *et al.*, 1981; Sorenson *et al.*, 1987; Frets *et al.*, 1990). Frets *et al.* (1990) interviewed 164 couples two to three years after genetic counselling about their reproductive decisions, and constructed a model that enabled identification of the reproductive decision in 96% of the cases according to three factors: reproductive outcome before genetic counselling, desire to have children, and interpretation of information gained from genetic counselling.

New developments in DNA analysis and increased options for prenatal and presymptomatic diagnosis may bring about a shift in the significance of factors influencing reproductive planning after genetic counselling. Frets and Niermeijer (1990) compared the literature in the 1970s, when the magnitude of the genetic risk was one of the decisive factors, with current literature, in which the weight in post-counselling reproduction planning shifted to factors such as interpretation of the risk as high or low and the desire to have children. It is becoming increasingly evident that the impact of genetic risks on decisions is mediated by cognitive processes and coping alternatives (Shiloh and Saxe, 1989; Sagi *et al.*, 1992), and integrated with factors similar to those operating in the general population (Welshimer and Earp, 1989).

The influence of genetic counselling on counsellees' decisions has also been studied. In a review of the literature, Kessler (1989) concluded that there is little evidence that genetic counselling had made any difference in the final decisions made by counsellees. In fact, a few studies finding such influences indicated that genetic counselling was followed by a net *increase* in plans to have children. Still, Kessler concluded, genetic counselling plays a role in the process of counsellees' decision-making by confirming or reinforcing their decision, and increasing their confidence in pre-intended decisions.

Relating factors assumed to influence decision-making to the decisions themselves can be misleading, because these factors are not necessarily identical to the reasons that people give for their choices. The discrepancy between direct and indirect measures of utilities is well known in the literature on decision-making (Brookhouse *et al.*., 1986). Spangler (1989) established that utilities measured directly by a questionnaire represent values (consciously held beliefs influenced by situational and social factors), whereas utilities measured indirectly through statistical analysis of choices represent motives (relatively stable and not necessarily conscious constructs). He found that the two measures together may predict choice behaviour better than either measure alone. Findings about genetic-related decisions may be better understood in terms of these notions: counsellees' direct reports about reasons for their choices presumably express their values, whereas their motives may be better assessed indirectly from their actual choices.

The literature on genetic-related decisions abounds with evidence of the difficulty in implying counsellees' values from their decisions. When asked to give reasons for their decisions, genetic counsellees often give different reasons for deciding similarly. For example, Black (1979) reported different reasons for the reproduction plans expressed by parents of a Down syndrome child compared with parents of a child with retardation of unknown cause. 11% of the former and 70% of the latter parents gave recurrence risks as a reason against further reproduction. The reason most often given against additional children by Down syndrome parents was that they had reached their desired family size, whereas parents of other retarded children cited the burden of caring for a retarded child and financial strain. When patients in an antenatal clinic who wished to enter a programme of testing for cystic fibrosis carriers were asked why, the majority said that avoiding the birth of a child with CF was the reason, but almost as many reported that they wanted the test because they were interested to know their carrier status (Mennie *et al.*, 1992). Two main classes of reasons were given by subjects asked about their decision to take the predictive test for Huntington's disease: reduction of anxiety and uncertainty associated with being at risk, and enhanced planning and decision-making about one's future (Meissen *et al..*, 1991; Tibben *et al.*, 1993). Future research will have to examine the status of these reasons as motives or values, their origins, and the implications for seeking genetic information and choosing a behavioural option. Psychological constructs such as information-seeking styles (Miller, 1980), desire for control (Burger, 1992), and health locus of control (Wallston, 1989), which have been applied successfully to other health settings, may prove especially relevant in this new field.

3.4   Relationship to the literature on risky decisions

Decision-making involving genetic risks can be clarified by looking at it in the context of risky decisions in general. Although the ties between decision theory and genetic decision-making have occasionally been considered (see, for example, Vlek, 1987), there is far too little collaborative work and the literature on genetic decision-making has remained largely non-theoretic. This is unfortunate for both disciplines. Genetic counselling could benefit from anchoring its rationale and techniques on theoretical models rather than on descriptive findings; and decision-making research, often accused of a lack of ecological validity (Edwards, 1990), could benefit from examining predictions in an uniquely structured real-life setting in which deciders are provided with factual information about risks.

Let us consider first how 'risk' and 'risky decision' are defined in genetic counselling and in the field of risky decision-making. Yates and Stone (1992) analysed risk in a variety of circumstances, and defined risk-taking problems as special kinds of decision problems, which are 'problematic' mainly because the relevant options entail other considerations besides risk. Thus, the worth of an alternative is a function of the risk and of other considerations that have attractive benefits as well as possible negative features. Yates and Stone proposed that 'risk' is characterised by three critical elements that degrade an alternative's worth interactively: potential losses, the significance of those losses, and the uncertainty of those losses. They noted that many measures and operational definitions of risk focus on only one of the risk elements, and that risk is not an objective feature of a decision alternative itself, but rather represents an interaction between the alternative and the risk taker. Risk is an inherently subjective construct: what is considered a loss is peculiar to the person concerned, as are the significance of that loss and its chance of occurring.

It is quite safe to say that all of Yates and Stone's notions, based mostly on experimental data, can be supported by findings in the literature on genetic risks. Cases in point are the fact that genetic risk is perceived by counsellees as a global concept, sometimes interpreted as severity (Lippman-Hand and Fraser, 1979*b*); as one among other considerations that complicate the decision (see, for example, Welshimer and Earp, 1989; Frets *et al.*, 1990); as subjective by nature, and relevant and used in decision-making only in so far as it is subjective (Shiloh and Saxe, 1989; Sagi *et al.*, 1992).

Most important for genetic counselling is the emphasis on the multidimensionality of the 'risk' concept. Some observers (Kaplan and Garrick, 1981, p. 25) claim that 'a single number is not a big enough concept to communicate risk'. This applies even to attempts to develop an 'overall risk' concept integrating the significance and uncertainty of losses into a general index. Such an index is limited, because (1) it uses probabilities for specific outcomes as representations of uncertainty, whereas uncertainty is a wider concept; and (2) it ignores the multi-dimensional nature of the significance of the negative outcome (Yates and Stone, 1992). Nevertheless, in the literature on genetic counselling, not only is the term 'risk' used as a single number, it is usually used in a most restricted definition: the probability of occurrence of a negative genetic outcome. Genetic counsellees, on the other hand, attach a more global and personal meaning to 'risk' (Lippman-Hand and Fraser, 1979*b*; Shiloh, 1994). This corresponds to the general tendency of professionals dealing with risks to focus on probabilities, in contrast to lay people whose meaning of 'risk' is more closely connected to the severity of the outcome (Teigen, 1988).

Einhorn and Hogarth (1986) noted that the real world of risk involves factors such as ambiguous probabilities, dependencies between probabilities and utilities, context and framing effects, and regret. Differences in cognitive frames attached to words were demonstrated as major obstacles to effective communication between doctors and patients (Evans *et al.*, 1986). The ambiguity of the word 'risk', and the restricted way in which it is usually used by genetic counsellors in contrast to counsellees, might be a major impediment to clear communication in genetic counselling. The process could benefit from more neutral terms such as 'chances' when referring to probabilities, and from clarifying the meaning of 'risk' with clients.

In addition to the general difficulties in decision-making under uncertainty, decisions involving risks are even more difficult. People often make poor choices in risky situations (Slovic *et al.*, 1976), because there is no such thing as an acceptable risk, and risk should always be rejected (Slovic, 1987). Fischhoff *et al.* (1981) identified the reasons that risk problems are so difficult as being: uncertainty about the definition of the problem, difficulty in evaluating facts, difficulty in evaluating values, uncertainty about the human factor (biased perceptions, motivations, prior experience, etc.), and difficulty in evaluating the quality of a decision. It seems that decisions about genetic risks involve all these complications and more. In the study by Frets *et al.* (1991) of reproductive decisions of 164 couples two to three years after genetic counselling, several factors were found to be independently and significantly associated with problems in the decision-making process: no post-counselling relief, anticipation of a high risk level, relatives' disapproval of the decision, a decision not to have children, and the presence of an affected child. These findings illustrate the complex interplay of cognitive, emotional and social elements that must be considered with regard to the burden of making decisions in the presence of genetic risks.

In risky decisions, like any decision, people would like to make the 'right' decision, or at least the 'optimal' one. Decades of research on this issue have yielded a normative, albeit controversial, theoretical framework that provides practical methods for analysing and making decisions. An 'optimal' decision, according to the normative approach, maximises 'expected utility', which is defined as the total of the products of probability and utility for each combination of act and event. The principle of 'expected utility maximization' (von Neumann and Morgenstern, 1947), which describes how people should ideally formulate their preferences among alternatives, was generalised in the 1950s to include subjective measures of probabilities (Savage, 1954). In the subjective unexpected utility (SEU) model, both probabilities and utilities (or losses) are defined subjectively. The model is, in theory, as suitable for risky decisions as for other decisions (Neumann and Polister, 1992). The practical value of these formulations was adopted for clinical decision-making in medicine (Weinstein and Fineberg, 1980). Normative decision-making theory was also acknowledged as a possible theoretical framework for decisions regarding genetic risks (Antley, 1979; Pauker and Pauker, 1987; Pitz, 1987). Some suggestions were made to reframe the normative models for genetic decisions in terms of minimisation of 'expected burden' rather than the original maximisation formulation (Côté, 1983).

Despite the general acceptance of the normative models as 'rational', there is wide agreement that these models have a limited value in describing how people actually make decisions. Criticism was also voiced about the applicability and appropriateness of normative quantitative approaches to describe genetic-related decisions (Lippman-Hand and Fraser, 1979c; Kessler, 1980; Beeson and Golbus, 1985). The expected utility concept of rationality asserts that people should make decisions without violating probability theory's rules. However, evidence is accumulating that we have limited abilities to process information, especially in situations of uncertainty when one is required to estimate and interpret probabilities correctly. Some researchers have pointed out that individuals act particularly irrationally when the probabilities of specific outcomes are very low (Kunreuther *et al.*, 1978). Judgement errors impair people's choices in many ways (Kahneman *et al.*, 1982). The cognitive process by which genetic counsellees subjectively appraise their genetic risk after being given an 'objective' risk figure by their counsellor was recently found to be subject to the same general heuristics and biases typical of judgement under uncertainty (Shiloh, 1994). In that analysis, major cognitive heuristics like 'representativeness', 'availability' and 'anchoring' (Tversky and Kahneman, 1974) were shown to influence genetic counsellees' perceptions of risk and decision after counselling. Experiments show that people may incorrectly use information or not use all the relevant information (Fischoff *et al.*, 1978), a finding documented in clinical medical decision-making as well (Wortman, 1972; Elstein *et al.*, 1978).

A factor that almost always accompanies and complicates counsellees' decision-making about genetic risks is stress (Schild, 1984). Janis and Mann (1977), in their conflict theory, used studies of historical and political decisions, clinical research, and experiments in which stress was manipulated, to test the deleterious effects of high levels of stress on decision-making and risk-taking. The conflict model maintains that extremely low stress and extremely intense stress give rise to defective decision patterns, whereas moderate levels of stress are more adaptive and enhance vigilant decision-making patterns.

In high-stress situations, like those typical of genetic-related decisions, two sub-optimal patterns of decision-making would be predicted by the conflict theory. The first is 'defensive avoidance', triggered by high conflict and pessimism about finding a good solution to the dilemma. In this case people become motivated to reduce the distressing state of high emotional arousal by procrastination, shifting responsibility to someone else, and the invention of fanciful rationalisations in support of one of the choice alternatives. A second 'defective' decision-making pattern, 'hypervigilance', arises when the decision maker believes that there is insufficient time to search for and evaluate solutions, and is frantically preoccupied with the threatened losses that seem to loom larger every minute. This situation may occur after receiving a prenatal test result indicating the existence of an affected fetus, which necessitates a fast decision about pregnancy continuation (Blumberg, 1984). The behaviour of a person in such a state may be marked by very high vacillation and emotionality, an anxious search for a way out of the dilemma, impulsiveness, overlooking of the full implications of a choice, reduced memory span, and simplistic repetitive thinking that in the most extreme form is equivalent to 'panic'. The literature on genetic counsellees' post-counselling reactions contains descriptions of behaviours that can be interpreted as the above 'defensive-avoidance' and 'hypervigilant' patterns (Lippman-Hand and Fraser, 1979b; Antley, 1979; Beeson and Golbus, 1985; Frets et al., 1991). Pregnant women who have to decide after prenatal diagnosis may be even more at risk for such maladaptive behaviour, owing to the general altered cognitive functioning in pregnancy (Condon and Ball, 1989).

Decisional processes involving genetic risks seem to relate more closely to behavioural–descriptive, than to normative, models of decision making. Simon (1979) argued that decisions, instead of obeying expected utility theory, follow concepts of 'bounded rationality'. Individuals reduce problems to a modest number of variables and make decisions that appear 'reasonable' rather than 'optimal'. An individual's search for a solution may terminate when an aspiration level has been achieved. In this way, people 'satisfice' (i.e., terminate the search and choose the alternative that meets the level of aspiration) rather than optimise. This analysis fits perfectly with the observations made by Lippman-Hand and Fraser (1979c) about the decision process of genetic counsellees.

Researchers and genetic counsellors who want to generate and test hypotheses, or explain observations about decisions related to genetic risks based on theoretical grounds, may find a rich field from which to draw. The following are a few examples. Prospect theory (Kahneman and Tversky, 1979), one of the most elaborate models of decisions under risk, outlines a few phenomena highly relevant to genetic decision-making: the 'certainty effect' (people's tendency to give too much weight to outcomes that are considered certain, relative to outcomes that are probable); the 'reflection effect' (the reversal of the certainty effect in the negative domain, whereby preference is given to a loss that is merely probable over a smaller loss that is certain); 'probabilistic insurance' (the dramatic difference between the value of reducing risk to zero compared with reducing it to a remote chance); 'simplification' (the tendency to ignore some small but non-zero probabilities and treat them as though they were impossible); 'value function' (that in a given situation, individuals evaluate outcomes in relation to personal 'reference points' that define which outcome will be treated as a gain or a loss). Although no known study on genetic decision-making has yet been designed to test these rules directly, findings in quite a few (e.g., Lippman-Hand and Fraser, 1979*a,b,c*) support them indirectly.

Other relevant literature is that dealing with the relation between risk perceptions, risk-taking and perceived control. A robust finding shows the acceptability of risks that are perceived to be controlled (Slovic, 1987). Considerable research indicates that people underestimate their personal probability of encountering negative events (see, for example, Weinstein, 1980; Taylor and Brown, 1988), which can be interpreted either as 'unrealistic optimism' (Weinstein, 1980) or in terms of 'illusion of control' (Langer, 1975). Recent research supports the 'control hypothesis' as the main mechanism underlying unrealistic optimism (McKenna, 1993). Perceived control was recently found to play a major role as an outcome of genetic counselling (S. Shiloh, M. Berkenstadt, N. Miran, M. Bat-Miriam-Katznetson and B. Goldman, unpublished manuscript). It was also found that genetic clients, unlike subjects in studies on other risks, tend to overestimate rather than underestimate their risks, thus exhibiting 'unrealistic pessimism' (Sagi *et al.*, 1992). Is this related to perceived lack of control, to other characteristics of genetic risk, to individual differences among counsellees, or to self-enhancement motivations? Many questions regarding the influences of behavioural (e.g. through prenatal diagnosis) and other (including illusory) means of control over genetic outcomes on genetic risk perceptions and decisions await investigation.

Lopes's (1987) elaborate theory on hope and fear may provide yet another perspective for explaining genetic risk decision-making data. According to this theory, risk-taking differences derive from weightings of two competing motivations: a desire for security (avoiding bad outcomes) and a desire for high return (approaching good outcomes). The proportional weights of these motivations distinguish between those who are risk-seeking and those who are risk-avoiding. However, the theory introduces another factor – 'aspiration level' – that reflects the opportunities at hand as well as the constraints imposed by the environment. This situational factor may influence individuals' risk-taking behaviour to change their typical risk-seeking or risk-avoiding tendencies. Predictions derived from this theoretical framework may help explain and predict individual differences in decisions entailing genetic risks, while explicitly integrating counsellees' personal traits with qualities of the genetic problem they have to solve.

Finally, there is a growing literature focused on peoples' decisions and behaviour regarding health risks. The main emphasis in models developed for health behaviour is on understanding why individuals take (or fail to take) protective actions or put their health at risk. The most widely used model has been the health belief model (HBM) (Maiman and Becker, 1974), which specifies the perceived threat of the negative outcome if one fails to take the action, and the perceived benefits and barriers associated with taking the action, as the major predictors of the likelihood of engaging in a specific health-promoting action. Protection motivation theory (Rogers, 1983) includes elements similar to the HBM, but focuses on fear as a major motivating factor in health behaviour. Later modifications of both theories (Maddux and Rogers, 1983; Rosenstock *et al.*, 1988) incorporated perceived self-efficacy that one can perform the preventive behaviour (Bandura, 1977) as another exploratory variable. Self-regulation theory (Leventhal *et al.*, 1984) provides yet another perspective, that includes affective responses to health threats in addition to cognitive processes. According to this theory, a distinction is made between coping with the danger (the objective threat) and coping with the fear that the communication about it arouses. Separate and partly independent information-processing systems influence these two types of response. The relevance of these theories for genetic-risk research is evident. At present, only the HBM has been used in studies aimed at predicting parents' intentions to use prenatal diagnosis (Loader *et al.*, 1991; Sagi *et al.*, 1992).

3.5   Genetic counselling for decision-making?

Whereas some view the facilitation of counsellees' decision-making as the main role of the genetic counsellor (Antley, 1979), others claim that the genetic counsellor should confine his/her role to information provision (Hsia, 1979). Reality, however, compels genetic counsellors to try to help counsellees make their decisions without advising them what to decide. Willingly or not, most genetic counsellors become decision-making counsellors. The literature for genetic counsellors on decision aids and other techniques for enhancing genetic counsellees' decision-making is surprisingly small, although one does find counselling tips in the concluding sections of many of the studies reviewed here (for example, Frets *et al.*, 1990).

A few comprehensive and explicit attempts to suggest theoretically based frameworks for decision-making counselling are available. There is also evidence, from studies with medical professionals in well-defined clinical problems, that using decision aids improves decisions (Polister, 1981). Quantitative analytical techniques based on normative decision-making theory have been introduced into genetic counselling (Humphreys and Berkeley, 1979; Pauker and Pauker, 1987; d'Ydewalle and Evers-Kiebooms, 1987; Pitz, 1987; Vlek, 1987; Scholz, 1992). Pauker and Pauker's (1987) analytical technique designed to help with the decision to use amniocentesis for prenatal diagnosis includes the presentation of possible outcomes of performing the test, choices between therapeutic abortion and the various risks of having an affected child, measuring attitudes on a utility scale, and recommending a decision that would yield the highest expected utility based on normative decision rules.

Other authors present 'softer', more clinically oriented techniques, such as a therapeutic model (Van Spijker, 1992) of decision process therapy in genetic counselling based on principles from brief psychotherapy. This model includes clarification of information, discerning patterns of behaviour and emotional aspects, assessment of options, modelling, rational–emotive techniques and modifying personal balance. Other methods applied in genetic counselling were the 'balance sheet technique' drawn from Janis (1982), aimed at reducing stress and defective decisional patterns according to the conflict theory (Leonard and Beck-Black, 1984); and 'structured scenarios' techniques for enhancing counsellees' decision-making process (Arnold and Winsor, 1984; Huys *et al.*, 1992).

3.6   Conclusions

This chapter has shown that genetic-related decisions share features that can be addressed by the general field of decision-making. Focusing attention on the different perspectives that the involved parties hold on genetic decisions clarifies possible sources of conflict and flaws in communication. Re-evaluation of the underlying assumptions and practical implications of the non-directive approach in genetic counselling points to the necessity of developing new and creative counselling techniques more suited to clients' needs. Better understanding of the role of uncertainty and perceptions of risk in the genetic context has proved to be a major issue, with implications for the different ways in which genetic risks are presented and their impact on decisions. The interaction between information seeking, individual differences in ways of coping, control issues and genetic decisions point to the need to personalise genetic counselling according to the counsellees' personal styles. The relevance to genetic counselling of theories about risk, decision-making, rationality, emotionality, stress and conflict raise basic and applied research questions, and point to ways to adapt general counselling techniques to genetic counselling. Genetic counselling must be tied more closely to the research and practice of health psychology. Relating the specific needs of genetic counsellees to health psychology, decision theory and general counselling techniques will benefit all worlds, and lay the foundation for the modern training of genetic counsellors.

3.7   References

Adam, S., Wiggins, S., Whyte, P., Bloch, M., Shokeir, M. H. K., Soltan, H., Meschino, W., Summers, A., Suchowersky, O., Welch, J. P., Huggins, M., Theilmann, J. and Hayden, M. R. (1993) Five year study of prenatal testing for Huntington's disease: demand, attitudes, and psychological assessment. *Journal of Medical Genetics*, **30**, 549–56.

Al-Jader, L. N., Goodchild, M. C., Ryley, H. C. and Happer, P. S. (1990) Attitudes of parents of cystic fibrosis children towards neonatal screening and antenatal diagnosis. *Clinical Genetics*, **38**, 460–5.

Antley, R. M. (1979) The genetic counselor as facilitator of the counselee's decision process. In A. M. Capron, M. Lappé, R. F. Murray, T. M. Powledge, S. B. Twiss and D. Bergsma (eds.), *Genetic Counseling: Facts, Values and Norms*, pp. 137–68. New York: Alan R. Liss, Inc., for the National Foundation – March of Dimes. (Birth Defects: Original Articles Series, vol. 15.)

Applebaum, E. G. and Firestein, S. K. (1983) *A Genetic Counseling Casebook*. New York: Free Press.

Arnold, J. R. and Winsor, E. J. T. (1984) The use of structured scenarios in genetic counseling. *Clinical Genetics*, **25**, 485–90.

Bandura, A. (1977) Self-efficacy: toward a unifying theory of behavior change. *Psychological Review*, **84**, 191–215.

Barwell, B. E. and Pollitt, R. J. (1987) Attitudes des parents vis-à-vis du diagnostic prénatal de la phénylcétonurie. *Archives of French Pediatrics*, **44**, 665–6.

Beeson, D. and Golbus, M. S. (1985) Decision making: whether or not to have prenatal diagnosis and abortion for X-linked conditions. *American Journal of Medical Genetics*, **20**, 107–14.

Black, R. B. (1979) The effects of diagnostic uncertainty and available options on perceptions of risk. In C. J. Epstein, C. J. R. Curry, S. Packman, S. Sherman and B. D. Hall (eds.), *Risk, Communication, and Decision Making in Genetic Counseling*, pp. 341–54. Alan R. Liss, Inc., for the National Foundation – March of Dimes. (Birth Defects: Original Article Series, vol. 15.)

Bloch, M., Fahy, M., Fox, S. and Hayden, M. R. (1989) Predictive testing for Huntington disease. II. Demographic characteristics, life-style patterns, attitudes, and psychosocial assessments of the first fifty-one test candidates. *American Journal of Medical Genetics*, **32**, 217–24.

Blumberg, B. (1984) The emotional implications of prenatal diagnosis. In A. E. H. Emery and I. Pullen (eds.), *Psychological Aspects of Genetic Counseling*. London: Academic Press.

Bringle, R. G. and Antley, R. M. (1980) Elaboration of the definition of genetic counseling into a model for counselee decision-making. *Social Biology*, **27**, 304–18.

Brookhouse, K. J., Guion, R. M. and Doherty, M. E. (1986) Social desirability bias as one source of the discrepancy between subjective weights and regression weights. *Organizational Behavior and Human Decision Process*, **37**, 316–28.

Burger, J. M. 1992) *Desire for Control: Personality, Social, and Clinical Perspectives*. New York: Plenum Press.

Carver, C. S, Scheier, M. F. and Weintraub, J. K. (1989) Assessing coping strategies: a theoretically based approach. *Journal of Personality and Social Psychology*, **56**, 267–83.

Condon, J. T. and Ball, S. B. (1989) Altered psychological functioning in pregnant women: an empirical investigation. *Journal of Psychosomatic Obstetrics and Gynecology*, **10**, 211–20.

Costakos, D., Abramson, R. K., Edwards, J. G., Rizzo, W. B. and Best, R. G. (1991) Attitudes toward presymptomatic testing and prenatal diagnosis for adrenoleukodystrophy among affected families. *American Journal of Medical Genetics*, **41**, 295–300.

Côté, G. B. (1983) Reproductive drive and genetic counseling. *Clinical Genetics*, **23**, 359–62.

Deci, E. L. and Ryan, R. M. (1985) *Intrinsic Motivation and Self-Determination in Human Behavior*. New York: Plenum Press.

Decruyenaere, M., Evers-Kiebooms, G. and Van den Berghe, H. (1993) Perception of predictive testing for Huntington's disease by young women: preferring uncertainty to certainty? *Journal of Medical Genetics*, **30**, 557–61.

d'Ydewalle, G. and Evers-Kiebooms, G. (1987) Experiments of genetic risk perception and decision making: explorative studies. In G. Evers-Kiebooms, J. J. Cassiman, H. Van den Berghe and G. d'Ydewalle (eds.), *Genetic Risk, Risk Perception, and Decision Making*, pp. 209–26. New York: Alan R. Liss, Inc., for the National Foundation – March of Dimes. (Birth Defects: Original Articles Series, vol. 23.)

Edwards, W. (1990) Unfinished tasks: a research agenda for behavioral decision theory. In R. M. Hogarth (ed.), *Insights in Decision Making*. Chicago: University of Chicago Press.

Einhorn, H. J. and Hogarth, R. M. (1986) Decision making under ambiguity. *Journal of Business*, **59**, 225–50.

Elstein, A. S., Shulman, L. S. and Sprafka, S. A. (1978) *Medical Problem Solving; An Analysis of Clinical Reasoning*. Cambridge, MA: Harvard University Press.

Evans, D. A., Block, M. R., Steinberg, E. R. and Penrose, A. M. (1986) Frames and heuristics in doctor–patient discourse. *Social Science in Medicine*, **22**, 1027–34.

Evers-Kiebooms, G. (1990) Predictive testing for Huntington's disease in Belgium. *Journal of Psychosomatic Obstetrics and Gynecology*, **11**, 61–72.

Evers-Kiebooms, G. and Van den Berghe (1979) Impact of genetic counseling: a review of published follow-up studies. *Clinical Genetics*, **15**, 465–74.

Evers-Kiebooms, G., Denayer, K. and Van den Berghe, H. (1990) A child with cystic fibrosis. II. Subsequent family planning decisions, reproduction and use of prenatal diagnosis. *Clinical Genetics*, **37**, 207–15.

Evers-Kiebooms, G., Swerts, A. and Van den Berghe, H. (1991) Partners of Huntington patients: implications of the disease and opinions about predictive testing and prenatal diagnosis. *Genetic Counseling*, **39**, 151–9.

Fischhoff, B., Lichtenstein, S., Slovic, P., Derby, S. and Keeney, R. (1981) *Acceptable Risk*. New York: Cambridge University Press.

Fischhoff, B., Slovic, P. and Lichtenstein, S. (1978) Fault trees: sensitivity of estimated failure probabilities to problem representation. *Journal of Experimental Psychology: Human Perception and Performance*, **4**, 330–4.

Fishbein, M. and Ajzen, I. (1975) *Belief, Attitude, Intention and Behavior: An Introduction to Theory and Research*. Reading, MA: Addison-Wesley Publishing Co.

Fisher, S. (1986) *Stress and Strategy*. London: Lawrence Erlbaum.

Fost, N. (1989) Commentary: guiding principles for prenatal diagnosis. *Prenatal Diagnosis*, **9**, 335–7.

Fraser, F. C. (1974) Genetic counseling. *American Journal of Human Genetics*, **15**, 1–10.

Frets, P. G. and Niermeijer, M. F. (1990) Reproductive planning after genetic counseling: a perspective from the last decade. *Clinical Genetics*, **38**, 295–306.

Frets, P. G., Duivenvoorden, J. H., Verhage, F., Ketzer, E. and Niermeijer, M. F. (1990) Model identifying the reproductive decision after genetic counseling. *American Journal of Medical Genetics*, **35**, 503–9.

Frets, P. G., Duivenvoorden, H. J., Verhage, F., Peters-Romeyn, B. M. T. and Niermeijer, M. F. (1991) Analysis of problems in making the reproductive decision after genetic counseling. *Journal of Medical Genetics*, **28**, 194–200.

Gamberini, M. R., Canella, R., Lucci, M., Vullo, C. and Barrai, I. (1991) Reproductive behavior of thalassemic couples segregating for Cooley anemia. *American Journal of Medical Genetics*, **38**, 103–6.

Hodginson, K. A., Kerzin-Storrar, J., Watters, E. A. and Harris, R. (1990) Adult polycystic kidney disease: knowledge, experience and attitudes to prenatal diagnosis. *Journal of Medical Genetics*, **27**, 552–8.

Hsia, Y. E. (1979) The genetic counselor as information giver. In C. J. Epstein, C. J. R. Curry, S. Packman, S. Sherman and B. D. Hall (eds.), *Risk, Communication, and Decision Making in Genetic Counseling*, pp. 169–86. New York: Alan R. Liss, Inc., for the National Foundation – March of Dimes. (Birth Defects: Original Articles Series, vol. 15.)

Humphreys, P. and Berkeley, D. (1979) Representing risks: supporting genetic counseling. In C. J. Epstein, C. J. R. Curry, S. Packman, S. Sherman and B. D. Hall (eds.), *Risk, Communication, and Decision Making in Genetic Counseling*, pp. 227–50. New York: Alan R. Liss, Inc., for the National Foundation – March of Dimes. (Birth Defects: Original Articles Series, vol. 15.)

Huys, J., Evers-Kiebooms, G. and d'Ydewalle, G. (1990) Framing biases in genetic risk perception. In J. P. Caverni, J. M. Fabre and M. Gonzalez (eds.), *Cognitive Biases*. Amsterdam: Elsevier Science Publishers (North Holland).

Huys, J. Evers-Kiebooms, G. and d'Ydewalle, G. (1992) Decision making in the context of genetic risk: the use of scenarios. In: G. Evers-Kiebooms, J. P. Fryns, J. J. Cassiman, and H. Van den Berghe (eds.) *Psychological Aspects of Genetic Counseling*, pp. 17–20. New York: Wiley–Liss, Inc., for the National Foundation – March of Dimes. (Birth Defects: Original Articles Series, vol. 28.)

Janis, I. L. (1982) *Counseling on Personal Decisions: Theory and Research on Short-term Helping Relationships*. New Haven, CN: Yale University Press.

Janis, I. L. and Mann, L. (1977) *Decision-making: A Psychological Analysis of Conflict*. New York: Free Press.

Kaplan, S. and Garrick, B. J. (1981) On the quantitative definition of risk. *Risk Analysis*, **1**, 11–27.

Karp, L. E. (1983) The terrible question. *American Journal of Medical Genetics*, **23**, 359–62.

Kahneman, D., Slovic, P. and Tversky, A. (eds.) (1982) *Judgment Under Uncertainty: Heuristics and Biases*. Cambridge: Cambridge University Press.

Kahneman, D. and Tversky, A. (1979) Prospect theory: an analysis of decision under risk. *Econometrica*, **47**, 263–91.

Kendzierski, D. (1990) Decision making versus decision implementation: an action control approach exercise adoption and adherence. *Journal of Applied Social Psychology*, **20**, 27–45.

Kessler, S. (1980) The psychological paradigm shift in genetic counseling. *Social Biology*, **27**, 167–85.

Kessler, S. (1989) Psychological aspects of genetic counseling: VI. A critical review of the literature dealing with education and reproduction. *American Journal of Medical Genetics*, **34**, 340–53.

Kessler, S., and Levine, E. K. (1987) Psychological aspects of genetic counseling. IV. The subjective assessment of probability. *American Journal of Medical Genetics*, **28**, 361–70.

Kessler, S., Field, T., Worth, L. and Mosbarger, H. (1987) Attitudes of persons at risk for Huntington's disease toward predictive testing. *American Journal of Medical Genetics*, **26**, 259–70.

Kraus, E. M. and Brettler, D. B. (1988) Assessment of reproductive risks and intentions by mothers of children with hemophilia. *American Journal of Medical Genetics*, **31**, 259–67.

Kunreuther, H., Ginsberg, R., Miller, L. *et al.* (1978) *Disaster Insurance Protection: Public Policy Lessons*. New York: Wiley.

Lajos, I. and Czeizel, A. (1987) Letter to the Editor: reproductive choices in hemophilic men and carriers. *American Journal of Medical Genetics*, **28**, 519–20.

Langer, E. J. (1975) The illusion of control. *Journal of Personality and Social Psychology*, **32**, 311–28.

Lave, L. B. (1987) Health and Safety risk analysis: information for better decisions. *Science*, **236**, 291–5.

Leonard, C. O. and Beck-Black, R. (1984) Decision-making dilemmas in genetic counseling. In J. O Weiss, B. A. Bernhardt and N. W. Paul (eds.), *Genetic Disorders and Birth Defects in Families and Society: Toward Interdisciplinary Understanding*, pp. 62–70. New York: Alan R. Liss, Inc., for the National Foundation – March of Dimes. (Birth Defects: Original Articles Series, vol. 20.)

Leventhal, H., Nerenz, D. R. and Steele, D. F. (1984) Illness representations and coping with health threats. In A. Baum and J. Singer (eds.), *A Handbook of Psychology and Health*. Hillsdale, NJ: Erlbaum.

Lippman, A. (1992) Led (astray) by genetic maps: the cartography of the human genome and health care. *Social Science in Medicine*, **35**, 1469–76.

Lippman, A. and Wilfond, B. S. (1992) Twice-told tales: stories about genetic disorders. *American Journal of Human Genetics*, **51**, 936–7.

Lippman-Hand, A. and Fraser, F. C. (1979a) Genetic counseling: provision and reception of information. *American Journal of Medical Genetics*, **3**, 113–27.

Lippman-Hand, A. and Fraser, F. C. (1979b) Genetic counseling – the postcounseling period. I. Parents' perceptions of uncertainty. *American Journal of Medical Genetics*, **4**, 51–71.

Lippman-Hand, A. and Fraser, F. C. (1979c) Genetic counseling: parents' responses to uncertainty. In C. J. Epstein, C. J. R. Curry, S. Packman, S. Sherman and B. D. Hall (eds.), *Risk, Communication, and Decision Making in Genetic Counseling*, pp. 325–39. New York: Alan R. Liss, Inc., for the National Foundation – March of Dimes. (Birth Defects: Original Articles Series, vol. 15.)

Loader, S., Sutra, C. J., Walden, M., Kozyra, A. and Rowley, P. T. (1991) Prenatal screening for hemoglobinopathies. II. Evaluation of counseling. *American Journal of Human Genetics*, **48**, 447–51.

Lopes, L. L. (1987) Between hope and fear: the psychology of risk. *Advances in Experimental Social Psychology*, **20**, 255–95.

Maddux, J. E. and Rogers, R. W. (1983) Protection motivation and self-efficacy: a revised theory of fear appeals and attitude change. *Journal of Experimental Social Psychology*, **19**, 469–79.

Maiman, L. A. and Becker. M. H. (1974) The health belief model: origins and correlates in psychological theory. *Health Education Monograph*, **2**, 336–53.

Markel, D. S., Young, A. B. and Penney, J. B. (1987) At-risk persons' attitudes toward presymptomatic and prenatal testing of Huntington's disease in Michigan. *American Journal of Medical Genetics*, **26**, 295–305.

Marteau, T. M. (1989) Framing of information: its influence upon decisions of doctors and patients. *British Journal of Social Psychology*, **28**, 89–94.

Marteau, T. M. (1991) Psychological aspects of prenatal testing for fetal abnormalities. *Irish Journal of Psychology*, **12**, 121–32.

Marteau, T. M., Plenicar, M. and Kidd, J. (1993) Obstetricians presenting amniocentesis to pregnant women: practice observed. *Journal of Reproductive and Infant Psychology*, **11**, 5–82.

Mastromauro, C., Myers, R. H. and Berkman, B. (1987) Attitudes toward presymptomatic testing in Huntington's disease. *American Journal of Medical Genetics*, **26**, 271–82.

McKenna, F. P. (1993) It won't happen to me: unrealistic optimism or illusion of control? *British Journal of Psychology*, **84**, 39–50.

Meissen, G. J. and Berchek, R. L. (1987) Intended use of predictive testing by those at risk for Huntington's disease. *American Journal of Medical Genetics*, **26**, 283–93.

Meissen, G. J., Mastromauro, C. A., Kiely, D. K., McNamara, D. S. and Myers, R. H. (1991) Understanding the decision to take the predictive test for Huntington's disease. *American Journal of Medical Genetics*, **39**, 404–10.

Mennie, M. E., Liston, W. A. and Brock, D. J. H. (1992) Prenatal cystic fibrosis carrier testing: designing an information leaflet to meet the specific needs of the target population. *Journal of Medical Genetics*, **29**, 308–12.

Meryash, D. L. and Abuelo, D. (1988) Counseling needs and attitudes toward prenatal diagnosis and abortion in fragile-X families. *Clinical Genetics*, **33**, 349–55.

Miller, S. M. (1980) Why having control reduces stress: if I can stop the roller coaster I don't want to get off. In M. Seligman and J. Garber (eds.), *Human Helplessness: Theory and Applications*. New York: Academic Press.

Miller, S. M., Brody, D. S. and Summerton, J. (1988) Styles of coping with threat: implications for health. *Journal of Personality and Social Psychology*, **54**, 345–53.

Miller, W. B. (1983) Chance, choice and the future of reproduction. *American Psychologist*, **38**, 1198–205.

Neumann, P. J. and Polister, P. E. (1992) Risk and optimality. In J. F. Yates (ed.), *Risk-Taking Behavior*. Chichester: John Wiley & Sons.

Pauker, S. G. and Pauker, S. P. (1987) Prescriptive models to support decision making in genetics. In G. Evers-Kiebooms, J. J. Cassiman, H. Van den Berghe and G. d'Ydewalle (eds.), *Genetic Risks, Risk Perception, and Decision Making*, pp. 279–96. New York: Alan R. Liss, Inc., for the National Foundation – March of Dimes. (Birth Defects: Original Articles Series, vol. 23.)

Pitz, G. F. (1987) Evaluating decision aiding technologies for genetic counseling. In G. Evers-Kiebooms, J. J. Cassiman, H. Van den Berghe and G. d'Ydewalle (eds.), *Genetic Risk, Risk Perception, and Decision Making*, pp. 251–78. New York: Alan R. Liss, Inc., for the National Foundation – March of Dimes. (Birth Defects: Original Articles Series, vol. 23.)

Polister, P. E. (1981) Decision analysis and clinical judgment: a re-evaluation. *Medical Decision Making*, **1**, 361–89.

Quaid, K. A., Brandt, J., Faden, R. R. and Folstein, S. E. (1989) Knowledge, attitudes, and the decision to be tested for Huntington's disease. *Clinical Genetics*, **36**, 431–8.

Reif, M. and Baitsch, H. (1985) Psychological issues in genetic counseling. *Human Genetics*, **70**, 193–9.

Roberts, L. (1990) To test or not to test? *Science*, **247**, 17–19.

Robinson, A., Bender, G. B. and Linden, M. G. (1989) Decisions following the intrauterine diagnosis of sex chromosome aneuploidy. *American Journal of Medical Genetics*, **34**, 552–4.

Rogers, R. W. (1983) Cognitive and physiological processes in attitude change: a revised theory of protection motivation. In J. Cacioppo and R. Petty (eds.), *Social Psychophysiology*. New York: Guilford Press.

Rosenstock, I. M., Strecher, V. J. and Becker, M. H. (1988) Social learning theory and the health belief model. *Health Education Quarterly*, **15**, 175–83.

Rowley, P. T., Lipkin, M. and Fisher, L. (1984) Screening and genetic counseling for beta-thalassemia trait in a population unselected for interest: comparison of three counseling methods. *American Journal of Human Genetics*, **36**, 677–89.

Sagi, M., Shiloh, S. and Cohen, T. (1992) Application of the health belief model in a study on parents' intentions to utilize prenatal diagnosis of cleft lip and/or palate. *American Journal of Medical Genetics*, **44**, 326–33.

Savage, L. J. (1954) *The Foundations of Statistics*. New York: Wiley.

Schild, S. (1984) Markers for stress and criteria for ongoing counseling. In J. O. Weiss, B. A. Bernhardt and N. W. Paul (eds.), *Genetic Disorders and Birth Defects in Families and Society: Toward Interdisciplinary Understanding*, pp. 107–13. New York: Alan R. Liss, Inc., for the National Foundation – March of Dimes. (Birth Defects: Original Articles Series, vol. 20.)

Schilirò, G., Romeo, M. A. and Mollica, F. (1988) Prenatal diagnosis of thalassemia: the viewpoint of patients. *Prenatal Diagnosis*, **8**, 231–3.

Scholz, C. (1992) On the interactive accomplishment of decision in genetic counseling before prenatal diagnosis. In G. Evers-Kiebooms, J. P. Fryns, J. J. Cassiman and H. Van den Berghe (eds.), *Psychological Aspects of Genetic Counseling*, pp. 47–56. New York: Wiley – Liss, Inc., for the National Foundation – March of Dimes. (Birth Defects: Original Articles Series, vol. 28.)

Shiloh, S. (1994) Heuristics and biases in health decision making: Their expression in genetic counseling. In L. Heath, R. S. Tindale, J. Edwards, E. J. Posavac, F. B. Bryant, E. Henderson–King, Y. Suarez-Balcazar and J. Myers (eds.), *Applications of Heuristics and Biases and Social Issues*. New York: Plenum Press.

Shiloh, S. and Sagi, M. (1989) Framing effects in the presentation of genetic recurrence risks. *American Journal of Medical Genetics*, **33**, 130–5.

Shiloh, S. and Saxe, L. (1989) Perception of recurrence risks by genetic counselees. *Psychology and Health*, **3**, 45–61.

Simon, H. A. (1979) Rational decision making in business organizations. *American Economic Review*, **69**, 493–513.

Sissine, F. J., Rosser, L., Steele, M. W., Marchese, S., Garver, K. L. and Berman, N. (1981) Statistical analysis of genetic counseling impacts: a multi-method approach to retrospective data. *Evaluation Review*, **5**, 745–57.

Slovic, P. (1987) Perception of risk. *Science*, **236**, 280–5.

Slovic, P., Fischhoff, B. and Lichtenstein, S. (1976) Cognitive processes and societal risk taking. In J. S. Carroll and J. W. Payne (eds.), *Cognition and Social Behavior*. Hillsdale, NJ: Erlbaum.

Sorenson, J. R. and Wertz, D. C. (1986) Couple agreement before and after genetic counseling. *American Journal of Medical Genetics*, **25**, 549–55.

Sorenson, J. R., Scotch, N., Swazey, J., Wertz, D. C. and Heeren, T. (1987) Reproductive plans of genetic counseling clients not eligible for prenatal diagnosis. *American Journal of Medical Genetics*, **28**, 345–52.

Spangler, W. D. (1989) Direct versus indirect measures of utilities. *Psychological Reports*, **64**, 307–18.

Steele, M. W., Rosser, L., Rodnan, J. B. and Bryce, M. (1986) Effects of sibship position on reproductive behavior of couples after the birth of a genetically handicapped child. *Clinical Genetics*, **30**, 328–34.

Taylor, S. E. and Brown, J. D. (1988) Illusion and well-being: a social psychological perspective on mental health. *Psychological Bulletin*, **103**, 193–210.

Teigen, K. H. (1988) The language of uncertainty. *Acta Psychologica*, **68**, 27–38.

Thompson, E. and Rothenberg, K. (1992) National Institutes of Health workshop statement: reproductive genetic testing: impacts on women. *American Journal of Human Genetics*, **51**, 1161–3.

Tibben, A., Frets, P. G., van de Kamp, J. J. P., Neirmeijer, M. F., Vegter-van der Vlis, M., Roos, R. A. C., van Ommen, G. J. B., Duivenvoorden, H. J. and Verhage, F. (1993) Presymptomatic DNA-testing for Huntington's disease: pretest attitudes and expectations of applicants and their partners in the Dutch program. *American Journal of Medical Genetics*, **48**, 10–16.

Tversky, A. and Kahneman, D. (1974) Judgement under uncertainty: heuristics and biases. *Science*, **185**, 1124–31.

Van Spijker, H. G. (1992) Support in decision making processes in the post-counseling period. In G. Evers-Kiebooms, J. P. Fryns, J. J. Cassiman and H. Van den Berghe (eds.), *Psychological Aspects of Genetic Counseling*, pp. 29–36. New York: Wiley–Liss, Inc., for the National Foundation – March of Dimes. (Birth Defects: Original Articles Series, vol. 28.)

Vlek, C. (1987) Risk assessment, risk perception and decision making about courses of action involving genetic risk: an overview of concepts and methods. In G. Evers-Kiebooms, J. J. Cassiman, H. Van den Berghe and G. d'Ydewalle (eds.), *Genetic Risk, Risk Perception, and Decision Making*, pp. 171–207. New York: Alan R. Liss, Inc., for the National Foundation – March of Dimes. (Birth Defects: Original Articles Series, vol. 23.)

von Neumann, J. and Morgenstern, O. (1947) *Theory of Games and Economic Behavior*, 2nd edn. Princeton, NJ: Princeton University Press.

Wallston, K. A. (1989) Assessment of control in health-care settings. In A. Steptoe and A. Appels (eds.), *Stress, Personal Control and Health*. Chichester: John Wiley & Sons.

Weinstein, M. C. and Fineberg, H. V. (1980) *Clinical Decision Analysis*. Philadelphia: W. B. Saunders Company.

Weinstein, N. D. (1980) Unrealistic optimism about future life events. *Journal of Personality and Social Psychology*, **39**, 806–20.

Welshimer, K. J. and Earp, J. A. L. (1989) Genetic counseling within the context of existing attitudes and beliefs. *Patient Education and Counseling*, **13**, 237–55.

Wertz, D. C., Janes, S. R., Rosenfield, J. M. and Erbe, R. W. (1992) Attitudes toward the prenatal diagnosis of cystic fibrosis: factors in decision making among affected families. *American Journal of Human Genetics*, **50**, 1077–85.

Wortman, P. (1972) Medical diagnosis: an information processing approach. *Computer Biomedical Research*, **5**, 318–28.

Yates, J. F. and Stone, E. R. (1992) The risk construct. In J. F. Yates (ed.), *Risk-Taking Behavior*. Chichester: John Wiley & Sons.

4

Genetic counselling: some issues of theory and practice

SUSAN MICHIE and THERESA MARTEAU

4.1 Introduction

With the rapid rate of discoveries in human genetics and their increasing clinical application, the demand for genetic counselling is increasing. We know little about what makes for effective or efficient genetic counselling. This chapter will focus upon the methodological issues that need to be considered if we are to further our understanding of the effective ingredients of genetic counselling. In doing so, we will necessarily touch upon decision-making in the counselling context. This is, however, dealt with in more detail in Chapter 3.

The attempt to evaluate genetic counselling requires a definition of its aims. Broadly speaking, genetic counselling is a communication process aimed at helping people with problems associated with genetic disorders or the risk of these in their family. Its most uncontroversial goal is to improve the quality of life of the families that seek such help (Twiss, 1979).

The central issue in genetic counselling has been described as 'the provision of "objective" information from the counsellor and its interpretation by a patient' (Shiloh and Saxe, 1989). Other definitions have included decision-making. One such example is: 'the essence of genetic counselling is the counsellor's ability to transmit genetic information about an inherited disorder of concern to the counsellee(s) so that it will be incorporated into decision making' (Falek, 1984). A widely quoted, and comprehensive, definition of genetic counselling was provided by Fraser (1974, p. 637).

> ... a communication process which deals with the human problems associated with the occurrence, or risk of occurrence, of a genetic disorder in a family. This process involves an attempt by one or more appropriately trained persons to help the individual or the family to

> i.   comprehend the medical facts, including the diagnosis, the probable course of the disorder and the available management;
> ii. appreciate the way heredity contributes to the disorder and the risk of recurrence in specified relatives;
> iii. understand the options for dealing with the risk of recurrence;
> iv. choose the course of action which seems appropriate to them in view of their risk and their family goals and act in accordance with that decision; and

v.   make the best possible adjustment to the disorder in an affected family member and/or to the risk of recurrence of that disorder.

While there is some consensus in the literature about the objectives of genetic counselling, we do not know the extent to which this is shared in theory or in practice by genetic counsellors. The effectiveness of genetic counselling has been evaluated primarily by means of outcome studies, with the major outcomes being educational effectiveness, reproductive intentions and behaviour, and risk assessment (see Kessler (1990) for a review).

In addition to defining genetic counselling in terms of its objectives, it has also been described in terms of its process. Non-directiveness is an approach that is emphasised as important in allowing people to make the best decisions for themselves (see, for example, Emery, 1984; Chadwick, 1993), and is a principle embraced by all professional bodies (Royal College of Physicians, 1989; Institute of Medicine, 1994). There is little research documenting what counsellors describe themselves as doing, or what they actually do, during the counselling process. Self-report studies suggest a marked variation in directiveness of counselling style between professional groups, and between geneticists of different nationalities (Marteau *et al.*, 1994*a,b*). The extent to which counselling is non-directive in practice is unclear from research findings. For example, Wertz *et al.* (1984) found that in 58% of almost 900 genetic counselling sessions, counsellors were not aware of the topic that the counsellee most wanted to discuss. Kessler interprets these findings to mean that

> genetic counsellors may not be giving sufficient attention to the major issues on the minds of counsellees and may be too strongly wedded to their own agendas so that room for adequate discussion of the counsellees' concerns may not be made available. One of the things we need to know more about is how counsellors set the counselling agenda and the degree to which counsellees can find an entree and receptiveness in presenting their individual questions and concerns.

There is some evidence to suggest that counsellors may influence the thinking of counsellees about certain things, such as risk magnitudes (Lubs, 1979; Wertz *et al.*, 1986).

If we are to determine the extent to which genetic counselling makes a difference to outcome, we need to establish whether the process of counselling changes the relationship between what is brought to counselling and the outcome. One of the methodological problems limiting evaluation of genetic counselling is the lack of controlled designs. Thus, it has been possible to describe change and stability in such outcome measures as patient knowledge, risk perception and reproductive intention, but not to attribute changes specifically to genetic counselling.

Studies are needed that include:

(1)   greater attention to the expectations and beliefs that patients and counsellors bring to the counselling consultations;
(2)   a broader range of outcome measures, for example, patients' experience of the consultation; understanding of, and commitment to, the information received; counsellors' perceptions of patients' concerns;
(3)   standardised measures of the input, process and outcomes of genetic counselling.

4.2   Objectives of genetic counselling

Although much has been written about genetic counselling, there have been few empirical studies, especially controlled trials. Research before 1979 has been reviewed and found to be poor methodologically and inconclusive in demonstrating the impact of genetic counselling (Evers-Kiebooms and Van den Berghe, 1979). Of 9000 articles on genetic counselling in the major genetic and obstetric journals appearing from 1985 to 1989, only 45 presented empirical data from studies dealing with psychological, social and ethical issues (Lippman, 1991). A review of the studies since 1979 has been carried out by Kessler (1990).

His three main conclusions concerning the educational aspects of genetic counselling were that:

(1)   genetic counselling is effective in educating counsellees about recurrence risks and diagnostic issues;
(2)   there is room for improvement in this area since many counsellees still leave counselling poorly informed, and
(3)   the educational agendas of counsellors and counsellees appear to be somewhat discordant.

The way in which risks are perceived may be influenced by the decisions made before counselling. Although there is evidence of some influence from genetic counselling upon perception of risk, in that discrepancies in risk perception between counsellors and counsellees are reduced after counselling, it is counsellees' perceptions of risk *before* counselling, rather than the risk they are given by counsellors, that correlate with reproductive intentions (Shiloh and Saxe, 1989).

The impact of genetic counselling on reproductive intention or decision-making has not been empirically demonstrated. Pre-counselling reproductive intentions appear to be the major determinant of post-counselling reproductive behaviour (see Chapter 3). In the study by Sorenson and colleagues, the effect of counselling appeared to be one of increased reproductive intention, particularly among those deemed to be at low risk. For example, before counselling 27% of those deemed to be at high risk intended a further pregnancy but after counselling 42% did, an increase of 56%.

4.3   Psychological models and genetic counselling

Most studies to date have been descriptive investigations of attitudes and reactions of counsellees to genetic counselling and prenatal diagnosis, and are weak theoretically and methodologically. Very few have used psychological models to guide question framing, data collection or interpretation of results. We shall consider some of the models that may be useful in shaping future research questions and studies.

Knowledge

The finding that those who have been through genetic counselling may have a relatively poor understanding of what they have been told has been explained in several ways: difficulty in grasping key concepts, intense emotion interfering with information processing, and counsellors' poor communication skills, including failure to use educational aids.

According to Ley's (1988) cognitive model, the two main reasons for failures in comprehension in medical contexts are that information is presented in too difficult a form to be understood, and even if understood, much will be forgotten. A further factor is that what is understood and remembered will be interpreted within the counsellee's framework of existing ideas. People often have their own theories about illnesses, leading to discrepancies between the intended and the received messages. People's pre-existing beliefs about inheritance, e.g. about their own 'proneness' and about the way that familial diseases are or are not 'passed on' to children, may influence the extent to which they understand the genetic information that is provided to them in counselling and may determine their attitude to genetic services.

New knowledge, especially where science or technology is involved, is assimilated into individuals' existing frameworks of knowledge and understanding by anchoring the unfamiliar to what is already familiar, and rendering abstract concepts into something more concrete. These concepts lie at the heart of social representation theory (Farr and Moscovici, 1984; see also Chapter 11). Knowledge acquisition is therefore a dynamic process that takes place within a social, rather than an individual, context. This is illustrated in a study of public representations of the Human Genome Project in Britain (Durant *et al.*, 1992). While there was a relatively low public awareness and knowledge of the Human Genome Project, representations of the project were dominated by its close association with the better-known fields of genetic engineering and genetic fingerprinting, which shaped the concerns that people had.

Risk perception

Risk perception is influenced by two key factors: the likelihood of an outcome and the perception of that outcome. These are further influenced by the nature of the risk and the presentation of information pertinent to that risk (Slovic, 1987). In genetic counselling, the primary focus of interest has been the communication of probabilistic information, to the neglect of information about the actual condition. The risk of occurrence or recurrence is not the most common reason given by clients for attending genetic counselling. This may explain why 54% of clients are not able to report the risk they have just been given in a counselling session (Sorenson *et al.*, 1981*b*).

Risk is perceived as higher if the adverse outcome is novel, unfamiliar, causes more damage, is perceived as uncontrollable, unpredictable, frightening, affects later generations and has delayed consequences (Slovic, 1987). Rare but very serious outcomes may be perceived as both higher in risk and less acceptable. A common finding in genetic counselling is that risk perception is inextricably bound up with the perceived burden of the disease or condition (Leonard *et al.*, 1972; Lippman-Hand and Fraser, 1979; Frets *et al.*, 1990; Kessler, 1990; Sagi *et al.*, 1992). This may explain the reported gender difference, that men learn probabilistic information better than women (Sorenson *et al.*, 1981*c*). Women may be more influenced by burden as the ones who are likely to have greater responsibility for a disabled child.

Parents' reproductive intentions after genetic counselling are associated with perceptions of risk, but not with the probability of the outcome (Shiloh and Saxe, 1989). There are consistent differences between the perceived risk and the probability. Several factors have been found to account for this. How a probability is presented effects how it is perceived and its influence on subsequent decisions (see, for example, Tversky and Kahneman, 1981). Risks may be perceived as higher when presented in genetic counselling as odds rather than as percentages (Kessler and Levine, 1987). A framing effect is also evident: genetic risks presented as the chance of having an unhealthy child were associated with higher intentions to use prenatal diagnosis than the same probabilities presented in terms of having a healthy child (Marteau, 1989).

Genetic counselling may influence counsellees' perceptions of risk, not only by the provision of new information, but also by the very process of discovering risks. Of women who showed changes in their reproductive plans and said that counselling had influenced them, the largest number reported that they had decided to have a child in the next two years because of their counselling (Sorenson *et al.*, 1981*b*). Of the remainder, most became uncertain, whereas a minority were discouraged reproductively by their counselling. This may occur because talking about risk desensitises people to risk, or it may be the result of a changed understanding of probabilities and their consequences. Alternatively, it may reflect the kinds of people who attend for counselling.

Implicit in many studies of genetic counselling is the notion that counsellors' perceptions of risk are based largely upon probabilities. Very little is known about how genetic counsellors convey risk information, and what determines this. In a study of obstetricians presenting prenatal testing to older women, risks were conveyed categorically (Marteau *et al.*, 1993), reflecting patients' perceptions of risks (Lippman-Hand and Fraser, 1979). So, for example, the same probability was described as a high or a low risk according to whether the outcome was the birth of a child with Down syndrome or a miscarriage following amniocentesis. Determining how genetic counsellors convey risks, and the causes and consequences of different presentations, is a relatively neglected topic, given that risk is at the core of genetic counselling.

Decision-making

The questions of 'What do people decide?', 'Who decides?' and 'How are decisions reached?' are addressed in Chapter 3. In this chapter, two related questions will be considered: first, what influences decision-making and, secondly, what effect genetic counselling has on decision-making.

Decision-making has been studied within the framework of laws of probability and normative models. One of the basic assumptions is that 'rational' decision-making involves the systematic weighing of alternatives and exploration of the consequences of each alternative to produce judgements and choices. Yet people rarely follow formal statistical rules in making decisions outside the laboratory. These observations generated studies aimed at finding out the 'errors and biases' of thinking that lead to 'irrational' or 'illogical' decisions (see, for example, Hogarth, 1981; Kahneman *et al.*, 1982).

In some cases, it may be that people are using normative decision-making processes, but in the context of questions and problems that are different to those defined by the researcher. It has also been recognised that non-normative thinking and decision-making may be far from 'irrational'. It may be highly adaptive, especially when decisions are being taken in situations that are changing, uncertain and dynamic.

Lippman-Hand and Fraser (1979) are among many who have questioned the appropriateness of the concept of rational decision-making in genetic counselling. In situations where risks and burdens cannot be defined unambiguously and a choice must be made within this uncertainty, the decision problem does not meet the requirements for mathematical decision-making models. There is some evidence that genetic counselling may confirm a decision already made before counselling. For example, 65% of 836 fertile women seeking genetic counselling reported that their main reason for doing so was to obtain information to help in deciding whether they should have a child (Wertz *et al.*, 1984). Yet the same study also found that pre-counselling reproductive intentions were the major determinant of post-counselling intentions.

Two of the most frequently used models of health behaviours within a decision or choice context have been the health belief model (HBM) developed by Rosenstock (1966), and decision analysis.

Health belief model

This model is described in Chapter 3. Within genetic counselling, Shiloh and Saxe (1989) argued that the factors identified as predictors of risk perception correspond to the three categories of the HBM: perceptions (prior expectations), modifying variables (reproductive intentions) and cues (percentage of affected children, neutrality of communication). Parents' intentions to use prenatal diagnosis have also been explained by the HBM (Sagi *et al.*, 1992). The variables studied explained 38% of the variance in intentions to undergo prenatal diagnosis for cleft lip and/or palate and 56% of the variance in intentions to terminate affected pregnancies. The HBM has also been found to predict partner testing in haemoglobinopathy screening (Rowley *et al.*, 1991).

The shortcomings of this model have been summarised by Schwarzer (1992). The main shortcoming of the HBM is the huge number of non-prioritised variables that could by hypothesised to affect health-related behaviour (over 80 have been catalogued by Becker and Maiman (1975)). Kessler (1990) criticised the HBM as a model for genetic counselling, because it assumes that risk perception motivates preventive behaviour. For many counsellees, he argued, evaluation of a risk is shaped by prior reproductive decisions, *not* vice versa.

In genetic counselling, in so far as parents perceive themselves as having options, their choices appear to centre on social meanings and imagined social consequences (for example, of living with a disabled child) rather than around biomedical information, such as risks of occurrence, or abstract ethical principles (Beeson and Golbus, 1985). Parents may not perceive themselves as engaging in a weighing of alternatives or as making decisions at all – 54% of a sample of 26 women and 16 of their partners had made clear decisions before counselling. Another study found that 56% of 528 clients followed up for six months reported that counselling had not influenced their reproductive plans (Sorenson *et al.*, 1981*b*). What is needed are controlled studies to determine the extent to which counselling does influence decisions, as opposed to what people say its influence is. Studies should also look at the decision-making of the couple where this is relevant, because partners may not agree on the best decision.

Decision analysis

This is one of the most prominent approaches to the empirical study of choice (see Raiffa, 1968; Weinstein and Fineberg, 1980; Keeney, 1982; Lindley, 1985). Decisions are seen as series of step by step choices, involving generating solutions and then choosing between them. The temporal stages are (1) recognition, (2) formulation, (3) alternative generation, (4) information search, (5) judgement or choice, (6) action, and (7) feedback. An example of this approach in genetic counselling is the decision tree of the amniocentesis decision generated by Pauker and Pauker (1979). Within this, risks are presented in both a positive and negative frame to minimise the effect of framing on decisions. The individual's values are incorporated with individualised risks in a logical framework to derive 'the best possible choice under conditions of certainty'.

This model is limited in being based most commonly upon the researcher's, rather than the decision-maker's, representation of the problem, the decision, the choices and the outcomes. The process of decision-making is seen as normative and based on a series of individual acts, and not as an interaction sequence. Three shortcomings are described by Vlek (1987). The first is that decision analysis can be rationally applied only to well-defined decision problems; the second is that decision theory holds no recipe for determining the balance between analysis and synthesis (aggregate judgement and evaluation) for a particular type of decision problem. Thirdly, decision analysis may lead to an infinite regress of ever more detailed justifications of original preferences and expectations.

To overcome these limitations, the model has been adapted by several researchers. Vlek (1987) has suggested that the goal should be an *acceptable* rather than an *optimal* decision and that this should be made by considering the possible consequences of each course of action, together with an assessment of future demands and abilities to cope with them. This leads to a weighing of the 'expected desirability' with the 'expected stress' of one option against another. Adaptations include a feedback loop within the linear chain of action in the rational choice model, so that participants reflect on previous choices and may take up issues seen as already settled during the preceding counselling conversation. Strategies of coping with stress and conflict, and counsellees' feelings, motives and intentions, have been introduced into the decision process (Black, 1982; Schild and Black, 1984; Pitz, 1987). Further work is needed to determine the clinical usefulness of this model (for a discussion of these issues, see Chapter 3).

4.4   Methodological issues in studying genetic counselling

Studies of other medical consultations suggest a framework that may have utility in the study of genetic counselling, involving the description of the input to, the process of, and the outcomes of counselling (Pendleton, 1983).

Input to genetic counselling

'Input' refers to what counsellors and patients bring to the genetic counselling. Counsellors vary in the amount of training and clinical experience and their qualifications, their style of counselling (including directiveness), and other individual differences such as their tolerance of uncertainty. These can be measured by questionnaires, standardised where they exist, or developed for the purpose. Style of counselling can be assessed by using standardised assessments of the consultation (see, for example, Roter, 1977). Counsellors also bring their particular expectations of the nature of the problem, the patient and/or family.

Counsellees also bring a set of expectations about what they think the problem is and what they hope to get out of the consultation. They vary in many important demographic aspects, e.g. cultural background, religion, age and reproductive stage. They also vary along many psychological and social dimensions: emotional state (e.g. anxiety and depression), beliefs and knowledge about the particular condition and genetics in general, decisions they and/or their family face, their resources for coping and support, and their family and general life situation. These are best measured by a combination of questionnaire and interviews, including both open questions and numerical rating scales. Initial results from an evaluation of genetic counselling being carried out by Theresa Marteau and colleagues suggest that results may be very different when questions are open-ended than when they include response options.

The process of counselling

The process of genetic counselling includes both its verbal content and the interaction within which that content is produced. There is little work studying the interactions within genetic counselling sessions that may facilitate, or hinder, the desired outcomes of the consultation. Psychologists have developed an expertise in measuring the emotions, cognitions and behaviours of individuals. No such general expertise exists in measuring interactions between individuals. The task we confront is how to analyse the interactions in ways that are, on the one hand, valid and reliable, and, on the other, meaningful, and feasible in terms of resources available for the task.

Interactions are important because of their relationship with outcome and their potential role in affecting improved outcomes (see, for example, Korsch *et al.*, 1968; Francis *et al.*, 1969; Roter and Hall, 1989). The system by which interactions are analysed is important in that the choice of system will influence the nature of the research findings. One area where they have been extensively studied is the consultation between patients and family doctors. Most studies of the process of the general practice (GP) consultation have analysed transcripts or audiotapes, with fewer using direct observation or videotapes. They have concentrated on affective tone and information exchange within verbal interactions, rather than non-verbal behaviour. The main outcome measures studied in GP consultations have been patient knowledge, satisfaction, compliance, responsiveness to treatment and reassurance.

According to Wasserman and Inui (1983) an ideal system of interaction analysis should:

(1) take account of salient dimensions of interpersonal communications, i.e. it should measure information transfer on multiple levels (e.g. content and relationship), by a variety of behaviours, within a sequence of interaction and within a context;

(2) deal with the distinctive and special character and issues of clinician–patient interactions, and

(3) lend itself to application within the healthcare system (e.g. clinical teaching and feedback).

The study of the genetic counselling process should include measures of the amount and content of communication, as well as its form. Below are detailed some of the aspects of consultations that appear to merit assessment.

Amount of communication

These are simple measures such as duration of the consultation and the proportion of time spent talking by each participant.

Content of communication

Interaction analysis has been criticised for failing to specify the content of the interaction, which may have a substantial impact on the pattern of communication and subsequent outcomes. In Wasserman and Inui's (1983) words: 'It is problematic to attempt to understand clinician–patient communications, while ignoring what the participants are talking about.'

Content can be described by the proportion of time spent on different aspects of the consultation, from broad categories, e.g. medical and non-medical, to much more specific and numerous categories. There have been many studies that have looked at both the quantity of different types of information and its presentation in relation to recall.

In their review of studies using this approach, Roter and Hall (1989) concluded that the minority of patients are given adequate information about the medicine they take or their medical condition. Patients forget a great deal of what they are told (on average, about 50%). This approach, however, fails to take account of the relative importance of different types of information for the patient.

A qualitative analysis was carried out by Lippman-Hand and Fraser (1979) on 30 transcripts of genetic counselling sessions. They found that the counselling process always included the collection of family history information, and explanation of the diagnosis where it could be established, a review of the hereditary pattern associated with this diagnosis, a statement of the associated recurrence rates, and an explanation of how these were determined. The amount of time spent dealing with psychological or social issues appears to be minimal (Sorenson *et al.*, 1981*b*).

A pilot study of our own of 36 tape-recorded genetic counselling consultations suggested the following as common categories in genetic counselling consultations: genetics, risk, condition, tests (process), tests (results), future action, decisions, emotional issues and areas of confusion or uncertainty. As has been found earlier, relatively little attention was paid to such social issues as the impact of genetic information upon other family members.

The interaction

Inui *et al.* (1982) compared three interactional analysis systems for their relative power to explain variance in the observed outcomes of medical consultations (patient knowledge, satisfaction, recall of prescribed medications and compliance). The systems compared were Bales's process analysis scheme (Bales, 1968), Roter's modification (Roter, 1977) and Stiles's Verbal Response Mode (Stiles, 1978). In all these systems, segments of speech are assigned to categories that describe a function of verbal behaviour.

None of the interactional systems was found to be ideally suited for describing the interaction at a relatively holistic level nor for describing the sequences of interaction. Categories are too 'atomistic' to capture and distinguish effectively such activities as doctor reassurance or patient efforts to raise issues for consideration, describe distressing symptoms or negotiate for services. There is also a problem with the Bales system of being forced to assign statements to one category rather than another, when both apply. The Stiles system avoids this problem, but also avoids dealing with content or context. This latter problem is successfully dealt with in Katz's Resource Exchange Analysis Scheme (Katz *et al.*, 1967), although it faces further problems of ambiguity and conflicts in classification.

Butler *et al.* (1992) described interaction in GP consultations in terms of the problems or agendas (implicit or explicit) that are brought to the consultation. By noting who initiates, follows or returns to agendas previously raised, and how these agendas are acted upon, this method is able to address the issue of control between doctor and patient. The method involves describing each 'floorholding', i.e. the period of time when one participant is speaking. Each floorholding is coded in terms of 10 agendas, 6 procedures and 6 processes. The system is unwieldy for rapid scoring and satisfactory reliability depends upon careful training.

Using this approach with 73 GP interviews, Butler and colleagues found that whereas both doctor and patient address physical agendas to a similar high degree, patients present emotional and social agendas to a greater extent than do doctors. This reflects the pattern of findings in Sorenson's study in 1981: whereas genetic counsellors tended to focus on diagnosis, aetiology and risk estimation, patients had a broader set of interests, including social, emotional and financial issues, as well as disease prognosis and treatment.

In their evaluation of an intervention aimed at increasing patient involvement in medical consultations, Greenfield *et al.* (1988) used a coding scheme that classified each verbal utterance as to whether it seeks to control the behaviour of the other, to communicate information, or to convey emotion. In addition to a total of 30 conversational codes, there were 8 indicators of interaction style. Using this method of analysing interactions, they found the intervention effective in increasing the number of controlling utterances by patients, the ratio of patient to doctor utterances, the proportion of the consultation reflecting 'interpersonal involvement' and reducing the portion directed by the doctor. The authors do not comment on how time-consuming their methodology is.

Kessler and Jacopini (1982) sought to apply quantitative methods to genetic counselling interactions. They used the Bales system to analyse one transcript. It took several weeks for two people to become trained and to achieve a satisfactory reliability in scoring. Previously qualitative analysis of this transcript had generated four hypotheses, three of which were confirmed by this quantitative analysis. The study showed that the counsellor focused on providing medical information and genetic facts rather than exploring personal meanings, attitudes and feelings. The male counsellor showed a high level of 'negativity' toward the wife but was supportive of the husband. The authors found that the Bales system was unable to evaluate the directiveness of the counsellor. This analysis took three weeks. The amount of work involved in this detailed analysis of *one* transcript clearly limits its usefulness in identifying the variation and range of interactions, and in generating sufficient data to test hypotheses.

Roter and Hall (1989) concluded that there are methodological limitations in all the systems of interactional analysis, and that there is no clear optimal approach to describing interaction sequence. Other investigators in this area (Inui and Carter, 1985) have stated that describing the communication process by the frequencies of verbal behaviour is analogous to describing *Hamlet* as a play in which the principal characters include ghosts, witches, lords, ladies, officers, soldiers, sailors, messengers and attendants – one of whom is already dead, one of whom dies by drowning, one by poisoned drink, two by poisoned sword, and one by sword and by drink!

Until we have more evidence about which aspects of interaction predict outcome, it is difficult to assess the pros and cons of the different methodologies that have been developed to study interaction. Because the communication process is highly complex, no one system can do justice to all its dimensions. One particular aspect of the process that requires development is a measure of directiveness. Given how central non-directiveness is to the identity of genetic counselling, development of reliable and valid measures of this would allow studies of the effect of directiveness upon outcomes. The most appropriate technique for analysis will depend in large part upon the particular research question.

The outcomes of counselling

The most commonly used cognitive outcome has been the amount of diagnostic and risk information that patients can remember (see, for example, Inui *et al.*, 1982; Ley *et al.*, 1988). Counting items of information begs the question as to what is sufficiently relevant or important to be defined as an item of information. This is usually defined by the researcher implicitly in the development of the methodology. Rarely has the value that counsellees attach to different kinds of information been assessed. Many studies reporting outcomes have blurred the critical distinction between description and evaluation. For example, a counsellee may recall only 5% of a consultation, but from the patient's perspective this 5% might have been extremely valuable.

It is important to look at both how much information is recalled **and** how that information is evaluated (see, for example, Tuckett *et al.*, 1985). This takes account of the ideas that a counsellor communicates and the way they are understood and evaluated by the patient. In the method developed by Tuckett and colleagues, success of communication is assessed along three dimensions:

(1) recall: how accurately patients recall what they are told;
(2) interpretation: how correctly patients make sense of it;
(3) commitment: how patients evaluate the doctors' ideas in the context of their own explanatory models.

Assessments of the patients' understanding are made by comparing what they might be supposed to know, based on a 'third party' judgement of what was said in their consultation, with their own account of the key points made by the doctor. Patients were interviewed at home and rating scales were devised to measure outcome.

This method involves judgements about what doctors actually say to patients and what it might mean. It makes explicit the process of determining what patients might be *supposed* to know as a result of their consultations, as a basis for comparison with what they *do* know. Such methods inevitably involve specifying values and therefore make apparent problems of validity that are only implicit in methods where such judgements are avoided.

Tuckett and colleagues found that this method can be used reliably to rate GP consultations. In their study of 328 patients, they found that only 10% failed to recall any of the key points discussed in consultation with their GPs. Of those who recalled the key points, 73% correctly made sense of at least one of the topics in the consultation. About a quarter of those correctly interpreting the points the doctor made to them were not committed to the doctor's point of view on at least one of the points. This was significantly more likely to be the case when patients were women, from a non-white ethnic group, over 60, or consulting about children under 10 years of age.

As well as the question of how information is remembered and understood, there are questions of how information *ought* to be conveyed and how decisions *ought* to be made. This begs the questions of 'What is a good decision?' and 'What is a good decision-facilitating counselling style?' One definition of 'a good decision' comes from

Shiloh and Saxe (1989): 'A good decision is one that is not discrepant with the person's value system.' If outcomes are to be assessed and models evaluated, we need some criteria of the validity of judgements and the quality of decisions made. Research needs to include the perspective of those counselled: what they hope to get from the process, what they *do* get from it, and what they regard as a 'good decision'. Framing research questions and evaluating empirical results using psychological models may help to bring the theory and practice of genetic counselling closer together.

Pilot study

The pilot study reported here was conducted in preparation for a study aimed at investigating the relationship between aspects of interaction within genetic counselling and outcomes.

Our aims in studying interactions within genetic counselling are:

(1) to further our understanding of the process by which patients remember, understand and evaluate the information they are given;

(2) to investigate the influence of prior expectations of counsellors and patients on the interaction, and the influence of expectations and interaction on outcome;

(3) to use our understanding of the above to evaluate the effectiveness of different ways of conducting genetic counselling.

Participants were all counsellees coming for their first visit to a genetics centre, having been referred either by their GP or a hospital doctor. In 11 consultations, the 7 different counsellors they were to see answered questions before the consultation about what they thought the counsellees wanted to get out of the session. 'Getting information' was seen as the main issue for patients in nine consultations, and 'diagnosis' was seen as the main issue in two. The 'information desired' concerned risks on seven occasions, tests on five occasions, the condition on three occasions, carrier status on two occasions and coping on two occasions (see Table 4.1).

The counsellees seen in these consultations had somewhat different views from those of the counsellors. Only four wanted information on risks, two wanted information on tests, one on the condition and one on carrier status. None mentioned coping. Of the 11 patients, 4 had hopes of the consultation that were matched by the perception of the counsellor. They all appeared satisfied with their experience (Table 4.2).

There was one patient who said that she did not expect much, and indeed did not get much, from the consultation. There were six who hoped for something not mentioned by the counsellor. Three wanted to clarify or reach a decision about further children, two wanted general advice and one wanted test results. Of these, two had their expectations met, two received other useful information and two said they got nothing out of the consultation.

Of these last two, one woman expressed a hope for test results, whereas the counsellor predicted she wanted information on the risk of recurrence. This was in spite of the referrer having said that the parents would like a further baby and would therefore like

Table 4.1. *Pre-consultation expectations of genetic counsellors and patients of what patients hoped to get from genetic counselling consultations*

	Counsellor (*n*=11)	Patient (*n*=11)
Diagnosis	2	0
Information about risks	7	4
Information about tests	5	2
Information about condition	3	1
Information about carrier status	2	1
Help with coping	2	0
Decision about future children	0	3
Advice	0	2
Reduce uncertainty	0	2
Treatment and management	0	2
Explanation	0	1
Research	0	1

Table 4.2. *Satisfaction of counsellees with counselling according to the extent to which pre-counselling expectations matched those of counsellors*

	Counsellor–counsellee expectations	
	'Satisfied'	'Not satisfied'
Matched	4	0
Mismatched	4	2
No expectations	0	1

to discuss the prospects. The counsellee judged the session to have been 'really disappointing' and said that she still didn't know what was happening. When she was asked how she was feeling, she replied, 'Confused. Want to go for another one, but not sure', suggesting that she had been, as the referrer had indicated, hoping to discuss the prospects of another child. The counsellor judged that the counsellee had got 'permission to get pregnant' from the consultation.

These findings are consistent with the findings of Sorenson and colleagues in the United States (Sorenson *et al.*, 1981*b*). They found that counsellees most frequently gave as their major reason for coming to counselling: 'to get information that will help me to make a decision about whether to have a child.' Yet counsellors are most likely to think that counsellees come to counselling to learn their risk (Sorenson *et al.*, 1981*a*). In another study, Sorenson *et al.* (1981*b*) found that when a counsellee's major interest in counselling was anything other than risk estimation or prognosis, this is not likely to be perceived by the counsellor. Counsellor interests focused on risk estimation, diagnosis and aetiology. Many patients had a broader set of interests, including other medical topics, such as treatment, feelings and financial considerations. Those using health services have been found to want information in a different order from that predicted by geneticists. In a study of the appropriate educational input for a cystic fibrosis screening programme, it was found that health service users would ask significantly earlier about the nature of the carrier test, and significantly later about the chance of having an affected child or about the effect that a child with cystic fibrosis would have on the family, than geneticists predicted (Myers *et al.*, 1994).

One of the reasons for looking at the interaction within these sessions is to determine the extent to which patients introduce their agenda into the consultation and the extent to which it is addressed by the counsellor. This is the next step.

4.5 Conclusion

The increasing potential for the developments of DNA technology to be applied in clinical practice is being paralleled by increasing constraints on the funding of health services. This necessitates a close look at and a scientific evaluation of the effectiveness and efficiency of both the services that are already offered and new services that become possible. Genetic counselling has emerged as an integral part of genetic services in Western medicine. Those with both a nursing and medical training have been developing skills for communicating complicated genetic and risk information to parents, discussing options and decisions with them and providing support for those experiencing emotional distress. Some genetic counsellors have had formal training in genetic counselling, others have learnt 'on the job'. If genetic counsellors are to achieve their goals as effectively as possible, and if the service is to develop in the most appropriate and efficient way, we need information on what aspects of the counselling process have what effects for whom. There are no short cuts to gathering this information through well-designed empirical studies using comprehensive and valid measures that can answer the critical questions. The results of such studies would not only inform the development of genetic services, but could be useful in other areas such as how to increase public understanding of the new genetics, and how to increase the effectiveness of counselling about other aspects of health.

4.6 Acknowledgements

Both authors of this chapter are supported by a grant from the Wellcome Trust, which is also supporting some of their research reported here.

4.7 Bibliography

Ajzen, I. and Fishbein, M. (1980) *Understanding Attitudes and Predicting Social Behavior*. Englewood Cliffs, NJ: Prentice-Hall.

Bales, R. (1968) Interaction process analysis. In D. L. Sills (ed.) *International Encyclopaedia of the Social Sciences*, vol. 7. London: Macmillan and The Free Press.

Bandura, A. (1986) *Social Foundations of Thought and Action: A Social Cognitive Theory*. Englewood Cliffs, NJ: Prentice-Hall.

Becker, M. H. (1979) Understanding patient compliance: the contributions of attitudes and other psychosocial factors. In S. J. Cohen (ed.), *New Directions in Patient Compliance*. Sexington, MA: Heath.

Becker, M. H. and Maiman, L. A. (1975) Socio-behavioral determinants of compliance with health and medical care recommendations. *Medical Care*, **13**, 10–24.

Beeson, D. and Golbus, M. (1985) Decision making: whether or not to have prenatal diagnosis and abortion for X-linked conditions. *American Journal of Medical Genetics*, **20**, 107–14.

Bishop, G. D. (1991) Understanding the understanding of illness. In J. A. Skelton and R. T. Croyle (eds.), *Mental Representation in Health and Illness*. New York: Springer-Verlag.

Black, R. B. (1982) Risk taking behavior: decision making in the face of genetic uncertainty. *Social Work in Health Care*, **7**, 11–25.

Brehmer, B. (1976) Social judgment theory and the analysis of interpersonal conflict. *Psychological Bulletin*, **83**, 985–1003.

Butler, N. M., Campion, P. D. and Cox, A. D. (1992) Exploration of doctor and patient agendas in general practice consultations. *Social Science and Medicine*, **35**, 1145–55.

Campion, P. D., Butler, N. M. and Cox, A. D. (1992) Principle agendas of doctors and patients in general practice consultations. *Family Practice*, **9**, 181–90.

Chadwick, R. F. (1993) What counts as success in genetic counselling? *Journal of Medical Ethics*, **19**, 43–6.

Croyle, R. (1992) Appraisal of health threats: cognition, motivation, and social comparison. *Cognitive Therapy and Research*, **16**, 165–182.

Croyle, R. and Barger, S. (1993) Illness cognition. In *International Review of Health Psychology*, **2**, 29–49.

Durant, J. R., Hansen, A., Bauer, M. and Gosling, A. (1992) *The Human Genome Project and the British Public*. EC Report, Brussels.

Edwards, W. (1961) Behavioral decision theory. *Annual Review of Psychology*, **12**, 473–98.

Eiser, J. R. and van der Pligt, J. (1988) *Attitudes and Decisions*. London: Routledge.

Emery, A. E. H. (1984) Introduction: the principles of genetic counselling. In A. E. H. Emery and I. Pullen (eds.), *Psychological Aspects of Genetic Counselling*. London: Academic Press.

Evers-Kiebooms, G. and Van den Berghe, H. (1979) The impact of genetic counselling: a review of published follow-up studies. *Clinical Genetics*, **15**, 465–74.

Falek, A. (1984) Sequential aspects of coping and other issues in decision making in genetic counselling. In A. E. H. Emery and I. M. Pullen (eds.), *Psychological Aspects of Genetic Counselling*. London: Academic Press.

Farr, R. and Moscovici, S. (eds.) (1984) *Social Representations*. Cambridge: Cambridge University Press.

Francis, V., Korsch, B. M. and Morris, M. J. (1969) Gaps in doctor–patient communication. Patient's response to medical advice. *New England Journal of Medicine*, **280**, 535–40.

Fraser, F. C. (1974) Genetic counselling. *American Journal of Human Genetics*, **26**, 636–59.

Frets, P. G., Duivenvoorden, H. J., Verhage, F., Niermeijer, M. F., Van de Berge, S. M. M. and Galjaard, H. (1990) Factors influencing the reproductive decision after genetic counseling. *American Journal of Medical Genetics*, **35**, 492–502.

Greenfield, S., Kaplan, S. H., Ware, J. E., Yano, E. M. and Harrison, J. L. P. (1988) Patient's participation in medical care: effects on blood sugar control and quality of life in diabetes. *Journal of General Internal Medicine*, **3**, 448–57.

Hogarth, R. M. (1981) Beyond discrete biases: functional and dysfunctional aspects of judgmental heuristics. *Psychological Bulletin*, **90**, 197–217.

Institute of Medicine (1994) *Assessing Genetic Risks. Implications for Health and Social Policy.* Washington, DC: National Academy Press.

Inui, T. S. and Carter, W. B. (1985) Problems and prospects for health services research on provider–patient communication. *Medical Care*, **209**, 521–38.

Inui, T. S., Carter, W. B., Kukull, W. A. and Haigh, V. H. (1982) Outcome-based doctor–patient interaction analysis. 1. Comparison of techniques. *Medical Care*, **20**, 535–49.

Janis, I. L. (1984) The patient as decision maker. In W. D. Gentry (ed.), *Handbook of Behavioral Medicine*. New York: Guilford.

Janis, I. L. and Mann, L. (1976) Coping with decisional conflict. *American Scientist*, **64**, 657–67.

Janz, N. K. and Becker, M. H. (1984) The health belief model: a decade later. *Health Educational Quarterly*, **11**, 1–47.

Kahneman, D., Slovic, P. and Tversky, A. (eds.) (1982) *Judgment under Uncertainty: Heuristics and Biases*. New York: Cambridge University Press.

Katz, E., Gurwitch, M., Tsiyona, P. and Dantet, B. (1967) Doctor–patient exchanges: a diagnostic approach to organisations and professions. *Human Relations*, **22**, 309.

Keeney, R. L. (1982) Decision analysis: an overview. *Operations Research*, **30**, 803–38.

Kessler, S. (1990) Current psychological issues in genetic counseling. *Journal of Psychosomatic Obstetrics and Gynecology*, **11**, 5–18.

Kessler, S. and Jacopini, G. (1982) Psychological aspects of genetic counseling. II. Quantitative analysis of a transcript of a genetic counseling session. *Americal Journal of Medical Genetics*, **12**, 421–35.

Kessler, S. and Levine, E. K. (1987) Psychological aspects of genetic counseling. IV. The subjective assessment of probability. *American Journal of Medical Genetics*, **28**, 361–70.

Korsch, B. M., Gozzi, E. K. and Francis, V. (1968) Gaps in doctor–patient communication. I. Doctor–patient interaction and satisfaction. *Paediatrics*, **42**, 855–71.

Kunda, Z. (1990) The case for motivated reasoning. *Psychological Bulletin*, **108**, 480–98.

Lazarus, R. S. and Folkman, S. (1984) *Stress, Appraisal and Coping*. New York: Springer-Verlag.

Leonard, C., Chase, G. and Childs, B. (1972) Genetic counseling: a consumer's view. *New England Journal of Medicine*, **287**, 433–39.

Leventhal, H. (1989) Emotional and behavioral processes. In M. Johnston and L. Wallace (eds.), *Stress and Medical Procedures*. Oxford: Oxford Science and Medical Publications.

Leventhal, H., Meyer, D. and Nerenz, D. R. (1980) The common sense representation of illness danger. In S. Rachman (ed.), *Medical Psychology*, vol. 2. New York: Pergamon Press.

Leventhal, H. and Nerenz, D. R. (1985) The assessment of illness cognition. In Karoly, P. (ed.), *Measurement Strategies in Health Psychology*. New York: John Wiley & Sons.

Ley, P. (1988) *Communicating with Patients*. London: Croom Helm.

Lindley, D. V. (1985) *Making Decisions*. New York: John Wiley & Sons.

Lippman, A. (1991) Research studies in applied human genetics: a quantitative analysis and critical review of recent literature. *American Journal of Medical Genetics*, **41**, 105–11.

Lippman-Hand, A. and Fraser, F. C. (1979) Genetic counseling – the postcounseling period. II. Making reproductive choices. *American Journal of Medical Genetics*, **4**, 73–87.

Lubs, M. (1979). Does genetic counseling influence risk attitudes and decision-making? *Birth Defects: Orig. Art. Serv.*, **15**, 355–67.

Marteau, T. M. (1989) Framing of information: its influence upon decisions of doctors and patients. *British Journal of Social Psychology*, **28**, 89–94.

Marteau, T., Drake, H. and Bobrow, M. (1994a) Counselling following diagnosis of a fetal abnormality: the differing approaches of obstetricians, clinical geneticists and genetic nurses. *Journal of Medical Genetics*, **31**, 864–7.

Marteau, T., Drake, H., Reid, M., Feijoo, M., Soares, M. Nippert, I., Nippert, P. and Bobrow, M. (1994*b*) Counselling following diagnosis of a fetal abnormality: a comparison between German, Portuguese and UK geneticists. *European Journal of Human Genetics*, **2**, 96–102.

Marteau, T. M., Kidd, J. and Plenicar, M. (1993) Obstetricians presenting amniocentesis to pregnant women: practice observed. *Journal of Reproductive and Infant Psychology*, **11**, 3–10.

Myers, M. F., Bernhardt, B. A., Tambor, E. S. and Holtzman, N. A. (1994) Involving consumers in the development of an educational program for cystic fibrosis carrier screening. *American Journal of Human Genetics*, **54**, 719–26.

Pauker, S. P. and Pauker, S. G. (1979) The amniocentesis decision: an explicit guide for parents. In C. J. Epstein, C. J. R. Curry, S. Packman, S. Sherman and B. D. Hall (eds.) *Risk, Communication, and Decision Making in Genetic Counseling*, pp. 289–324. New York: Alan R. Liss, Inc. (Birth Defects: Original Articles Series, vol. 15, no. 5C.)

Pearn, J. H. (1973) Patients' subjective interpretation of risks offered in genetic counselling. *Journal of Medical Genetics*, **10**, 129–34.

Pendleton, D. (1983) Doctor–patient communication: a review. In D. Pendleton and J. Hasler (eds.), *Doctor–Patient Communication*, pp. 5–53. New York: Academic Press.

Pitz, G. F. (1987) Evaluating decision aiding technologies for genetic counseling. In G. Evers-Kiebooms, J. J. Cassiman, H. Van den Berghe and T. d'Ydewalle (eds.), *Genetic Risk, Risk Perception, and Decision Making*, pp. 251–78. New York: Alan R. Liss, Inc. for the March of Dimes Birth Defects Foundation. (Birth Defects: Original Article Series, vol. 23, no. 2.)

Raiffa, H. (1968) *Decision Analysis: Introductory Lectures on Choices under Uncertainty*. Boston, MA: Addison-Wesley.

Rippetoe, P. A. and Rogers, P. W. (1987) Effects of components of protection–motivation theory on adaptive and maladaptive coping with a health threat. *Journal of Personality and Social Psychology*, **52**, 596–604.

Rosenstock, I. M. (1966) Why people use health services. *Millbank Memorial Fund Quarterly*, **44**, 94–127.

Rosenstock, I. M. (1974) The Health Belief Model and preventive health behaviour. *Health Education Monographs*, **2**, 354–86.

Roter, D. L. (1977) Patient participation in the patient–provider interaction: the effects of patient question-asking on the quality of interaction, satisfaction and compliance. *Health Education Monographs*, **5**, 281–315.

Roter, D. L. and Hall, J. H. (1989) Studies of doctor–patient interaction. *Annual Reviews of Public Health*, **10**, 163–80.

Rotter, J. B. (1954) *Social Learning and Clinical Psychology*. Englewood Cliffs, NJ: Prentice-Hall.

Rowley, P. T., Loader, S., Sutera, C. J., Walden, M. and Kozyra, A. (1991) Prenatal screening for hemoglobinopathies. III. Applicability of the Health Belief Model. *American Journal of Human Genetics*, **48**, 452–9.

Royal College of Physicians (1989) *Prenatal Diagnosis and Genetic Screening: Community and Service Implications*. London: The Royal College of Physicians.

Sagi, M., Shiloh, S. and Cohen, T. (1992) Application of the health belief model in a study on parents' intentions to utilize prenatal diagnosis of cleft lip and/or palate. *American Journal of Medical Genetics*, **44**, 326–33.

Sarafino, E. P. (1990) *Health Psychology: Biopsychosocial Interactions*. Chichester: John Wiley & Sons.

Schwarzer, R. (1992) Self-efficacy in the adoption and maintenance of health behaviours:

theoretical approaches and a new model. In R. Schwarzer (ed.), *Self-efficacy: Thought Control of Action*. London: Hemisphere Publishing Corporation.

Schild, S. and Black, R. B. (1984) *Social Work and Genetics. A Guide for Practice*. New York: Haworth.

Shafir, E. and Tversky, A. (1992) Thinking through uncertainty: nonconsequential reasoning and choice. *Cognitive Psychology*, **24**, 449–74.

Shiloh, S. and Saxe, L. (1989) Perception of recurrence risks by genetic counselees. *Psychology and Health*, **3**, 45–61.

Slovic, P. (1987) Perception of risk. *Science*, **236**, 280–5.

Sorenson, J. R., Kavenagh, C. M. and Mucatel, M. (1981*a*) Client learning of risk and diagnosis in genetic counseling. *Birth Defects: Orig. Art. Ser.* **27**(1), 215–88.

Sorenson, J. R., Scotch, N. A. and Swazey, J. P. (1981b) *Reproductive Pasts, Reproductive Futures, Genetic Counseling and its Effectiveness*. New York: Alan R. Liss, Inc.

Stiles, W. (1978) Verbal response modes and dimensions of interpersonal roles: a method of discourse analysis. *Journal of Personality and Social Psychology*, **36**, 693–703.

Taylor, S. E. (1983) Adjustment to threatening events: a theory of cognitive adaptation. *American Psychologist*, **38**, 1161–73.

Tuckett, D. A., Boulton, M. and Olson, C. (1985). A new approach to the measurement of patients' understanding of what they are told in medical consultations. *Journal of Health and Social Behaviour*, **26**, 27–38.

Turk, D. C., Rudy, T. E. and Salovey, P. (1986) Implicit models of illness. *Journal of Behavioral Medicine*, **9**, 453–74.

Tversky, A. and Kahneman, D. (1974) Judgment under uncertainty: heuristics and biases. *Science*, **185**, 1124–31.

Tversky, A. and Kahneman, D. (1981) The framing of decisions and the psychology of choice. *Science*, **211**, 453–8.

Twiss, S. B. (1979) Problems of social justice in applied human genetics. In A. M. Capron, R. F. Lappe, T. M. Murray, T. M. Powledge and D. Bergsma (eds.), *Genetic Counseling: Facts, Values and Norms*. New York: Alan R. Liss, Inc.

Vlek, C. (1987) Risk assessment, risk perception and decision making about courses of action involving genetic risk: an overview of concepts and methods. In G. Evers-Kiebooms, J. J. Cassiman, H. Van den Berghe and G. d'Ydewalle (eds.), *Genetic Risk, Risk Perception and Decision-Making*, pp. 171–207. New York: Alan R. Liss, Inc. for the March of Dimes Birth Defects Foundation. (Birth Defects: Original Article Series, vol. 23.)

Wasserman, R. C. and Inui, T. S. (1983) Systematic analysis of clinician–patient interactions: a critique of recent approaches with suggestions for future research. *Medical Care*, **21**, 279–93.

Weinstein, M. C. and Fineberg, H. V. (1980) *Clinical Decision Analysis*. Philadelphia: Saunders.

Wertz, D., Sorenson, J. and Heeren, T. (1984) Genetic counseling and reproductive uncertainty. *American Journal of Medical Genetics*, **18**, 79–88.

Wertz, D., Sorenson, J. and Heeren, T. (1986) Clients' interpretation of risks provided in genetic counseling. *American Journal of Human Genetics*, **39**, 253–64.

Yates, J. F. and Stone, E. R. (1992) Risk appraisal. In J. H. Yates (ed.), *Risk-taking Behaviour*. Chichester: John Wiley & Sons.

5

Evaluating carrier testing: objectives and outcomes

THERESA MARTEAU and ELIZABETH ANIONWU

Population screening to detect carriers of several recessive conditions is now possible. Such screening can be conducted preconceptually, antenatally or neonatally. This raises the question of which programmes should be implemented, and on what basis these decisions should be made. Professional organisations in many countries are responding to these vexed questions by producing reports that set criteria to be met by all genetic screening programmes (Health Council of The Netherlands, 1989; Royal College of Physicians, 1989; Nuffield Council on Bioethics, 1993; Andrews *et al.*, 1994). These reports emphasise the importance of patient autonomy in the decision whether or not to undergo a genetic test. The need to do more good than harm is also stressed. The purpose of this chapter is to consider the extent to which these principles are considered in evaluations of carrier screening programmes. Ways of realising the objectives of genetic screening programmes will then be discussed.

The number and type of screening programmes vary both between and within countries. The most frequently available programmes are for the common recessive conditions (see Table 5.1).

5.1 The objectives of carrier testing

To evaluate any clinical service, it is necessary to define its objective(s). Most often the objective is to improve one or more health outcomes. In genetic screening for recessive conditions, there is frequently no health improvement for the affected individual: the most common intervention is the offer of termination for affected pregnancies. To avoid past abuses of genetics as exemplified by Nazi Germany, there is a desire in many countries to ensure that decisions to terminate an abnormal pregnancy are well informed but not influenced by the views of others, particularly the state or other official bodies. The primary objective of genetic screening is usually seen to be providing information to allow informed choices, particularly concerning reproduction. This perspective is reflected in the guidelines regarding how genetic screening services should be implemented, such as those in the reports referenced above. Although carrier testing may result in many people opting to have pregnancies tested and affected ones terminated, this is not deemed to be an objective. Rather, this is seen as a consequence of the objective of the programme: informed decision-making, as

Table 5.1. *Examples of some carrier frequencies in various ethnic groups*

Condition	Ethnic group	Estimated carrier frequency
Cystic fibrosis[a]	White British	1 in 25
	Pakistanis	1 in 50[b]
Tay–Sachs disease[c]	Ashkenazi Jews	1 in 25
	Non-Jewish	1 in 250
Haemoglobinopathies[d]		
Beta-thalassaemia trait	Cypriots	1 in 7
	Asians	1 in 10–30
	Chinese	1 in 30
	Afro-Caribbeans	1 in 50
	White British	1 in 1000
Sickle cell trait	Afro-Caribbeans	1 in 10
	West Africans	1 in 4
	Cypriots	1 in 100
	Pakistanis, Indians	1 in 100
C trait	Afro-Caribbeans	1 in 30
	Ghanaians	Up to 1 in 6
D trait	Pakistanis, Indians	1 in 100
	White British	1 in 1000

Notes:
[a]   Watson *et al.* (1991).
[b]   Personal communication from Professor Bernadette Modell (University College London) and Dr Martin Schwarz (Royal Manchester Children's Hospital).
[c]   Ellis (1991).
[d]   Department of Health (1993).

stated in a recent US report (Andrews *et al.*, 1994, p. viii) on health and policy implications for genetic screening:

> Our committee stressed the importance of autonomous decision making by individuals and by the family even if the development of a genetic disease might be the outcome. We believe that in a society such as ours, autonomy far outweighs any public health considerations. In practice, most informed couples do, in fact, select various courses of action that lead to a lower frequency of genetic disease.

It is against these objectives that carrier and other genetic screening programmes need to be evaluated. But rarely, if ever, does this happen. Evaluations of screening programmes usually fail to assess the extent to which decisions about screening are based on good information. There have been no evaluations of screening that have reported the proportion of those offered screening that have made an informed choice. Rather, outcomes are reported as the proportion undergoing screening, diagnostic tests and terminations of affected pregnancies. There is thus discordance between the espoused objective of screening and the outcomes used in evaluating screening.

5.2 The outcomes of carrier testing

In describing the consequences of carrier testing, several outcomes are relevant: the decision-making process, and the cognitive, emotional, behavioural and social outcomes.

The process of decision-making

Although many acknowledge that the decision whether or not to participate in a genetic screening programme should be an informed one, there have been few attempts to define this, and even fewer that have tried to operationalise it.

Recent guidance on informed consent from the Department of Health in the UK states:

> Patients are entitled to receive sufficient information in a way they can understand about the proposed treatments, the possible alternatives, and any substantial risks, so that they can make a balanced judgement. *(Department of Health, 1990)*

Although this is perhaps a good starting point, it begs many questions. What is a balanced judgement? What understanding do people need or want in order to make a balanced judgement about whether to participate in a screening programme? What is sufficient information, and how is it most effectively provided?

One approach to defining informed consent is to determine how much knowledge people have when making a decision: the greater the knowledge, the more informed the decision. Another approach is to assume that knowledge, while necessary, is not a sufficient criterion. Shiloh and Saxe (1989) suggested that an informed decision is one that is not discrepant with a person's value system. It is not clear how this is assessed. It may be more useful to think about decision-making as a continuum, marked at one end by decisions about which the individual has no recollection of having made, and at the other by decisions that have involved the use of information in systematically weighing up the pros and cons of each option before deciding.

Many of those invited to participate in population-based screening programmes are making decisions based on very little understanding of the test being offered. In one study, one third of pregnant women who had been offered serum screening to detect spina bifida could not correctly recall whether they had undergone the test (Marteau *et al.*, 1988). Many people also do not know the meaning of their test results (Smith *et al.*, 1994). Understanding about the condition for which screening is being offered may also be poor (Hampton *et al.*, 1974). When asked what they understood about cystic fibrosis (CF), carriers recently detected in a population-based screening programme had very little information. Most overestimated the seriousness of the condition, thinking that death in early childhood was the norm, as opposed to a life expectancy that extends into early or mid-adulthood (Bekker *et al.*, 1994).

In population-based prenatal screening in the UK, uninformed decision-making is more common in ethnic minorities, particularly non-English speakers. In a survey of women using the maternity services in the Tower Hamlets district of London, fewer than 10% of non-English speaking women knew that they were having a test for a congenital abnormality (Akin-Deko, 1991). In a separate survey, only 30% of non-English speaking women who underwent serum screening for Down syndrome knew that they were having a blood test for congenital abnormalities (cited in Parsons *et al.*, 1992).

Many factors will influence how informed any decisions are, including how the screening programme is organised, the type of information provided, and the way in which it is given. Determining the type and amount of information that people require to make an informed decision is often considered a matter of clinical judgement (Wald and Law, 1992; Tobias and Souhami, 1993). Rarely is it an issue addressed by research.

One of the main factors influencing decisions about carrier testing and termination of pregnancy is the perceived seriousness or burden of a condition (for a review, see Marteau and Slack, 1992). Yet it can be one of the more difficult pieces of information to convey, particularly for conditions that are very variable. Many people do not have personal experience of the conditions for which carrier testing is being offered, and so are very reliant upon health professionals to inform them. However, health professionals vary in the information they provide. Different clinical experiences with a condition lead to different views on the nature of the condition (Christensen-Szalanski *et al.*, 1983; Marteau and Baum, 1984). What information should be given to present a 'fair' or 'balanced' view of the condition? In the recent UK screening programmes for CF, the literature prepared for those offered carrier testing was criticised as being unduly pessimistic about the life expectancy of a child born with CF (Britton and Knox, 1991). Uptake of screening for CF was far lower in the Baltimore programme than in the equivalent UK programmes (Tamber *et al.*, 1994). One reason for this might have been the more positive view of CF portrayed in the leaflets used in Baltimore. This view was based on discussions with families and individuals affected by CF. In order to achieve a representative view, information about the conditions needs to be based on the views of a wide range of health professionals involved in caring for people affected by the condition, together with the views of individuals and families affected by the condition.

The situations that encourage uptake of testing may be different from those that encourage informed decision-making. More people undergo tests when they are offered at the same time as the information is provided (Bekker *et al.*, 1993). Separating the provision of information from the provision of a test, although resulting in lower uptake, may result in better-informed decisions (Lorenz *et al.*, 1985; Tamber *et al.*, 1994).

Decisions about testing

The answer to the question, 'What is the uptake of carrier testing for any particular condition?' must be, 'It depends.' It depends upon the population offered testing and upon how the test is offered. Uptake of screening is higher among pregnant than non-pregnant populations, owing to the salience of the threat of disease in unborn children at that time (Richards and Green, 1993). But even among pregnant populations, uptake of the same screening test will vary. For carrier testing of cystic fibrosis (CF) in pregnant women, uptake rates in the UK varied across programmes from 70% (Mennie *et al.*, 1992) to 98% (Harris *et al.*, 1993). Among non-pregnant groups, variation in uptake can be even greater. In the recent CF programmes, uptake varied from 4% (Bekker *et al.*, 1993) to 86% (Watson *et al.*, 1991).

Outcomes of carrier testing

How people respond to undergoing carrier testing may depend upon many factors, including their expectations of receiving a negative or positive test result, the test results they actually receive, and their understanding of those results. These may be modified by pre-test counselling, and mediated by the individual's ethnic and cultural background. The relative importance of each of these factors is largely unknown. Responses may also be influenced by the stage in life when tests are conducted. Carrier screening may for example be associated with more distress in parents when conducted on their children neonatally rather than later. Similarly, there may be less distress if carrier status is determined before rather than during pregnancy. Below are summarised some of the outcomes of carrier detection.

Emotional outcomes

The emotional consequences of carrier detection are those most widely investigated. Of these, the most frequently studied has been anxiety. People may become anxious when they discover that they are carriers for a genetic disease (Childs *et al.*, 1976; Zeesman *et al.*, 1984). In recent population-based screening programmes for CF carriers, anxiety levels rose when told they were carriers, but fell to pre-existing levels shortly after (Mennie *et al.*, 1992; Bekker *et al.*, 1994). Among pregnant women, anxiety was reduced upon learning that their partners were not carriers (Mennie *et al.*, 1992).

However, not all people are made anxious by learning that they are carriers of a recessive condition. Among those who were tested as part of a population-based screening programme for cystic fibrosis (Bekker *et al.*, 1994) were 22 relatives of people with cystic fibrosis (Bekker *et al.*, 1994). Those with a known family history of cystic fibrosis did not become more anxious when they found out that they carried the gene. This contrasts with the raised anxiety following carrier detection in those without a family history (Figure 5.1). Prior knowledge and expectations may be factors influencing these responses. Among the detected carriers with a relative with CF, none reported being surprised at their test result, in contrast with 8 of the 14 carriers detected who did not have a relative with CF.

The emotional effects of being found to be a carrier have been less systematically documented for couples in which both are carriers.

Cognitive outcomes

The impact of genetic testing upon how people think has been less extensively studied than the impact upon emotions. However, research on people's cognitive responses to testing is a necessary adjunct to understanding their emotional responses. For example,

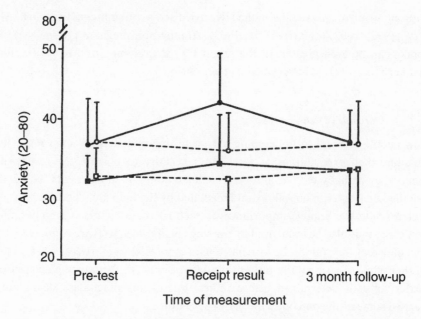

Figure 5.1 Anxiety in those receiving positive and negative test results in population-based screening and cascade screening. O- -O, Negative, population-based sample (*n*=467); □- -□, negative, cascade screening (*n*=14); ●—●, carriers, population-based sample (*n*=14); ■—■, carriers, cascade screening (*n*=8).

much concern was evident among parents of children found to be sickle cell carriers. In part this arose from beliefs about carrier status that were unfounded. For example, 43% of the parents thought of their child as having a disease, with 66% of parents thinking that their children needed dietary supplements to stay healthy (Hampton *et al.*, 1974).

How people understand their results is likely to influence how they respond to them. There was much concern that people would find negative results on CF population-based carrier testing too complex to understand. There are many mutations of the CF gene, 85% of which are routinely screened for. This means that those getting a negative test result have a residual risk of about 1:130 of being a carrier for CF. Although over 90% of those receiving a low-risk result understood that there was a residual risk, there was no evidence that this residual risk caused any residual concerns. Three months after undergoing population-based carrier testing for cystic fibrosis, nearly a fifth incorrectly believed that they were at no risk of being a carrier, and over a third of those with a positive result believed that they probably, but not definitely, carried the gene (Bekker *et al.*, 1994). While some forgetting is expected after the provision of complex medical information, the forgetting was selective and the recall of the meaning of the results optimistic. These findings may reflect the way that people process information. They may also reflect the way that people protect themselves from threatening information (for discussion, see Bekker *et al.*, 1994).

Those found to carry a gene for the recessive condition of Tay–Sachs disease had a less optimistic view of their future health compared with those screened and not found to be carriers (Marteau *et al.*, 1992). Carrier detection had taken place one to three years earlier. In contrast with the results of this study, no such effects have been found among those screened for cystic fibrosis carrier status (Bekker *et al.*, 1994; Meidzybrodzka *et al.*, 1995). These latter studies, however, have only documented initial reactions to screening. It remains to be determined whether pessimism about one's future health is a long-term consequence of learning that one is a carrier for a genetic condition. An alternative explanation for the lack of an effect of carrier status upon perceptions of future health concerns the amount and quality of counselling provided: less counselling was provided in the Tay–Sachs screening programmes than in the cystic fibrosis ones.

Reproductive behaviour

The primary purpose of carrier detection is to offer people information about their risks of having a child with a genetic condition, and the reproductive options available to them. These options include having children regardless of the risk, avoiding a child with a partner with the same recessive gene, undergoing prenatal testing with a view to termination of an affected fetus, and not having children. Few, if any studies, have assessed people's knowledge of the reproductive options available to them, particularly those found to be carriers. Certainly some women are not even aware that they have undergone testing. This is particularly so in testing for haemoglobinopathies. Historically these tests were part of routine blood investigations, undertaken to detect clinical conditions such as sickle cell disorders. In the process, carriers of haemoglobin disorders were identified, but the women were not always informed either that they had undergone a form of genetic screening or of the result (Nuffield Council on Bioethics, 1993).

We also know relatively little about how parents perceive their reproductive options. It would appear that carrier testing usually has little effect upon the selection of a partner. Following the introduction of sickle cell screening in Greece, no evidence was found that screening had affected the partners that people chose (Stamatoyannopoulos, 1974). Among some Jewish groups, however, the results of carrier testing for Tay–Sachs disease are used to influence choice of marital partner (Cornwell, 1994). Young people involved in such schemes are routinely tested to see if they carry a gene for Tay–Sachs. Their results are not divulged, but held on a register administered by a Rabbi. If a couple wish to marry, they apply to a Rabbi for approval. If they are both carriers for Tay–Sachs, approval would not be granted. In theory, this system circumvents the possibility of stigma attached to being a carrier, and avoids dilemmas of prenatal diagnosis. How well this system works in practice is unknown (Kleinman, 1992; for further discussion of this scheme, see Chapter 12).

Carrier detection can only affect decisions about whether and how to have children if done before pregnancy. Decisions about having children, however, are more strongly influenced by the desire to have children than by any perceived risk of a genetic condition (Frets, 1990; Shiloh, Chapter 3 in this volume). So, for example, the majority of couples with a risk greater than 15% of having a child with a genetic condition opt to have children. But whereas knowledge of carrier status does not affect most people's decisions to have children, it does affect the use that people make of prenatal diagnosis: for example, the majority of couples who undergo testing and are both found to carry a gene for thalassaemia use prenatal diagnosis (Modell *et al.*, 1980; Knott *et al.*, 1986). Prenatal diagnosis is more likely to be requested the earlier the pregnancy: Petrou *et al.* (1992) reported that of the 170 couples at risk of having a child with sickle cell disease, 50% requested prenatal diagnosis; 82% of mothers seen in the first trimester of pregnancy requested prenatal diagnosis compared with 49% of those seen in the second trimester. It was noted that there was a difference in the uptake of prenatal diagnosis for sickle cell in respect to ethnic origin, with more people of African than Afro-Caribbean origin choosing this option. Few studies, if any, have considered how carriers actually make decisions about the use of prenatal testing.

There has also been little research on the consequences of these decisions, or on the relationship between the process and outcomes of decisions. As with decisions to undergo testing, there is widespread agreement that decisions about whether to continue with an affected pregnancy should be made by the woman or the couple, informed and supported but not influenced by the views of health professionals. There is some circumstantial evidence to suggest that these conditions are not always met. In two uncontrolled studies, the proportion of women who terminated pregnancies affected by a sex chromosome anomaly was significantly higher among those consulting only an obstetrician than among those additionally consulting geneticists or paediatricians (Holmes-Siedle *et al.*, 1987; Robinson, 1989). These differences may reflect differences in the information given or the style in which it was given, or differences between women consulting different groups of health professionals. Support for the explanation that these differences reflect differences in counselling style comes from a recent study in which obstetricians described their own counselling style as significantly more directive than geneticists did (Marteau *et al.*, 1994). Further support for obstetricians being directive in counselling comes from observational studies of their counselling (Marteau *et al.*, 1993) and women's reports of how they were actually counselled (Farrant, 1985). An alternative explanation for these findings from uncontrolled studies is that parents who were more reluctant to terminate their affected pregnancies consulted health professionals other than obstetricians.

Both descriptive and controlled studies are needed to determine how people make decisions, and the outcomes of different decisions. This can provide the background against which to plan and evaluate different counselling approaches.

Social consequences

Concerns have been expressed about the possible adverse social and economic consequences of carrier detection (Lippman, 1991; see also Chapters 14 and 16). The nature and extent of these will vary between cultures, depending in part upon the history of screening, pre-existing social status of at-risk groups, and the attitudes of insurance companies and the way that health care is financed.

Selective carrier screening for conditions such as the haemoglobinopathies and

Tay–Sachs disease is based on the premise that there are high-risk populations in particular ethnic groups (Modell and Modell, 1990). Determining which ethnic groups are
at risk of being carriers for conditions such as cystic fibrosis, Tay–Sachs disease, thalassaemia and sickle cell disorders is complex. At present, with a negative family history
for a variety of autosomal recessive disorders an individual's ethnic background is
more important in defining a priori risk (e.g. CF in Caucasians, sickle cell in African
Americans). In the future, however, racial and ethnic status may not be as important a
risk predictor; this has already happened in Tay–Sachs disease, where more births of
infants with Tay–Sachs disease now occur in non-Jews than among Jews (Andrews *et
al.*, 1994). A similar situation has also been reported in Britain, with Ellis (1991) noting
that of the 10 children born each year in Britain with Tay–Sachs disease, 7 occur within
the non-Jewish community. The view that cystic fibrosis only occurs in whites is still
prevalent despite evidence to the contrary (Goodchild *et al.*, 1974; Neale and Kaur,
1993) and has led to delayed diagnosis in children from ethnic minorities (Spencer *et
al.*, 1993). When is a risk high enough to justify the offer of screening? Table 5.1 highlights the range of carrier frequencies of these four conditions within different ethnic
groups. If a screening programme selectively identifies a person from a high-risk ethnic
group as a carrier and she or he has a partner from a low-risk group, the latter should be
offered screening. Failure to do so has led to the birth of an affected child where the
white father was eventually tested and found to be a carrier for a haemoglobinopathy.
The local health authority made a payment to the family of £175 000 in an out-of-court
settlement (Younge, 1990).

The provision of screening services for the haemoglobinopathies in the UK has been
patchy and of variable quality (Anionwu, 1993). In the UK, universal screening for
haemoglobinopathies in pregnancy and in the neonatal period is now recommended
where the at risk populations exceed 15% of the total antenatal population
(Department of Health, 1993). These screening programmes will also identify carriers
in 'low-risk' ethnic groups, but very little is known about the impact that such information may have. Universal neonatal screening for sickle cell disease has been undertaken
for several decades in some states within the USA, and the prevalence of the condition
and the carrier state in various ethnic groups has been reported (Sickle Cell Disease
Guideline Panel, 1993). The mean prevalence of sickle cell trait per 100 000 population
among blacks was 6550, Hispanic (total) 579, Hispanic (Eastern States) 3040 and
Whites 258. A study of parents of newborns identified with sickle cell trait through a
neonatal screening programme in Rochester, NY, noted: 'Not all individuals with sickle
cell trait are black. In this group, 11% were Hispanic, Italian, or of other ethnic groups,
some of whom were surprised to learn that they carry a sickle cell gene' (Rowley, 1989).
There have been similar findings in Florida, with Mack (1989) commenting, 'The
common belief that they [Hb S and Hb C] occur only in blacks results in a reluctance to
consider the diagnosis or to accept it when made, because to do so implies racial inter-
mixing.'

There has been much greater discussion about the possibility that genetic screening
may create racial stigma within ethnic groups seen to be at higher risk (see Chapter 14).

In screening programmes conducted over 20 years ago, there was some evidence of stigma, felt or experienced, among those found to be carriers. So, for example, in Cyprus, those found to be carriers of sickle cell disease were far less likely to tell their friends that they had been screened than were those who had been screened and found not to be carriers (Stamatoyannopoulos, 1974). More recent studies in the UK of those found to be carriers of cystic fibrosis found no reluctance among carriers to tell others of their test results (Watson *et al.*, 1992).

Failure to find any social stigma in the recent studies may reflect a change in social attitudes towards genetic disease, or a difference between the way the programmes were organised. Alternatively, they may reflect differences between two conditions that differ in their ability to confer stigma. Sickle cell is predominantly found in black people, who already experience discrimination. Carrying a gene that predisposes to disease may reinforce negative attitudes. Some support for this latter view comes from two studies examining the perceptions of carriers held by non-carriers and carriers. Those who had participated in a screening programme for sickle cell trait were asked to describe what it felt like to be a carrier. Non-carriers perceived being a carrier in a more negative light than did carriers, seeing them as less happy, less healthy and less active than carriers (Wooldridge and Murray, 1989). Using a similar methodology, Evers-Kiebooms found that carriers and non-carriers of the gene for cystic fibrosis had similar and fairly positive perceptions of being a carrier (Evers-Kiebooms *et al.*, 1994). But carriers attributed more negative feelings to other carriers than to themselves. The authors argue that the cognitive bias of thinking oneself better than others (illusory superiority) may counterbalance any felt stigmatising effects of being a carrier.

One explanation for the differences in findings between these studies is that people discriminate between being a carrier for cystic fibrosis and for sickle cell disease. But the studies differ in many other respects that may also affect the results, for example, sample selection and culture (Europe compared with North America). Future studies need designs that allow cross-cultural comparisons of responses to different conditions.

Most concerns about discrimination against carriers of genetic conditions in Europe and North America revolve around insurance cover. Fears are widespread that genetic testing may be made mandatory for some insurance schemes (Wilfond and Fost, 1990; Billings *et al.*, 1992). Further concerns are highlighted in Billings's report of problems with insurance companies encountered by carriers. Several of the cases illustrate discrimination by employers and insurance companies. Part of the debate hinges around the perceived control and preventability of genetic conditions. Whereas some argue that genetic make-up is beyond an individual's control and hence something for which they should not be discriminated against, others argue that the birth of children with recessive conditions is potentially preventable, and hence deserving of higher premiums. Many countries are considering legislation to prevent such discrimination (see Chapter 9).

5.3 Evaluating carrier testing: outstanding research questions

Evaluations of carrier screening programmes rarely extend beyond a consideration of uptake of the service, and of the immediate psychological consequences of carrier detection. To consider how the objectives of the programmes may be met requires the main outcomes to be defined and operationalised for use in research.

Informed decisions

From studies of persuasive communication, we know that people make decisions in one of two ways. The first involves 'systematic processing' of information, with both sides of the argument being evaluated, resulting in attitudes that are relatively enduring and predictive of behaviour. The second uses 'heuristic processing' and involves the use of simple rules-of-thumb such as 'experts can be trusted'. Information is processed peripherally, leading to an attitude that is liable to change and is not predictive of behaviour.

In some situations, people prefer to make decisions about their health care heuristically rather than systematically. For example, following the diagnosis of cancer, about half of all patients prefer decisions about their treatment to be made by doctors rather than themselves (Degner and Sloan, 1992). In a study of non-patients who were asked what questions they would ask in trying to decide whether they wanted carrier testing for cystic fibrosis, the majority made a decision after having only one or two questions answered (Myers *et al.*, 1994). It was the minority that informed themselves about the nature, risks and benefits of the test before making a final decision.

The nature of carrier testing may be especially conducive to the use of heuristic rather than systematic processing of information. Outcomes may be unfamiliar, preferences ill-defined and events unknown. In addition there may be well-established values associated with medical technology. There is some evidence that decisions in this context are associated with the use of heuristic rather than systematic reasoning (Tallman and Gray, 1990). However, awareness of all the possible consequences of a decision may make people feel more confident that they made the correct decision, and may make them more psychologically prepared for their test results. There is some evidence from laboratory studies that appropriate confidence increases as the amount of cognitive processing on a problem requiring a choice increases (Sniezak *et al.*, 1990).

More resources may be required to facilitate people's processing of information centrally, rather than their use of peripheral cues and heuristics. The latter, for example, could be achieved by the usual information leaflet, whereas the former may require some more active and structured provision of different types of information alongside suggestions about how to integrate these in the process of making a decision. The use of interactive videos may be one method of achieving this. Research is needed to evaluate the outcomes of these two approaches to decision-making to determine whether extra resources do lead to extra benefits for those undergoing carrier testing.

Understanding responses

Most studies of carrier testing are designed in such a way that any differences between carriers and non-carriers are attributed to their test results. The possible effects of how screening is organised cannot be determined from such uncontrolled studies. Similarly, most studies do not consider the possible effect of individual differences such as gender, ethnicity or coping style upon responses to screening. These are important in understanding responses to screening, as a first step to determining the most effective and efficient ways of providing such services.

Men and women respond differently to the offer of carrier testing to their test results (see Chapter 12): women are more likely to respond to an invitation for carrier screening. In one study, they were twice as likely as men to respond to an invitation for carrier status detection (Bekker *et al.*, 1993). Possibly many men see carrier testing as associated with reproduction and childbirth and hence more the responsibility of women. But when offered screening when already attending a clinic, men were significantly more likely to accept the offer of testing. This suggests that social influences may plan an important role in determining responses of men to the offer of testing.

Longer-term consequences of carrier detection

Evaluations of screening programmes are largely confined to assessments of the immediate consequences. A full evaluation of any screening programme requires consideration of its longer-term consequences. There are several groups of people whose longer-term responses to screening merit more study: women undergoing terminations of affected pregnancies after carrier detection and prenatal diagnosis, and women who give birth to children with different screening histories. Such histories would include having declined screening, not having been offered screening, accepting screening but declining a diagnostic test, and receiving a false negative result on screening.

There have been several follow-up studies of women who have undergone terminations following detection of Down syndrome or neural tube anomalies (see Chapter 6). Terminating a pregnancy for a fetal abnormality causes acute grief as well as relief (Lloyd and Laurence, 1985; Black, 1989). In a two-year follow-up of couples who had undergone terminations of pregnancies affected by fetal abnormalities, 20% of the 84 women interviewed still experienced regular bouts of crying, sadness and irritability (White-Van Mourik *et al.*, 1992). Iles (1989) found that women often experienced guilt at 'destroying a life' and at 'opting out of' rearing a disabled child. Women undergoing terminations for pregnancies affected by spina bifida, a condition compatible with life, had poorer psychological outcomes than women whose babies had anecephaly, and hence did not survive much beyond birth. Extrapolating from these findings, women who undergo terminations for CF, for example, a condition compatible with leading a fulfilling life into adulthood (Walters *et al.*, 1993) may have poorer psychological outcomes. Loader *et al.* (1991) interviewed women who had been at risk for having a child with a haemoglobinopathy. Half the women accepted the offer of prenatal diagnosis.

Of four affected pregnancies, just one was terminated. Of 26 women interviewed, 18 were happy with the decisions they had made about the use of prenatal diagnosis. Although women were generally very satisfied with the service they had received, the authors raise the question of whether the screening programme was a good use of scarce resources, given that 18 907 people were screened, to detect four affected pregnancies of which just one was terminated.

Even with the most comprehensive carrier screening programmes, some parents will give birth to children with conditions for which carrier testing and prenatal diagnosis were available This is for two main reasons: most carrier tests will not detect all mutations; and not all parents will undergo screening or prenatal diagnosis, some declining and others not being offered testing. There have been no formal studies of the impact of giving birth to a child that could have been detected prenatally. In a recent study, we found that geneticists, obstetricians and the general public were more blaming towards mothers who give birth to a child with Down syndrome, having declined screening, than they were towards mothers not offered screening who gave birth to affected children (Marteau and Drake, 1995). Evidence from a preliminary study suggests that some parents do blame health professionals for failing to detect their affected children during pregnancy. This was evident both for those receiving false negative results, and those who had not been offered screening (Hall, Bobrow and Marteau, submitted for publication). In four of the six families interviewed who had received a negative result in pregnancy screening, one or both of the parents blamed health professionals for failing to detect their affected child prenatally. In 7 of the 19 families not offered screening, at least one parent blamed health professionals for not having detected the affected child. None of the four families who had declined screening blamed health professionals.

There is a substantial body of evidence showing that blaming others for misfortune is frequently associated with poorer adjustment (Taylor *et al.*, 1984; Tennen *et al.*, 1986; Shapp *et al.*, 1992). The pattern of results from this pilot study suggests that the availability of prenatal screening may be adversely affecting adjustment to caring for affected children in those parents who did not make a decision to eschew such tests. This is an area that requires more research both to understand the processes behind these reactions, and to determine whether adjustment can be facilitated through prenatal or postnatal interventions.

Concluding comment

Population-based screening services are organised and evaluated in a way that suggests that the primary objective is carrier detection and termination of affected pregnancies. Although the importance of informed decision-making is frequently acknowledged in screening programmes, this is not reflected in evaluations of screening programmes. The need to correct this is urgent.

5.4   Acknowledgement

Theresa Marteau is supported by the Wellcome Trust, UK.

5.5   References

Akin-Deko, A. L. S. (1991) *Maternity Services Liaison Scheme: Working for the Community*. London: Maternity Services Liaison Scheme.

Andrews, L. B., Fullarton, J. E., Holtzman, N. A. and Motulsky, A. G. (eds.) (1994) *Assessing Genetic Risks. Implications for Health and Social Policy*. Washington, DC: National Academy Press.

Anionwu, E. N. (1993) Sickle cell and thalassaemia: community experiences and official response. In W. I. U. Ahmad (ed.), *'Race' and Health in Contemporary Britain*. Buckingham: Open University Press.

Bekker, H., Dennis, G., Modell, M., Bobrow, M. and Marteau, T. (1994) The impact of screening for carriers of cystic fibrosis. *Journal of Medical Genetics*, **31**, 364–8.

Bekker, H., Modell, M., Dennis, G., Silver, A., Mathew, C., Bobrow, M. and Marteau, T. M. (1993) Uptake of cystic fibrosis carrier testing in primary care: supply push or demand pull? *British Medical Journal*, **306**, 1584–6.

Billings, P. R., Kohn, M. A., De Cuevas, M. Beckwith, J., Alper, J. S. and Natowicz, M. R. (1992) Discrimination as a consequence of genetic testing. *American Journal of Human Genetics*, **50**, 476–82.

Black, R. B. (1989) A 1 and 6 month follow-up of prenatal diagnosis patients who lost pregnancies. *Prenatal Diagnosis*, **9**, 795–804.

Britton, J. and Knox, A. J. (1991) Screening for cystic fibrosis. *Lancet*, **338**, 1524.

Childs, B., Gordis, L., Kaback, M. M. and Kazazian, H. H. (1976) Tay–Sachs screening: social and psychological impact. *American Journal of Human Genetics*, **28**, 550–8.

Christensen-Szalanski, J. J. J., Beck, D. E., Christensen-Szalanski, C. M. and Koepell, T. D. (1983) Effects of expertise and experience on risk judgements. *Journal of Applied Psychology*, **68**, 278–84.

Cornwell, T. (1994) Jewish marriage makers embrace testing for genetic disease. *Observer*, 6 March, p. 27.

Degner, L. F. and Sloan, J. A. (1992) Decision making during serious illness: what role do patients really want to play? *Journal of Clinical Epidemiology*, **45**, 941–50.

Department of Health (1990) *Patient Consent to Examination or Treatment*. (HC (90) 22). London: Department of Health.

Department of Health (1993) *Standing Medical Advisory Committee Working Party Report on Sickle Cell, Thalassaemia and other Haemoglobinopathies*. London: HMSO.

Ellis, I. (1991) Carrier screening for Tay–Sachs. *Nursing*, **4**, 16–18.

Evers-Kiebooms, G., Denayer, L., Welkenhuysen, M., Cassiman, J. J. and Van den Berghe, H. A. (1994) The stigmatizing effect of the carrier status for cystic fibrosis? *Clinical Genetics*, **46**, 336–43.

Farrant, W. (1985) 'Who's for amniocentesis?' The politics of prenatal screening. In H. Homans (ed.), *The Sexual Politics of Reproduction*, pp. 96–177. London: Gower.

Frets, P. G. (1990) The reproductive decision after genetic counselling. Ph.D. thesis, Erasmus University, The Netherlands.

Goodchild, M. C., Insley, J., Rushton, D. I. and Gaze, H. (1974) Cystic fibrosis in 3 Pakistani children. *Archives of Disease in Childhood*, **49**, 739–41.

Hall, S., Bobrow, M. and Marteau, T. M. Parents' attributions for the birth of a child with Down syndrome: A pilot study. Submitted for publication.

Hampton, M. L., Anderson, J., Lavizzo, B. S. and Bergman, A. B. (1974) Sickle cell 'nondisease'. *American Journal of Diseases of Children*, **128**, 58–61.

Harris, H., Scotcher, D., Hartley, N., Wallace, A., Craufurd, D. and Harris, R. (1993) Cystic fibrosis carrier testing in early pregnancy by general practitioners. *British Medical Journal*, **306**, 1580–3.

Health Council of The Netherlands (1989) *Heredity, Science and Society*. The Hague.

Holmes-Siedle, M., Rynanen, M. and Lindenbaum, H. (1987) Parental decisions regarding termination of pregnancy following prenatal detection of sex chromosome abnormality. *Prenatal Diagnosis*, **7**, 239–44.

Iles, S. (1989) The loss of early pregnancy. *Ballière's Obstetrics & Gynaecology*, **3**, 769–90.

Kleinman, M. (1992) An alternative program for the prevention of Tay–Sachs disease. In B. Bonné-Tamir and A. Adam (eds.), *Genetic Diversity Among Jews*, pp. 346–8. Oxford University Press.

Knott, P. D., Ward, R. H. T. and Lucas, M. K. (1986) Effect of chorionic villus sampling and early pregnancy counselling on uptake of prenatal diagnosis. *British Medical Journal*, **293**, 479–80.

Lippman, A. (1991) Prenatal genetic testing and screening: constructing needs and reinforcing inequities. *American Journal of Law and Medicine*, **17**, 15–50.

Lloyd, J. and Lawrence, K. M. (1985) Sequelae and support after termination of pregnancy for fetal malformation. *British Medical Journal*, **290**, 907–9.

Loader, S., Sutera, J., Segelman, S., Kozyra, A. and Rowley, P. T. (1991) Prenatal hemoglobinopathy screening. IV. Follow-up of women at risk for a child with a clinically significant hemoglobinopathy. *American Journal of Human Genetics*, **49**, 1292–9.

Lorenz, R. P., Botti, J. J., Schmidt, C. M. and Ladda, R. L. (1985) Encouraging patients to undergo prenatal genetic counseling before the day of amniocentesis: its effect on the use of amniocentesis. *Journal of Reproductive Medicine*, **30**, 933–5.

Mack, A. K. (1989) Florida's experience with newborn screening. *Pediatrics*, **63**, 861–3.

Marteau, T. M. and Baum, J. D. (1984) Doctors' views on diabetes. *Archives of Diseases in Childhood*, **54**, 566–70.

Marteau, T. M. and Drake, H. (1995) Attributions for disability: the influence of genetic screening. *Social Science and Medicine*, **40**, 1127–32.

Marteau, T. M. and Slack, J. (1992) Psychological implications of prenatal diagnosis for patients and health professionals. In D. J. H. Brock, C. H. Rodeck and M. A. Ferguson-Smith (eds.), *Prenatal Diagnosis and Screening*, pp. 663–73. London: Churchill Livingstone.

Marteau, T. M., Drake, H. and Bobrow, M. (1994) Counselling following diagnosis of a fetal abnormality: the differing approaches of obstetricians, clinical geneticists and genetic nurses. *Journal of Medical Genetics*, **31**, 864–7.

Marteau, T. M., Johnston, M., Plenicar, M., Shaw, R. W. and Slack, J. (1988) Development of a self-administered questionnaire to measure women's knowledge of prenatal screening and diagnostic tests. *Journal of Psychosomatic Research*, **32**, 403–8.

Marteau, T. M., Plenicar, M. and Kidd, J. (1993) Obstetricians presenting amniocentesis to pregnant women: practice observed. *Journal of Reproductive and Infant Psychology*, **11**, 3–10.

Marteau, T. M., van Duijn, M. and Ellis, I. (1992) Effects of genetic screening on perceptions of health: a pilot study. *Journal of Medical Genetics*, **29**, 24–6.

Meidzybrodzka, A., Hall, M., Mollison, J., Templeton, A., Russell, I., Dean, J., Kelly, K., Marteau, T. and Haites, N. (1995) Antenatal carrier screening for cystic fibrosis: randomised trial of step-wise vs. couple screening. *British Medical Journal*, **310**, 353–7.

Mennie, M. E., Gilfillan, A., Compton, M., Curtis, L., Liston, W. A., Pullen, I. *et al.* (1992) Prenatal screening for cystic fibrosis. *Lancet*, **340**, 214–16.

Modell, M. and Modell, B. (1990) Genetic screening for ethnic minorities. *British Medical Journal*, **300**, 1702–4.

Modell, B., Ward, R. H. T. and Fairweather, D. V. I. (1980) Effect of introducing antenatal diagnosis on reproductive behaviour of families at risk for thalassaemia major. *British Medical Journal*, **280**, 1347–50.

Myers, M. F., Bernhardt, B. A., Tambor, E. S. and Holtzman, N. A. (1994) Involving customers in the development of an educational program for cystic fibrosis carrier screening. *American Journal of Human Genetics*, **54**, 719–26.

Neale, T. and Kaur, G. (1993) *Cystic Fibrosis and Asian Families*. Bromley, Kent: Cystic Fibrosis Research Trust.

Nuffield Council on Bioethics (1993) *Genetic Screening: Ethical Issues*. London: Nuffield Foundation.

Parsons, L., Richards, J., and Garlick, R. (1992) Screening for Down's syndrome. *British Medical Journal*, **305**, 1228.

Petrou, M., Brugiatelli, M., Ward, R. H. T. and Modell, B. (1992) Factors affecting the uptake of prenatal diagnosis for sickle cell disease. *Journal of Medical Genetics*, **29**, 820–3.

Richards, M. and Green, J. (1993) Attitudes towards prenatal screening for fetal abnormality and detection of carriers of genetic disease: a discussion paper. *Journal of Reproductive and Infant Psychology*, **11**, 49–56.

Robinson, A. (1989) Decisions following the intrauterine diagnosis of sex chromosome aneuploidy. *American Journal of Medical Genetics*, **34**, 552–4.

Rowley, P. T. (1989) Parental receptivity to neonatal sickle cell trait identification. *Pediatrics*, **63**(5.2), 891–3.

Royal College of Physicians (1989) *Prenatal Diagnosis and Genetic Screening. Community and Service Implications*. London: The Royal College of Physicians.

Shapp, L. C., Thurman, S. K. and DuCette, J. P. (1992) The relationship of attributions and personal well-being in parents of preschool children with disabilities. *Journal of Early Intervention*, **14**(4), 295–303.

Shiloh, S. and Saxe, L. (1989) Perception of risk in genetic counseling. *Psychology and Health*, **3**, 45–61.

Sickle Cell Disease Guideline Panel (1993) *Sickle Cell Disease: Screening, Diagnosis, Management, and Counseling in Newborns and Infants. (Clinical Practice Guideline* no. 6.) U.S. Department of Health and Human Services.

Smith, D., Shaw, R. W. and Marteau, T. (1994) Lack of knowledge in health professionals: a barrier to providing information to patients. *Quality in Health Care*, **3**, 75–8.

Sniezak, J. A., Paese, P. W. and Switzer, F. S. III (1990) The effect of choosing on confidence in choice. *Organizational Behavior and Human Decision Processes*, **46**, 264–82.

Spencer, D. A., Venkataraman, M. and Weller, P. H. (1993) Delayed diagnosis of cystic fibrosis in children from ethnic minorities. *Lancet*, **342**, 238.

Stamatoyannopoulos, G. (1974) Problems of screening and counseling in the hemoglobinopathies. In A. G. Motulsky and W. Lenz (eds.), *Birth Defects*, pp. 268–76. Amsterdam: Excerpta Medica.

Tallman, I. and Gray, L. N. (1990) Choices, decision, and problem-solving. *Annual Review of Sociology*, **16**, 405–33.

Tambor, E. S., Bernhardt, B. A., Chase, G. A., Faden, R. R., Geller, G., Hofman, K. J. and Holtzman, N. A. (1994) Offering cystic fibrosis carrier screening to an HMO population: factors associated with utilization. *American Journal of Human Genetics*, **55**, 626–37.

Taylor, S. E., Lichtman, R. R. and Wood, J. V. (1984) Attributions, beliefs about control, and adjustment to breast cancer. *Journal of Personality and Social Psychology*, **46**, 489–502.

Tennen, H., Affleck, G. and Gershman, K. (1986) Self-blame among parents of infants with perinatal complications: the role of self-protective motives. *Journal of Personality and Social Psychology*, **50**, 690–6.

Tobias, J. and Souhami, R. (1993) Fully informed consent can be needlessly cruel. *British Medical Journal*, **307**, 1199–201.

Wald, N. and Law, M. (1992) Screening, ethics and the law. *British Medical Journal*, **305**, 892.

Walters, S., Britton, J. and Hodson, M. E. (1993) Demographic and social characteristics of adults with cystic fibrosis in the United Kingdom. *British Medical Journal*, **306**, 549–52.

Watson, E. K., Mayall, E., Chapple, J., Harrington, K., Williams, C. and Williamson, R. (1991) Screening for carriers of cystic fibrosis through primary health care services. *British Medical Journal*, **303**, 504–7.

Watson, E. K., Mayall, E. S., Lamb, J., Chapple, J. and Williamson, R. (1992) Psychological and social consequences of community carrier screening programmes for cystic fibrosis. *Lancet*, **340**, 217–20.

White-van Mourik, M. C. A., Connor, J. K. and Ferguson-Smith, M. A. (1990) Patient care before and after termination of pregnancy for neural tube defects. *Prenatal Diagnosis*, **10**, 497–505.

Wilfond, B. S. and Fost, N. (1990) The cystic fibrosis gene: medical and social implications for heterozygote detection. *Journal of the American Medical Association*, **263**(20), 2777–82.

Wooldridge, E. Q. and Murray, R. F. (1989) The health orientation scale: a measurement of feelings about sickle cell trait. *Social Biology*, **35**, 123–36.

Younge, P. (1990) £175,000 for ruined life. *The Voice*, no. 372 (24 July), pp. 1–3.

Zeesman, S., Clow, C. L., Cartier, L. and Scriver, C. R. (1984) A private view of heterozygosity: eight-year follow-up study on carriers of the Tay Sachs gene detected by high school screening in Montreal. *American Journal of Medical Genetics*, **18**, 769–78.

6

Psychosocial aspects of prenatal screening and diagnosis

JOSEPHINE GREEN and HELEN STATHAM

The antenatal clinic has been the scene of much of the routine screening carried out by the medical profession, and much of what we know about people's reaction to routine screening has been derived from this setting. At present, few of the tests offered to pregnant women owe their existence to the new genetics. However, increasingly such tests will be possible and it is expected that a major application of the new genetics will be determining the genetic status of a fetus. This may involve new techniques such as direct sampling of fetal cells from maternal blood. In the immediate future, however, it will use the same techniques for obtaining information about the fetus as are currently used in prenatal testing (e.g. amniocentesis and chorionic villus sampling), even if subsequent laboratory techniques will be different.

In this chapter we shall describe the tests commonly used in pregnancy to detect fetal abnormality, summarise what is known of their psychosocial benefits and hazards and consider likely future developments including carrier screening for recessive disorders during pregnancy (see also Chapters 3 and 5). First, however, we shall discuss the important distinction between 'screening' and 'diagnosis'.

6.1 Screening versus diagnosis

Most genetic disorders are extremely rare, of the order of one in a number of thousand pregnancies, and there are a great many of them. It would not be feasible to test every pregnancy for every known disorder, even if this were thought desirable. This has always been the issue with diagnostic tests in pregnancy: the costs, both financial and otherwise, make it impracticable to give them to everybody. It is for this reason that we have **screening** tests that select out high-risk groups for whom the financial and psychological costs of diagnostic testing are thought to be justified.

Pregnant women are screened in various ways to identify sub-groups who are more likely to be carrying an abnormal fetus than the rest of the pregnant population. These women, once defined as **at risk**, can be offered diagnostic tests. A **risk factor** can be anything that has been observed to have an association with fetal anomaly that is higher than average. Those that are identifiable at the booking interview include maternal age, ethnic group and family history. Because screening tests cannot indicate that the baby definitely is or is not affected, it is in the nature of such tests that they 'get it wrong', i.e.

some people with a screening result indicating **high risk** have babies that are unaffected (false positive), and some with a **low risk** do in fact have affected babies (false negative). Cut-offs are usually chosen to minimise the number of false negatives, but that often means a high proportion of false positives (Wald, 1984). Uncertainty is therefore inherent in the process.

6.2 Serum screening

Since the late 1970s, the measurement of maternal serum alpha-fetoprotein in maternal blood (MSAFP) has been widely used in the UK to screen for neural tube defects (NTDs). Its introduction into routine use in the USA has been much slower, with only a small number of states adopting it as standard care and only California mandating that all pregnant women be offered the test (Press and Browner, 1993). The incidence of neural tube defects (NTDs) has, in fact, been falling worldwide, even in places where there has been no screening and selective abortion (Stone *et al.*, 1988). For this reason, and because ultrasound was becoming more sensitive and specific, a number of UK hospitals were starting to abandon mass MSAFP screening during the late 1980s, especially in places with a low incidence of NTDs. However, this changed with the discovery that low levels of MSAFP were associated with Down syndrome (Merkatz *et al.*, 1984) and the UK has seen a renewed interest in serum screening. Serum screening is usually carried out at 16 weeks and results are generally available within one week. Accuracy depends critically on gestational age being correctly known, because the level of MSAFP increases by about 19% per week in the second trimester (Wald and Cuckle, 1984). Women whose serum screening result indicates an increased risk of NTDs or Down syndrome will then be offered a diagnostic test: usually ultrasound for NTDs and amniocentesis for Down syndrome.

6.3 Ultrasound scanning

Ultrasound scanning is in a different category from other techniques for investigating fetal well-being. First, it gives instant results: the process is happening there and then. Secondly, it does not fit easily into a classification of 'screening versus diagnosis' as we have defined them. In the UK and much of Europe it is used on virtually everybody, irrespective of risk status, 'just to make sure that the baby's all right'. Even where routine use is not official policy, e.g. in the USA, the majority of pregnant women are still scanned at least once during pregnancy. In these circumstances it is being used as a screening technique. It may reveal findings that are of limited significance in their own right but are known to be associated with certain syndromes. In these cases the scan findings will be the cue for further diagnostic investigations. This is increasingly happening for chromosomal disorders (Nicolaides *et al.*, 1992). However, scans may also give definitive information, e.g. about structural abnormalities, number of fetuses, and fetal death, so in this sense they are diagnostic.

6.4 Diagnostic tests

To detect genetic abnormalities it is usually necessary to have a sample of fetal cells. This is most commonly obtained from amniotic fluid, from which fetal cells may be cultured. This procedure – amniocentesis – is generally carried out between 16 and 20 weeks of pregnancy. Results normally take 3 to 4 weeks because cells have to be cultured, although for some single-gene disorders results can now be made available within 48 hours by using polymerase chain reaction (PCR) techniques. Chorionic villus sampling (CVS) is an alternative, and less widely available, means of obtaining fetal cells. The villi can be sampled through the cervix or through the abdomen; the transabdominal route is tending to be preferred (see, for example, Elias *et al.*, 1989; Brambati *et al.*, 1990). The major advantage of CVS over amniocentesis is that it can be performed at a much earlier stage of pregnancy, usually between nine and eleven weeks. In addition, results are available more quickly because chromosome results can be obtained without the need for cells to be cultured. However, compared with amniocentesis, there is a greater likelihood of needing a repeat test. A further disadvantage is that it is only of relevance to women whose risk status is established early in pregnancy, and it is therefore not of use as an adjunct to screening programmes that take place during the second trimester. Both tests are associated with a risk of spontaneous abortion: about 1 in 150 for amniocentesis and about 1 in 50 for CVS.

Further details of the tests currently in use may be found in Green & Statham (1993).

6.5 Accepting testing

Routine testing

Richards and Green (1993) have argued that the system of care offered to pregnant women in Britain is taken up almost universally and with little questioning because women wish to do the best for their babies; they follow procedures presented to them as desirable by doctors, midwives and in the baby books they read. In the Cambridge Prenatal Screening Study (Green *et al.*, 1994), a study of 1824 pregnant women from 9 hospitals in 4 English Health Regions, 97% agreed with the statement 'regular antenatal check ups are essential for the health of mother and baby'. Most women have little appreciation of the possible consequences of the screening that is part of a series of routines that constitute antenatal care. Postnatally, 27% of women in a study in a London teaching hospital did not know that they had undergone maternal serum screening for neural tube defects (Marteau *et al.*, 1988).

Tests that are routine are assumed to be necessary and appropriate. Reassurance is derived from knowing that the pregnancy is being monitored and that if there were anything wrong it would be detected and dealt with. It is this, as much as the receipt of test results, that makes routine care reassuring. Therefore a woman in the Cambridge study, commenting on what she had particularly liked about her antenatal care, said:

> 'Seeing a doctor and he not saying there is anything wrong makes you feel better'
> *(Statham* et al., *1993*a, *p. 61).*

The 'routineness' of the package of antenatal care is at the same time a source of comfort and a potential source of threat. A woman who had had an abnormality detected on a routine ultrasound scan said:

> 'somehow, when you go for a routine scan for size, you can't believe that they'll find something so dreadful' *(Statham, 1994).*

One of the characteristics of tests that are routine is that women can feel that somebody else has already made the decision for them about the test's appropriateness. Thorpe *et al.* (1993) asked mothers of newborn children about their views of ultrasound scanning in a non-routine context: cerebral scanning of neonates. Many mothers had misgivings about this which they then had to accommodate within their framework of acceptance of antenatal scanning:

> 'If it was necessary I would have [consented] – if it was routine – no problem' *(Thorpe* et al., *1993).*

Thus, the acceptance of routines *because* they are routine means that pregnant women do not necessarily make an informed decision to undergo screening and diagnostic tests. They may not see it as appropriate to make a decision, they may not be given information on which to make a decision, may not read or understand information they are given, or may not know they are having a test. For example, a woman who received an abnormal result on a Down screening test said that she had not read the information given her about that test because she had assumed it was for spina bifida, as in a previous pregnancy (Statham and Green, 1993).

The decision to be tested

One of the first studies of women who were undergoing prenatal screening (Farrant, 1985) highlighted an important distinction between their viewpoint and that of those delivering the service: obstetricians sought to detect abnormality whereas mothers sought reassurance that there was no abnormality. It is likely that the desire for – and expectation of – reassurance underlie the decisions many women make about screening during pregnancy.

Many factors may influence a woman's decision about testing: attitudes to abortion, perception of risk, the perceived burden of having a disabled child, attitudes to risks of the test itself (Green *et al.*, 1992). Perceived risk of abnormality, rather than objective probability, may be particularly important (Marteau *et al.*, 1991). The very existence of tests for fetal abnormality can create pressures to use the technology (Green, 1990). Tymstra (1989) has suggested that 'anticipated decision regret' can be an important motivational force; a particular choice is made partly to avoid future regret at not having made that choice: 'I would never forgive myself if . . .'

The way in which staff present tests to pregnant women may also have an impact on the decisions they make: Marteau *et al.* (1993) have observed consultations in which amniocentesis was discussed with women. Similar numerical risks for Down syndrome and miscarriage were presented as high and low respectively. A midwife whom we knew to be opposed to serum screening for Down syndrome told us that she knew she gave non-directive counselling because only 50% of her mothers had the test compared with 70–80% of women booked with her colleagues.

Attitudes to abortion

The aim of prenatal diagnosis is to determine whether or not a fetus has an abnormality. It is unlikely, however, that prenatal diagnosis would have been resourced in the way it is, just to give parents reassurance, or if most parents with an abnormality chose to continue the pregnancy. The 'enormous potential for the avoidance of serious genetic disease and congenital malformation' (Weatherall, 1992) can only be realised in most cases if women who conceive fetuses with such a genetic disease or malformation terminate the pregnancy.

There seems to be agreement across a number of studies (Breslau, 1987; Faden *et al.*, 1987; Green *et al.*, 1993*a*) that around 30% of women say that they would not consider termination on grounds of fetal abnormality. However, attitude to abortion does not necessarily predict uptake of prenatal testing. Most studies where both have been measured have found a higher proportion of their sample thinking they might use prenatal testing than thinking that they might abort. Evers-Kiebooms *et al.* (1993) conducted a study of attitudes to a range of disorders among a non-pregnant sample attending a Belgian Adult Education Centre. Whereas 66% of the sample reported they would use prenatal testing for Down syndrome, only 40% thought that they would terminate an affected pregnancy. In one of our studies (Green *et al.*, 1993*a*), 29% of the women who had routine serum screening for NTDs had said they would *not* consider terminating a pregnancy on the grounds of a strong chance that the baby would be handicapped. Only a small number of women in this study took active steps to avoid screening tests; for a sizeable group, disinclination to terminate was not seen as a reason to avoid screening. Three reasons were suggested for this apparent mismatch: women did not know they were having screening tests or did not understand that abortion was the only 'treatment'; women wanted the information to prepare for the birth of a handicapped child; women wanted reassurance and, even if they were ambivalent about testing, the most likely outcome of the tests was that they would be told that the baby was healthy. The possibility of being faced with a moral dilemma should an abnormality be detected can be seen then as a potential hazard of testing, but not an overriding reason for not having a test.

High-risk couples

The situation is likely to be very different for people who embark on a pregnancy knowing that they are at above-normal risk. They are likely to have first-hand experience of the condition for which their fetus has increased risk and to know about the various test procedures and their costs and benefits. They may have thought about whether or not to have prenatal diagnosis and will probably have thought about whether or not to terminate the pregnancy in the event of a positive result. Such decisions about diagnosis and termination may be particularly difficult: they may have a child living with (and perhaps dying of) a disorder. To consider terminating a pregnancy when another fetus is found to be affected with the same disorder may seem to be rejection of that existing child. For the carrier mother of a child with a recessive disorder in the Cambridge study (Green *et al.*, 1992) the non-availability of prenatal diagnosis made her decision to become pregnant easier; she avoided putting herself into the position of saying to her six-year-old disabled son, 'We've got another one like you so we're getting rid of it.'

A number of studies of people who know themselves to be carriers of genetic disorders have shown that, as with groups who are not at risk, more are likely to consider using prenatal testing than to consider abortion. For different disorders and for the same disorder in different studies, reported attitudes to termination of affected pregnancies are very varied. For example, 4–8% of those at risk for autosomal dominant polycystic kidney disease said that they would terminate an affected pregnancy (Sujansky *et al.*, 1990), compared with 28% of those at risk for fragile-X syndrome (Meryash and Abuelo, 1988). Some of these differences may be due to differences in the severity and burden of the disorder. However, differences are even greater between different groups of parents at risk from the same disease, cystic fibrosis: 17–20% in an American study said they would terminate (Wertz *et al.*, 1991), 65% in Belgium (Evers-Kiebooms *et al.*, 1990) and 52% in Wales (Al-Jader *et al.*, 1990). Further research is needed to explain these differences. They are likely to relate to cultural and temporal differences, as well as methodological ones. For example, the American study was carried out at a time of considerable optimism about the long-term prognosis for a child with cystic fibrosis, because of the recent discovery of the gene (Wertz *et al.*, 1991). This study did find that willingness to terminate was related to level of education but this was not found by Al-Jader *et al.* (1990). Gender differences have not been systematically explored in this area but the available evidence suggests that attitudes are not related to gender (Wertz *et al.*, 1991; Snowdon and Green, 1994).

Care must be taken in generalising from hypothetical statements to actual behaviour. In a study of a population at risk for Huntington's disease, 43% said that they would use prenatal diagnosis in the event of pregnancy, and 40% of these said that they would definitely terminate, while 35% were uncertain (Adam *et al.*, 1993). In practice, only 7 out of 38 who became pregnant (18%) eventually used prenatal diagnosis. A major problem in interpreting studies of high-risk groups is in establishing an appropriate baseline: those to whom prenatal diagnosis and/or termination are unacceptable are probably less likely to become pregnant. Therefore rates of use of prenatal diagnosis and subsequent termination may be an overestimate of the acceptability of these techniques. The decisions made in practice by those having abnormalities detected will be discussed later in this chapter.

6.6 Women's experiences of screening

The major hazard of screening tests is the high level of anxiety associated with false positive results, which was first reported by Farrant (1980) and Fearn *et al.* (1982). This has been confirmed in all subsequent studies (Berne-Fromell *et al.*, 1983; Burton *et al.*, 1985; Marteau *et al.*, 1992; Green *et al.*, 1994). Fearn and colleagues found that the anxiety persisted even when women were told that there was not a problem after all. The most severe levels of anxiety were found in women who, having gone on to have amniocentesis, were not told the results. They were just told to assume that all was well if they did not hear to the contrary. A recent survey of 357 consultant obstetricians in England and Wales (Green, 1995) reveals that a small number (2%) still follow such a policy, although, thankfully, most do not.

Most published studies have been concerned with MSAFP screening for NTDs, but there have been a small number of reports of women's experiences of positive results from serum screening for Down syndrome. These indicate levels of anxiety at least as high as those associated with NTD screening (Evans *et al.*, 1988; Abuelo *et al.*, 1991; Keenan *et al.*, 1991; Marteau *et al.*, 1992; Roelofsen *et al.*, 1993; Statham and Green, 1993). The State Anxiety scores (Spielberger *et al.*, 1970) reported by Keenan and colleagues for women with *low* MSAFP levels (i.e. high risk for Down syndrome) are much higher than the comparable scores reported by Burton *et al.* (1985) for women with *high* MSAFP levels (i.e. high risk for NTDs). A number of authors (e.g. Abuelo *et al.*, 1991) have specifically drawn attention to the higher anxiety of women selected as high risk for Down syndrome on the basis of a serum screening test in contrast to those with the identical numerical risk based on age alone. The extent of this in Abuelo's study was that 'several women with low AFPs refused to participate in our study because they stated that they were "too nervous"' (p. 384).

Women's experience of ultrasound scanning

Studies of women's experiences of ultrasound from the early 1980s, e.g. Milne and Rich (1981) and the King's College Hospital study (Campbell *et al.*, 1982; Reading and Cox, 1982; Reading *et al.*, 1982) showed that what women liked about scans was a moving image that was interpreted for them. This has been confirmed by subsequent studies (Field *et al.*, 1985; Reading and Platt, 1985; Cox *et al.*, 1987). Women who are having scans that are potentially diagnostic (after raised MSAFP) seem to be just as enthusiastic about them as women for whom they are routine (Hunter *et al.*, 1987; Tsoi and Hunter, 1987; Tsoi *et al.*, 1987). For most British women, 'seeing' the baby on scan is a high spot of the pregnancy, an event to be shared with the baby's father and siblings (Green *et al.*, 1992). There is now an expectation that the ultrasound image will be deciphered for them and features of the fetal anatomy pointed out, with consequent disappointment when this does not occur (see, for example, Jacoby, 1988). Women do not view scans as a threat that might give them bad news (like amniocentesis), but as a benign procedure that allows them to see their baby and confirm that it is healthy. Their assumption that the procedure is benign is, of course, encouraged by the fact that scans are routine and given to everyone.

A telling comment from the study by Thorpe *et al.* (1993) of neonatal cerebral scanning was, 'This scan looks for things that are wrong whereas the scan in pregnancy checks to see that the baby is OK.' This distinction between 'checking that everything is OK' and 'looking for abnormalities' is probably one of the keys to women's enthusiasm for scans.

Surprisingly little is known about the experiences of women whose scans are not reassuring, particularly those whose results are 'false positives' (Griffiths and Gough, 1985; Madarikan *et al.*, 1990). This is an increasing area of concern for scan operators and neonatal paediatricians.

Experiences of diagnostic tests

At present, women who have diagnostic tests fall into four main groups: those who have had a previous handicapped child, those who have not but have a family history of a genetic disorder, women of 'advanced maternal age' (usually 35 or over), and women who have had positive screening tests. In the future it may be that the second category will be expanded to include those who have been found to carry a genetic disorder *without* there being a family history where appropriate tests are available, e.g. cystic fibrosis carrier couples. Until recently the largest of these categories was 'advanced maternal age' (7–8% of pregnancies in the UK), and this is therefore the group of women about whom most is known. It is in fact clear from the literature that these different groups do react to the process of prenatal testing in different ways, with those having the test on grounds of age alone tending to be less anxious and more likely to be reassured by a negative result than those having prenatal diagnosis for other reasons (see, for example, Chervin *et al.*, 1977; Beeson and Golbus, 1979; Evers-Kiebooms *et al.*, 1988). It is also clear that women having diagnostic tests as a result of positive screening tests are especially anxious (see, for example, Farrant, 1985; Statham and Green, 1993) and this is a point that needs to be remembered when new screening tests are introduced.

Studies of women's experiences of amniocentesis have been appearing since the mid-1970s, whereas those on CVS date only from the late 1980s. The amniocentesis literature shows some consistent findings, despite considerable methodological variation: worries about potential miscarriage and the stressful nature of the 3–4-week waiting period for results, followed, usually, by a drop in anxiety once normal results are given. The first controlled study comparing women having amniocentesis with women who were not (Fava *et al.*, 1982, 1983) found no differences between the groups by mid-pregnancy and therefore concluded that amniocentesis does not allay anxiety. Three subsequent studies (Phipps and Zinn, 1986; Marteau, 1991; Statham *et al.*, 1993a) have also failed to find significantly lower anxiety scores for women who have had negative amniocentesis results compared with untested controls. The latter two studies have concluded that differences between groups are likely to relate to pre-existing characteristics, which is a recurring theme in this area (see Green, 1990).

The assumption has always been that CVS would be less stressful for women than amniocentesis for three reasons: (1) they do not have to wait so long to have the test done; (2) they do not have to wait so long for results (1 week, compared with 3 or 4); (3) termination of pregnancy, if indicated, can be carried out in the first trimester and thus by dilatation and curettage rather than by inducing labour at 20 weeks or more. The first two assumptions are, on the whole, supported by the literature. McCormack *et al.* (1990), for example, in a retrospective study of 152 women who had had CVS, found high acceptability and willingness to accept higher miscarriage rates as a trade off for the benefits of earlier and quicker results. This was especially true for women at known risk of passing on a genetic disorder. However, although many women may be willing, in principle, to accept the higher miscarriage rate, it may not be so easy to accept in practice. Robinson *et al.* (1991) found that women who miscarried after CVS experienced a great deal of guilt and blamed their loss on their selfish desire for an earlier test result. This is an important point that should be taken into account when counselling women for prenatal diagnosis.

Data are lacking with respect to the third assumption: that earlier terminations are less distressing. The only relevant studies (Black, 1989; Robinson *et al.*, 1991) are too small and methodologically confused to allow conclusions to be drawn. However, Richard *et al.* (1992) did find that women terminating after early diagnosis by CVS tended to come for fewer follow-up visits than women who had had later diagnoses. A study by Iles (1989) also suggests that terminations late in the second trimester are associated with higher psychiatric morbidity than those earlier in the second trimester (for a further discussion, see Green *et al.*, 1992; Statham, 1992, 1994).

The first study comparing experiences of CVS and amniocentesis involved 61 women enrolled in the Canadian randomised controlled trial (Spencer and Cox, 1987, 1988). Robinson *et al.* (1988) published a separate study, based on another 54 women from the same trial. Both studies indicated that CVS evoked less anxiety over a shorter period of time than amniocentesis. There is also the suggestion that women delay becoming attached to their fetus until after they have had a negative result, whether at 15 weeks with CVS, or at 21 weeks with amniocentesis (Spencer and Cox, 1987, 1988; Caccia *et al.*, 1991). This is what Rothman (1986) has called the 'tentative pregnancy'. Kolker (1989) suggests that positive feelings after normal test results are a function of the anxiety created by the test itself: had the test not raised anxieties, there would be no need for reassurance. This interpretation is consistent with the findings from the amniocentesis studies, quoted above, that found no net benefit to tested women compared with those who were not tested.

Findings from randomised controlled trials raise another important issue: the extent to which people are undergoing procedures that they have actively chosen. A variety of data from disparate areas all support what psychologists might expect, namely that people have more 'loyalty' to something that they have chosen for themselves and are thus less likely to show adverse psychological effects. This was shown directly in a study by Verjaal *et al.* (1982). This can be a problem with randomised trials, because women will not necessarily be having the treatment that they would have chosen. In the

Canadian randomised controlled trial of amniocentesis and CVS, for example, CVS was not available to women outside the trial. Any woman wanting it therefore had to enrol in the trial and hope that she was randomised accordingly. Women who preferred amniocentesis would be less likely to join the trial because they could guarantee their choice by abstaining. Thus, in Spencer and Cox's study the negative feelings about amniocentesis were likely to come from women who had not only not chosen it, but who preferred CVS initially. In contrast, another study (Tunis *et al.*, 1990), which compared 30 women who had chosen to have amniocentesis with 151 who had chosen CVS, found virtually no differences between the groups on a range of psychological measures.

6.7 Diagnosis of abnormality

Rothman (1986) suggests that prenatal testing encourages women to view their babies as commodities that may be rejected if found to be substandard. Pregnancies thus come to be regarded as 'tentative', that is, the mother holds back from relating to her fetus until after appropriate tests have revealed a healthy baby. Although, as we have seen, this is true for some women, it is likely that the reason they do this is to try to protect themselves from the pain they know will be associated with the news of an abnormal baby, rather than for consumerist reasons. There is no evidence that such a strategy does in fact lessen the trauma of a positive result.

The discovery that a fetus has an abnormality may be seen either as a hazard or as a benefit of prenatal screening and diagnosis, but it is always devastating for parents. It is usually unexpected, for as the report of the Royal College of Physicians (1989) stated: 'most infants with congenital malformations and chromosomal disorders are born to healthy young women with no identifiable risk factor.' A possible consequence of the new genetics is that more individuals will be identified as carriers of genetic disorders before any pregnancy following population screening programs or following the birth of an affected child to a family member. Thus, we can expect that more people will start a pregnancy knowing that they are at increased risk, although it is also possible that some people may refrain from becoming pregnant under these circumstances. As the number of genetic disorders that can be detected during pregnancy continues to grow, more couples will also face decisions about terminating affected pregnancies.

Making the decision to terminate an affected pregnancy

A number of factors may contribute to the decision to terminate an affected pregnancy, including: cultural and individual attitudes to abortion, the nature of the abnormality, when it is detected, the certainty of the diagnosis and prognosis, parental age and reproductive history, and who counsels the couple. These factors cannot always be disentangled from each other.

Clayton-Smith *et al.* (1989) reported that 'most' of 37 women undergoing amniocentesis initially opted for abortion when told of a sex chromosome aneuploidy. After counselling, 23 chose to continue the pregnancy. Where the prognosis is poor, however, most women choose termination. In a study by Drugan *et al.* (1990), 93% of women terminated pregnancies after the discovery of chromosomal abnormalities with severe consequences, compared with 27% where the prognosis was uncertain. Timing of the diagnosis did not influence the decision for either category of chromosome defect in this study, and this has also been reported for ultrasound-detected abnormalities (Pryde *et al.*, 1992). However, in a study by Verp *et al.* (1988), parents were more likely to terminate early in pregnancy after diagnosis with CVS (97.6%) than later after diagnosis with amniocentesis (78.1%).

The decision to terminate a pregnancy is different in a number of ways when the baby is affected with a lethal condition. Jorgensen *et al.* (1985) found that the doctor saying the baby would not survive was helpful in decision-making. In some cases, however, continuing a pregnancy where the baby will not survive may be a better option for women who are intrinsically opposed to abortion (Watkins, 1989).

In a study of pregnant women (Faden *et al.*, 1987), respondents were asked about their hypothetical willingness to consider abortion as both the severity and certainty of a diagnosis of fetal abnormality increased. The greatest difference in willingness to consider abortion was between 95% and 100% certainty, a difference that may not be considered significant by doctors and a degree of certainty that cannot be guaranteed for those whose prenatal diagnosis is based on linkage analysis (See Chapter 2).

Holmes-Siedle *et al.* (1987) found that parental age and reproductive history were significantly associated with the decision to terminate. This study also showed that the profession of the person counselling the parents was important; parents were more likely to terminate following counselling by a general obstetrician than by a geneticist. Marteau *et al.* (1994) have shown recently that obstetricians report themselves to be more likely to counsel directively than clinical geneticists, who were themselves more likely to be directive than genetic nurses.

Not all women may feel that they have a real choice about terminating an affected pregnancy. For some the socio-economic realities of caring for a disabled child deprive them of any real option, but for others there is an imperative generated by the testing process itself. A positive screening result creates uncertainties that most people feel the need to resolve.

> I had a choice in that no one would have *forced* me to have it [amniocentesis], but psychologically I did not have any choice in that if I hadn't had it I couldn't have gone through with the pregnancy   *(Farrant, 1985, p. 111).*

If, eventually, they are found to have a baby with a handicapping condition they need to make sense of having been through the anxieties created by the screening by acting on the result.

> If they'd handed her to me and said she was Down's I'd have been upset but I'd have got on with it; but once you've got into the testing trap you have to get to the end *(Statham & Green, 1993, p. 175).*

Even if there is no overt pressure on a woman to terminate an affected pregnancy it will be the expected course of action because no other action (as opposed to inaction) is possible. In addition, women may be subject to direct pressures. Over one third of a sample of obstetricians in England and Wales said that they generally require a woman to agree to terminate an affected pregnancy before proceeding with prenatal diagnosis, and 13% agreed with the statement, 'The state should not be expected to pay for the specialised care of a child with a severe handicap where the parents had declined the offer of prenatal diagnosis of the handicap (Green, 1995). This suggests that women who make an active choice to proceed with affected pregnancies (or who choose to avoid testing) may experience more negative attitudes than those who give birth to such children unwittingly.

Another small group of women for whom no real choice is available are those who would choose termination but for whom it is unavailable because of the time at which the abnormality is detected. In most countries termination of pregnancy is illegal after 24 weeks' gestation. The law in England and Wales has recently been amended by the 1992 Human Fertilization and Embryology Act such that terminations on the grounds of 'substantial risk that if the child were born it would suffer from such physical and mental abnormalities as to be seriously handicapped' can be performed without time limit, as also in France. However, a recent British survey (Green, 1993), suggested that obstetricians are either ignorant of the law or are interpreting it such that terminations after 24 weeks must be for more serious conditions than at earlier gestations.

Termination of pregnancy

A number of studies have described the psychological responses of women to termina-tion for fetal abnormality, although nearly all are retrospective and lacking in compari-son groups. The main issues that are important for parents were addressed by Blumberg *et al.* (1975), one of the earliest studies. Post-termination depression was found in both parents, with guilt at being responsible for the decision-making and loss of self esteem because of the conception of a handicapped child. These authors recog-nised the importance of the pregnancy being wanted. Subsequent studies confirmed these findings (Donnai *et al.*, 1981; Leschot *et al.*, 1982).

As stillbirth, neonatal death and the birth of a handicapped child were recognised as distressing events for parents, so Lloyd and Laurence (1985) and Jorgensen *et al.* (1985) likened the experiences of women undergoing abortion for fetal abnormality to those of women experiencing a perinatal death. Jorgensen commented, 'A decisive difference between stillbirth and abortion of a malformed fetus is that in the case of abortion, the parents have actively decided to terminate the pregnancy, thereby causing the death of a living fetus. The fact that the fetus was malformed might result in even stronger self-accusations and feelings of guilt.' Iles (1989) showed a better psychiatric outcome for women a few months after terminations for lethal conditions.

Current literature from support groups in the UK and the USA (Minnick *et al.*, 1990; SATFA, 1994) in which women and men describe their experiences, confirms that feelings first described 20 years ago remain unchanged. A recent study, which inter-viewed parents in the west of Scotland two years after their terminations (White-van

Mourik *et al.*, 1992), also confirmed the persistence of distress in many parents. Two particularly vulnerable groups were identified: those who were young or immature, and those with secondary infertility.

Further discussion of these studies can be found in Green *et al.* (1992), Statham (1992) and Statham (1994).

6.8 The social context of screening

One of the most important lessons to be learnt from the literature is that reactions to tests change over time. This can be seen in each of the major areas that we have considered so far. Amniocentesis, for example, when it first became available, was greeted very positively and all those having it said that they would have it again and recommend it to their friends (see, for example, Robinson *et al.*, 1975). It was offering women something that they had not had before, whose disadvantages had not as yet received much publicity and furthermore most of the women having it at this early stage were those who had actively sought it. In contrast, by the late 1980s, when CVS was at its height of popularity, amniocentesis was being viewed much more negatively. In the study by Spencer and Cox (1988), for example, every one of the 31 women planning further pregnancies said that they would want CVS rather than amniocentesis next time. It may well be that the more recent negative publicity concerning the risk of limb reduction after CVS (Firth *et al.*, 1994) will have changed this again.

Ultrasound offers a somewhat different example of changes over time, partly because its introduction into routine use in the UK was so rapid that there was little opportunity to study women's reactions in the early phase. The only study comparing the views of women to whom routine scanning was available with those to whom it was not (Hyde, 1986) showed that women were much less positive than they are today, especially those to whom it was not being routinely offered. Hyde's study draws attention to a more general point, what Porter and Macintyre (1984) have termed 'what is must be best'. The very existence of a particular test makes an implicit statement that the test is worth having. People assume that they would not be offered a procedure if it were not of proven benefit, and conversely that if a test is not on offer it is because it is not worth having. It may be that users of health services have become more cynical in the past decade but we have certainly seen recent evidence in the context of testing for the breast cancer susceptibility gene *BRCA1*:

> I would assume from the knowledge so far that if they were prepared to offer me the test they would then be prepared to come up with a treatment that would either contain it, kill it or monitor it so that it didn't get to the stage of killing me *(Richards* et al.*, 1995)*.

In fact it cannot be assumed that the offer of a test means that the doctor thinks that it is clinically useful. A study by Ennis *et al.* (1991) showed that many obstetricians in the UK routinely use tests that they consider to be inaccurate. In another recent survey of consultant obstetricians in England and Wales, 35% said that they were carrying out screening/diagnostic procedures because of outside pressures rather than because they considered them to be clinically valuable (Green, 1994).

The other change that may happen over time is that a procedure becomes taken for granted as part of the antenatal package, and, as we have seen with MSAFP screening for NTDs, some women do not know that they are having it (Marteau *et al.*, 1988). This is even more apparent with tests that have been around for longer; for example, Green *et al.* (1993*b*) found that only 23% of pregnant women knew that their blood had been tested for syphilis, and the same study (Statham *et al.*, 1993*b*) found similar ignorance about which neonatal tests had been carried out. Once a test is completely taken for granted it apparently makes no difference to women whether they are informed of a negative result as opposed to simply assuming that all is well (Green *et al.*, 1994). However, once a test has reached this stage it has lost its power to reassure.

The recent introduction of serum screening for Down syndrome provides us with an interesting case study with which to examine the pitfalls of introducing new tests. The first problem was that the test was replacing something that was already there; in fact it was replacing two things, and that has been the first source of confusion. On the one hand it was replacing serum screening for neural tube defects. Having the test was just the same: both involved taking blood at around 16 weeks and measuring levels of MSAFP. Therefore women thought that they knew about it. However, it was also replacing an existing form of screening for Down syndrome: maternal age. The relationship between age and Down syndrome was something that women and midwives felt they knew about: older women had amniocentesis. It was never really viewed as screening, and thus the majority of older women (those without babies with Down syndrome) were not being seen as 'false positives', just as the young women who gave birth to babies with Down syndrome were not seen as 'false negatives'.

The introduction of a new screening method challenged old ideas by drawing attention to the fact that any pregnant woman could be carrying a Down syndrome fetus. Some older women felt deprived at not having greater access to amniocentesis than younger women, and the belief that they had grown up with – that they were at increased risk because of their age – had to be rethought. At the same time, however, there was strong consumer demand for serum screening for Down syndrome from many younger women who resented their exclusion from testing, and also from older women who wished to avoid amniocentesis.

The next problem was that the test was difficult to understand. This was partly because the details of the test (what was measured and what cut-offs were used) varied from place to place. However, a more fundamental difficulty was a confusion in the minds of both women and health workers between screening and diagnosis. The test was seen as a poor test because it had false positives and false negatives, even though it performed relatively well on these criteria compared with maternal age. (The positive predictive value of the triple test is 1 in 68, which is about twice that of age.) Both women and health workers seem to have difficulty extracting meaningful information from 'one-in-something' risk figures. It is likely that the other factor that makes serum screening so much more alarming to women is the fact that it is based on specific information about this particular pregnancy, not just membership of an impersonal risk group. Finally, because the biochemical basis of the test is not understandable to the non-specialist, the news that the baby does not, after all, have Down syndrome may not be totally reassuring, but rather leave parents asking, 'If the baby doesn't have Down syndrome, what does it have?' (Statham and Green, 1993). These lingering doubts produced by the belief that 'there's no smoke without fire' have been reported in screening programmes in other contexts also (for example, by Rothenberg and Sills, 1968; Bodegard *et al.*, 1983).

6.9 Issues for the future

The contributions of the new genetics to testing in pregnancy in the future are likely to be in the areas of both screening and diagnosis. The techniques of the new genetics are already of relevance in diagnostic testing where a couple is known to carry a particular genetic disorder. PCR techniques can be used to speed up the process of detecting a specific gene mutation, so that results can be available within hours instead of weeks. At present, known carriers of single-gene defects are a minority of those having prenatal diagnosis, but this is a group that might be expected to increase in the future.

Currently, most carriers of recessive disorders do not learn that they are carriers until they have had at least one affected child. Population testing for heterozygotes for common mutations (see Chapter 5) could change this. Such testing could, in principle, happen at any stage of the life-cycle. Testing once a pregnancy is ongoing is arguably too late, but the counter-argument is that pregnant women are the most cost-effective target group, and that uptake of heterozygote testing is dramatically higher during pregnancy than at other times (Watson *et al.*, 1991; Bekker *et al.*, 1993; Harris *et al.*, 1993; Mennie *et al.*, 1993).

Although uptake of cystic fibrosis carrier testing is much higher during pregnancy, that does not mean to say that that makes it a better option. When such testing is offered during pregnancy, even if at the very first GP consultation, there are inevitably constraints on decision-making that would not exist at other times. The first, obviously, is that the option of avoiding pregnancy is not available to the couple. The second is that they may have to consider whether they wish to terminate an affected pregnancy. Thirdly, if they wish to exercise that option, they will have to have all necessary tests and make their decision very quickly. In the recent study of obstetricians' attitudes towards prenatal screening and diagnosis, 13% said that they would not be prepared to recommend termination of pregnancy on grounds of cystic fibrosis at *any* gestational age, and only 8% would do so beyond 24 weeks (Green, 1995).

As we have said, offering any test carries with it a message that the test is worth having. With tests offered in pregnancy that are specific to a particular disorder, for example Down syndrome or cystic fibrosis, the message is that the disorder in question is serious enough to justify termination of pregnancy. One concern is that those being offered these tests probably know very little about the disorder that they are being invited to avoid, so that this message may be unduly influential. As studies of MSAFP testing (e.g. Farrant, 1985) have shown, it is very difficult to get off the rollercoaster once embarked. Once a woman discovers that she is a cystic fibrosis carrier, she will almost certainly want to know if her partner is too, if only to make sense of having been tested in the first place. If he is, then of course she will want the baby tested for the same reason. To reach the stage of a positive diagnosis on the baby and then not to act on this information raises the question of why one has undergone all the previous testing. Unfortunately the only 'action' available is to terminate the pregnancy. These circumstances may result in couples making decisions that they would not have made had they known more about the disorder or been given more time to consider options. The report of parents requesting termination of fetuses who, like them, are healthy carriers of cystic fibrosis (Brambati *et al.*, 1993) raises particular questions in this respect.

We have to be careful in assuming that any one screening test can be used as a model for the introduction of another. This is clearly seen with the example of MSAFP for testing for NTDs/Down syndrome that we have already discussed. As well as the question of historical and cultural context, each test is likely to have some idiosyncratic feature that creates a new set of problems. With cystic fibrosis, as with other recessive disorders, there are a number of issues associated with obtaining samples from both parents (see Chapter 5). There is also the problem that there are a large number of mutations and it is practicable to test for only the few most common. Thus, although the test can tell someone that they definitely are a carrier, it cannot tell them that they definitely are not. This is a problem that we may expect to see with tests for other genetic conditions with many mutations.

Fragile-X syndrome is the second most common genetic cause of mental retardation after Down syndrome. It primarily affects boys, although one third of carrier females show some intellectual impairment (Davies, 1989). Mass screening in early pregnancy has been suggested and is already technically possible (Palomaki, 1994). Like cystic fibrosis testing, this can be done very simply with a mouthwash sample, and its simplicity will probably help to ensure high uptake. Overall, an estimated 1 in 500 pregnant women would be found to have a fragile-X mutation (25% full mutation and 75% premutation). Prenatal diagnosis could then be used to discover which fetuses had inherited the full mutation. The idiosyncrasy of this test, however, is that it is of different value for male and female fetuses. Whereas all males with the full mutation will show characteristic mental retardation, this will be true for only some of the females: at present it is not possible to know which.

Another issue that we may need to tackle in the future is that of prenatal testing for late onset conditions. So far this has only arisen for the relatively small number of people who know that they are at high risk of passing on Huntington's disease and one or two other late onset disorders for which the gene has been located. As we have seen, there are considerable dilemmas involved for parents in deciding to terminate a pregnancy even if they know that their child would not be able to lead a normal life. For late onset conditions, the dilemma is that the child would have a normal childhood but, at some point in adulthood, perhaps not until middle age, would develop an incurable life-threatening illness. Fortunately, Huntington's disease affects very few people.

Breast cancer, on the other hand, affects one woman in 12. Current developments in cancer genetics open up the possibility of prenatal detection of breast cancer suscepti- bility. People could be faced with the option of terminating a pregnancy because their unborn daughter has an 80% chance of developing breast cancer.

It has been argued (Brambati *et al.*, 1993) that if a woman is to be subjected to the risks of an invasive procedure such as amniocentesis or CVS, there is a moral obliga- tion to obtain the maximum information from the sample. This has led to routine examination of such samples to determine cystic fibrosis carrier status for women undergoing diagnostic testing for chromosomal abnormalities on grounds of maternal age. With the techniques currently available, most laboratories would feel that, what- ever the morality of this position, they could not afford to do this. However, technical advances will almost certainly make it possible to automate laboratory procedures, thereby opening up the possibility of a wider range of routine testing of such samples. This will pose serious counselling problems (Elias and Annas, 1994).

It is already clear that women are inadequately counselled about the relatively small number of conditions for which they are tested. The evidence of Marteau *et al.* (1993) suggests that women only receive a small proportion of the information that they might be thought to need in order to make informed decisions about testing. In Green's (1994) survey of obstetricians, 45% said that they had inadequate resources for counselling women about serum screening for Down syndrome. If the system cannot cope with the counselling needs of this one test, it is unlikely that women will receive adequate coun- selling if, in the future, it becomes feasible to carry out batteries of checks for a variety of genetic disorders. As the recent report *Genetic Screening: Ethical Issues* (Nuffield Council on Bioethics, 1993) observed with regard to prenatal testing:

> The practical difficulties relating to consent and counselling and the psychological con- sequences do not appear to have been given sufficient attention.

There are four main points that emerge from the literature on women's experiences of prenatal screening and diagnosis. First, women are likely to undergo tests for reassur- ance. As Richards and Green (1993) have argued, if there is no perceived need for reas- surance, people will be less inclined to have screening tests. Secondly, positive screening results are associated with considerable anxiety, and people are not always reassured by subsequent negative results. This is probably inevitable: if a test has the power to reas- sure, the absence of that reassurance must cause anxiety. Thirdly, women find waiting for the results of diagnostic tests very stressful, especially if they did not enter preg- nancy expecting to have such tests. Fourthly, communication – both what is communi- cated and how it is communicated – is very important. This is not to say that all problems associated with screening can be solved by optimal communication, but it is still a prerequisite for any screening programme that is to be based on informed consent.

6.10 References

Abuelo, D. N., Hopmann, M. R., Barsel-Bowers, G. and Goldstein, A. (1991) Anxiety in women with low maternal serum alpha-fetoprotein screening results. *Prenatal Diagnosis*, **11**, 381–5.

Adam, S., Wiggins, S., Whyte, P., Block, M., Shokeir, M. H. K., Soltan, H., Meschino, W., Summers, A., Suchowersky, O., Welch, J. P., Huggins, M. H., Theilmann, J. and Hayden, M. R. (1993) Five year study of prenatal testing for Huntington's disease: demand, attitudes, and psychological assessment. *Journal of Medical Genetics*, **30**, 549–56.

Al-Jader, L. N., Goodchild, M. C., Ryley, H. C. and Harper, P. S. (1990) Attitudes of parents of cystic fibrosis children towards neonatal screening and antenatal diagnosis. *Clinical Genetics*, **38**, 460–5.

Beeson, D. and Golbus, M. S. (1979) Anxiety engendered by amniocentesis. In C. J. Epstein, C. J. R. Curry, S. Packman, S. Sherman and B. D. Hall (eds.), *Risk, Communication, and Decision Making in Genetic Counseling*, pp. 191–7. New York: Alan R. Liss, Inc., for the National Foundation – March of Dimes. (Birth Defects: Original Articles Series, vol. 15.)

Bekker, H., Modell, M., Denniss, G., Silver, A., Mathew, C., Bobrow, M. and Marteau, T. (1993) Uptake of cystic fibrosis testing in primary care: supply push or demand pull? *British Medical Journal*, **306**, 1584–6.

Berne-Fromel, K., Kjessler, B. and Josefson, G. (1983) Anxiety concerning fetal malformation in women who accept or refuse alpha-fetoprotein screening in pregnancy. *Journal of Psychosomatic Obstetrics and Gynecology*, **2**, 94–7.

Black, R. B. (1989) A 1 and 6 month follow-up of prenatal diagnosis patients who lost pregnancies. *Prenatal Diagnosis*, **9**, 795–804.

Blumberg, B. D., Golbus, M. S. and Hanson, K. H. (1975) The psychological sequelae of abortion performed for a genetic indication. *American Journal of Obstetrics and Gynecology*, **122**, 799–808.

Bodegard, G., Fyro, K. and Larsson, A. (1983) Psychological reaction in 102 families with a newborn who has a falsely positive screening test for congenital hypothyroidism. *Acta Paediatrica Scandinavica, Supplement*, **304**, 3–21.

Brambati, B., Lanzani, A. and Tului, L. (1990) Transabdominal and transcervical chorionic villus sampling: efficiency and risk evaluation of 2,411 cases. *American Journal of Medical Genetics*, **35**, 160–4.

Brambati, B., Tului, L., Fattore, S. and Ferec, C. (1993) First-trimester fetal screening of cystic fibrosis in low-risk population. *Lancet*, **342**, 624.

Breslau, N. (1987) Abortion of defective fetuses: attitudes of mothers of congenitally impaired children. *Journal of Marriage and the Family*, **49**, 839–45.

Burton, B. K., Dillard, R. G. and Clark, E. N. (1985) The psychological impact of false positive elevations of maternal serum α-fetoprotein. *American Journal of Obstetrics and Gynecology*, **151**, 77–82.

Caccia, N., Johnson, J. M., Robinson, G. E. and Barna, T. (1991) Impact of prenatal testing on maternal–fetal bonding: chorionic villus sampling versus amniocentesis. *American Journal of Obstetrics and Gynecology*, **165**, 1122–5.

Campbell, S., Reading, A. E., Cox, D. N., Sledmore, C. M., Mooney, R., Chudleigh, P., Beedle, J. and Ruddick, H. (1982) Ultrasound scanning in pregnancy. *Journal of Psychosomatic Obstetrics and Gynecology*, **1**, 57–61.

Chervin, A., Farnsworth, P. B., Freedman, W. L., Duncan, P. A. and Shaprio, L. R. (1977)

Amniocentesis for prenatal diagnosis. *New York State Journal of Medicine*, August, 1406–8.

Clayton-Smith, J., Andrews, T. and Donnai, D. (1989) Genetic counselling and parental decisions following antenatal diagnosis of sex chromosome aneuploidies. *Journal of Obstetrics and Gynaecology*, **10**, 5–7.

Cox, D. N., Wittmann, B. K., Hess, M., Ross, A. G., Lind, J. and Lindahl, S. (1987) The psychological impact of diagnostic ultrasound *Obstetrics and Gynecology*, **70**, 673–6.

Davies, K. (1989) Study throws doubt on site of 'manic depression gene'. *New Scientist*, 18 November, p. 20.

Donnai, P., Charles, N. and Harris, R. (1981) Attitudes of patients after 'genetic' termination of pregnancy. *British Medical Journal*, **282**, 621–2.

Drugan, A., Greb, A., Johnson, M. P., Krivchenia, E. L., Uhlmann, W. R., Moghissi, K. S. and Evans, M. I. (1990) Determinants of parental decisions to abort for chromosome abnormalities. *Prenatal Diagnosis*, **10**, 483–90.

Elias, S. and Annas, G. J. (1994) Generic consent for genetic screening. *New England Journal of Medicine*, **330**, 1611–13.

Elias, S., Simpson, J. L., Shulman, L. P., Emerson, D., Tharapel, A. and Seely, L. (1989) Transabdominal chorionic villus sampling for first-trimester prenatal diagnosis. *American Journal of Obstetrics and Gynecology*, **160**, 879–86.

Ennis, M., Clark, A. and Grudzinskas, J. G. (1991) Change in obstetric practice in response to fear of litigation in the British Isles. *Lancet*, **338**, 616–18.

Evans, M. I., Bottoms, S. F., Carlucci, T., Grant, J., Belsky, R. L., Solyom, A. E., Quigg, M. H. and LaFerla, J. J. (1988) Determinants of altered anxiety after abnormal maternal serum alpha-fetoprotein screening. *American Journal of Obstetrics and Gynecology*, **159**, 1501–4.

Evers-Kiebooms, G., Denayer, L. and Van den Berghe, H. (1990) A child with cystic fibrosis. II. Subsequent family planning decisions, reproduction and use of prenatal diagnosis. *Clinical Genetics*, **37**, 207–15.

Evers-Kiebooms, G., Denayer, L., Decruyenaere, M. and Van den Berghe, H. (1993) Community attitudes towards prenatal testing for congenital handicap. *Journal of Reproductive and Infant Psychology*, **11**, 21–30.

Evers-Kiebooms, G., Swerts, A. and Van den Berghe, H. (1988) Psychological aspects of amniocentesis: anxiety feelings in three different risk groups. *Clinical Genetics*, **33**, 196–206.

Faden, R. R., Chwalow, A. J., Quaid, K., Crane, J. P. and McNellis, D. (1987) Prenatal screening and pregnant women's attitudes toward the abortion of defective fetuses. *American Journal of Public Health*, **77**, 288–90.

Farrant, W. (1980) Stress after amniocentesis for high serum alpha-feto-protein concentrations. *British Medical Journal*, **281**, 452.

Farrant, W. (1985) 'Who's for amniocentesis?' The politics of prenatal screening. In H. Homans (ed.), *The Sexual Politics of Reproduction*, pp. 96–177. London: Gower.

Fava, G. A., Kellner, R., Michelacci, L., Trombini, G., Pathak, D., Orlandi, C. and Bocicelli, L. (1982) Psychological reactions to amniocentesis: a controlled study. *American Journal of Obstetrics and Gynecology*, **143**, 509–13.

Fava, G. A., Trombini, G., Michelacci, L., Linder, J. R., Pathak, K. and Bovicelli, L. (1983) Hostility in women before and after amniocentesis. *Journal of Reproductive Medicine*, **28**, 29–34.

Fearn, J., Hibbard, B. M., Laurence, K. M., Roberts, A. and Robinson, J. O. (1982) Screening for neural-tube defects and maternal anxiety. *British Journal of Obstetrics and Gynaecology*, **89**, 218–21.

Field, T., Sandberg, D., Quetal, T. A., Garcia, R. and Rosario, M. (1985) Effects of ultrasound feedback on pregnancy anxiety, fetal activity, and neonatal outcome. *Obstetrics and Gynecology*, **66**, 525–8.

Firth, H. V., Boyd, P. A., Chamberlain, P. R., MacKenzie, I. Z., Morriss-Kay, G. M. and Huson, S. M. (1994) Analysis of limb reduction defects in babies exposed to chorionic villus sampling. *Lancet*, **343**, 1069–72.

Green, J. M. (1990) Calming or harming?: a critical review of psychological effects of fetal diagnosis on pregnant women. *Galton Institute Occasional Papers*, Second series, no. 2.

Green, J. (1993) Ethics and late termination of pregnancy. *Lancet*, **342**, 1179.

Green, J. M. (1994) Serum screening for Down's syndrome: the experiences of obstetricians in England and Wales. *British Medical Journal*, **309**, 769–72.

Green, J. M. (1995) Obstetricians' views on prenatal diagnosis and termination of pregnancy: 1980 compared with 1993. *British Journal of Obstetrics and Gynaecology*, **102**, 228–32.

Green, J. and Statham, H. (1993) Testing for fetal abnormality in routine antenatal care. *Midwifery*, **9**, 124–35.

Green, J. M., Snowdon, C. and Statham, H. (1993*a*) Pregnant women's attitudes to abortion and prenatal screening. *Journal of Reproductive and Infant Psychology*, **11**, 31–9.

Green, J. M., Statham, H. and Snowdon, C. (1992) Screening for fetal abnormalities: attitudes and experiences. In T. Chard and M. P. M. Richards (eds.), *Obstetrics in the 1990s: Current Controversies*. London: McKeith Press.

Green, J. M., Statham, H. and Snowdon, C. (1993*b*) Women's knowledge of prenatal screening tests. I. Relationships with hospital screening policy and demographic factors. *Journal of Reproductive and Infant Psychology*, **11**, 11–20.

Green, J. M., Statham, H. and Snowdon, C. (1994) *Pregnancy: A Testing Time.* (Report of the Cambridge Prenatal Screening Study.) Centre of Family Research, University of Cambridge.

Griffiths, D. M. and Gough, M. H. (1985) Dilemmas after ultrasonic diagnosis of fetal abnormality. *Lancet*, **1**, 623–4.

Harris, H., Scotcher, D., Hartley, N., Wallace, A., Craufurd, D. and Harris, R. (1993) Cystic fibrosis carrier testing in early pregnancy by general practitioners. *British Medical Journal*, **306**, 1580–3.

Holmes-Siedle, M., Ryynanen, M. and Lindenbaum, R. H. (1987) Parental decisions regarding termination of pregnancy following prenatal detection of sex chromosome abnormality. *Prenatal Diagnosis*, **7**, 239–44.

Hunter, M. S., Tsoi, M. M., Pearce, M., Chudleigh, P. and Campbell, S. (1987) Ultrasound scanning in women with raised serum alpha feto protein: long term psychological effects. *Journal of Psychosomatic Obstetrics and Gynecology*, **6**, 25–31.

Hyde, B. (1986) An interview study of pregnant women's attitude to ultrasound scanning. *Social Science and Medicine*, **22**, 587–92.

Iles, S. (1989) The loss of early pregnancy. *Baillière's Clinical Obstetrics and Gynaecology*, **3**, 769–90.

Jacoby, A. (1988) Mothers' views about information and advice in pregnancy and childbirth: findings from a national study. *Midwifery*, **4**, 103–10.

Jorgensen, C., Uddenberg, N. and Ursing, Z. (1985) Ultrasound diagnosis of fetal malformation in the second trimester: the psychological reactions of the women. *Journal of Psychosomatic Obstetrics and Gynecology*, **4**, 31–40.

Keenan, K. L., Basso, D., Goldkrand, J., Butler, W. J. (1991) Low level of maternal serum alpha-

160   *Josephine Green and Helen Statham*

fetoprotein: its associated anxiety and the effects of genetic counselling. *American Journal of Obstetrics and Gynecology*, **164**, 54–6.

Kolker, A. (1989) Advances in prenatal diagnosis: social–psychological and policy issues. *International Journal of Technology Assessment in Health Care*, **5**, 601–617.

Leschot, N. J., Verjaal, M. and Treffers, (1982) Therapeutic abortion on genetic grounds. *Journal of Psychosomatic Obstetrics and Gynecology*, **1**(2), 47–56.

Lloyd, J. and Laurence, K. M. (1985) Sequelae and support after termination of pregnancy for fetal malformation. *British Medical Journal*, **290**, 907–9.

Madarikan, B. A., Tew, B. and Lari, J. (1990) Maternal response to anomalies detected by antenatal ultrasonography. *British Journal of Clinical Practice*, **44**, 587–9.

Marteau, T. M. (1991) Psychological aspects of prenatal testing for fetal abnormalities. *Irish Journal of Psychology*, **12**, 121–32.

Marteau, T. M., Cook, R. Kidd, J. Michie, S., Johnston, M., Slack, J. and Shaw, R. W. (1992) The psychological effects of false positive results in prenatal screening for fetal abnormality: a prospective study. *Prenatal Diagnosis*, **12**, 205–14.

Marteau, T. M., Drake, H. and Bobrow, M. (1994) Counselling after diagnosis of fetal abnormality: the differing approaches of obstetricians, clinical geneticists and genetic nurses. *Journal of Medical Genetics* **31**, 864–7.

Marteau, T. M., Johnston, M., Plenicar, M., Shaw, R. W. and Slack, J. (1988) Development of a self-administered questionnaire to measure women's knowledge of prenatal screening and diagnostic tests. *Journal of Psychosomatic Research*, **32**, 403–8.

Marteau, T. M. Kidd, J., Cook, R., Michie, S., Johnston, M., Slack, J., Shaw, R. W. (1991) Perceived risk not actual risk predicts uptake of amniocentesis. *British Journal of Obstetrics and Gynaecology*, **98**, 282–6.

Marteau, T. M., Plenicar, M. and Kidd, J. (1993) Obstetricians presenting amniocentesis to pregnant women: practice observed. *Journal of Reproductive and Infant Psychology*, **11**, 3–10.

McCormack, M. J., Rylance, M. E., Newton, J. *et al.* (1990) Patients' attitudes following chorionic villus sampling. *Prenatal Diagnosis*, **10**, 253–5.

Mennie, M. E., Compton, M. E., Gilfillan, A., Liston, W. A., Pullen, I., Whyte, D. A. and Brock, D. J. H. (1993) Prenatal screening for cystic fibrosis: psychological effects on carriers and their partners. *Journal of Medical Genetics*, **30**, 543–8.

Merkatz, I. R., Nitowski, H. M., Macri, J. N. and Johnson, W. E. (1984) An association between low maternal serum alpha feto-protein and fetal chromosomal abnormalities. *American Journal of Obstetrics and Gynecology*, **148**, 866–94.

Meryash, D. L. and Abuelo, D. (1988) Counseling needs and attitudes toward prenatal diagnosis and abortion in fragile-X families. *Clinical Genetics*, **33**, 349–55.

Milne, L. S. and Rich, U. J. (1981) Cognitive and affective aspects of the responses of pregnant women to sonography. *Maternal-Child Nursing Journal*, **10**, 15–39.

Minnick, M. A., Delp, K. J., Ciotti, M. C. (eds.) (1990) *Parents Faced with Grief. A Time to Decide, a Time to Heal*. East Lansing, MI: Pineapple Press.

Nicolaides, K. H., Snijders, R. J. M., Gosden, C. M., Berry, C. and Campbell, S. (1992) Ultrasonographically detectable markers of fetal chromosomal abnormalities. *Lancet*, **340**, 704–7.

Nuffield Council on Bioethics (1993) *Genetic Screening: Ethical Issues*. London: Nuffield Council on Bioethics.

Palomaki, G. E. (1994) Population based prenatal screening for the fragile X syndrome. *Journal of Medical Screening*, **1**, 65–72.

Phipps, S. and Zinn, A. B. (1986) Psychological response to amniocentesis. I. Mood state and adaptation to pregnancy. *American Journal of Medical Genetics*, **25**, 131–42.

Porter, M. and Macintyre, S. (1984) What is, must be best: a research note on conservative or deferential responses to antenatal care provision. *Social Science and Medicine*, **19** (11), 1197–200.

Press, N. A. and Browner, C. H. (1993) 'Collective fictions': similarities in reasons for accepting maternal serum alpha-fetoprotein screening among women of diverse ethnic and social class backgrounds. *Fetal Diagnosis and Therapy*, **8**, 97–106.

Pryde, P. G., Isada, N., Hallak, M., Johnson, M. P., Odgers, A. E. and Evans, K. I. (1992) Determinants of parental decision to abort or continue after non-aneuploid ultrasound-detected fetal abnormalities. *Obstetrics and Gynecology*, **80**, 52–6.

Reading, A. E. and Cox, D. N. (1982) The effects of ultrasound examination on maternal anxiety levels. *Journal of Behavioral Medicine*, **5**, 237–47.

Reading, A. E. and Platt, L. D. (1985) Impact of fetal testing on maternal anxiety. *Journal of Reproductive Medicine*, **30** (12), 907–10.

Reading, A. E., Cox, D. N., Campbell, S. (1982) Ultrasound scanning in pregnancy. The psychological effect of fetal feedback. *Ultrasound Medical Bulletin*, **8**, 323–4.

Richard, R., van Zonder, H., Verjaal, M. and Leschot, N. J. (1992) Termination of pregnancy after first trimester CVS: a need for supportive counselling. Paper presented to the 3rd European Meeting on 'Psychosocial Aspects of Genetics', Nottingham, September.

Richards, M. P. M. and Green, J. M. (1993) Attitudes toward prenatal screening for fetal abnormality and detection of carriers of genetic disease: a discussion paper. *Journal of Reproductive and Infant Psychology*, **11**, 49–56.

Richards, M. P. M., Hallowell, N., Green, J. M., Murton, F. and Statham, H. (1995) Counselling families with hereditary breast and ovarian cancer: a psychosocial perspective. *Journal of Genetic Counselling* (in press).

Robinson, G. E., Carr, M. L., Olmsted, M. P. and Wright, C. (1991) Psychological reactions to pregnancy loss after prenatal diagnostic testing: preliminary results. *Journal of Psychosomatic Obstetrics and Gynecology*, **12**, 181–92.

Robinson, G. E., Garner, D. M., Olmstead, M. P., Shime, J., Hutton, E. and Crawford, B. M. (1988) Anxiety reduction after chorionic villus sampling and genetic amniocentesis. *American Journal of Obstetrics and Gynecology*, **159**, 953–6.

Robinson, J., Tennes, K. and Robinson, A. (1975) Amniocentesis: its impact on mothers and infants. A 1-year follow-up study. *Clinical Genetics*, **8**, 97–106.

Roelofsen, E. E. C., Kamerbeek, L. I. and Tymstra, T. J. (1993) Chances and choices. Psycho-social consequences of maternal serum screening. A report from The Netherlands. *Journal of Reprodutive and Infant Psychology*, **11**, 41–7.

Rothenberg, M. B. and Sills, E. M. (1968) Iatrogenesis: the PKU anxiety syndrome. *Journal of the American Academy of Child Psychology*, **7**, 689–92.

Rothman, B. K. (1986) *The Tentative Pregnancy: Prenatal Diagnosis and the Future of Motherhood*. New York: Viking.

Royal College of Physicians (1989) *Prenatal Diagnosis and Genetic Screening*. London: Royal College of Physicians.

SATFA (1994) *SATFA News*, April. London: Support around Termination for Abnormality.

Snowdon, C. M. and Green, J. M. (1994) *Attitudes to and Experiences of New Reproductive Technologies of Carriers of Recessive Disorders*. Centre of Family Research, University of Cambridge.

Spencer, J. W. and Cox, D. N. (1987) Emotional responses of pregnant women to chorionic villi sampling or amniocentesis. *American Journal of Obstetrics and Gynecology*, **157**, 1155–60.

Spencer, J. W. and Cox, D. N. (1988) A comparison of chorionic villi sampling and amniocentesis: acceptability of procedure and maternal attachment to pregnancy. *Obstetrics and Gynecology*, **72**, 714–18.

Spielberger, C. D., Gorsuch, R. L. and Lushene, R. E. (1970) *The State-Trait Anxiety Inventory*. Palo Alto, CA: Consulting Psychologists Press.

Statham, H. (1992) Professional understanding and parents' experience of termination. In D. J. Brock, C. H. Rodeck and M. A. Ferguson-Smith (eds.), *Prenatal Diagnosis and Screening*. London: Churchill Livingstone.

Statham, H. (1994) The parents' reaction to termination of pregnancy for fetal abnormality: from a mother's point of view. In L. Abramsky and J. Chapple (eds.), *Prenatal Diagnosis: The Human Side*. London: Chapman & Hall.

Statham, H. and Green, J. (1993) Serum screening for Down's syndrome: some women's experiences. *British Medical Journal*, **307**, 174–6.

Statham, H., Green, J. and Snowdon, C. (1993*a*) Psychological and social aspects of screening for fetal abnormality during routine antenatal care. In S. Robinson, A. Thomson and V. Tickner (eds.), *Research and the Midwife Conference Proceedings for 1992*. Manchester School of Nursing Studies, University of Manchester.

Statham, H., Green, J. and Snowdon, C. (1993*b*) Mothers' consent to screening newborn babies for disease. *British Medical Journal*, **306**, 858–9.

Stone, D. H., Smalls, M. J., Rosenberg, K. and Wormersley, J. (1988) Screening for congenital neural tube defects in a high-risk area: an epidemiological perspective. *Journal of Epidemiology and Community Health*, **42**, 271–3.

Sujansky, E., Kreutzer, S. B., Johnson, A. M., Lezotte, D. C., Schier, R. W. and Gabour, P. A. (1990) Attitudes of at-risk and affected individuals regarding presymptomatic testing for autosomal dominant polycystic kidney disease. *American Journal of Medical Genetics*, **35**, 510–15.

Thorpe, K., Harker, L., Pike, A. and Marlow, N. (1993) Women's views of ultrasonography: a comparison of women's experiences of antenatal ultrasound screening with cerebral ultrasound of their newborn infant. *Social Science and Medicine*, **36**, 311–15.

Tsoi, M. M. and Hunter, M. (1987) Ultrasound scanning in pregnancy: consumer reactions. *Journal of Reproductive and Infant Psychology*, **5**, 43–8.

Tsoi, M. M., Hunter, M., Pearce, M., Chudleigh, P. and Campbell, S. (1987) Ultrasound scanning in women with raised serum alpha feto protein: short term psychological effects. *Journal of Psychosomatic Research*, **31**, 35–9.

Tunis, S. L., Golbus, M. S., Copeland, K. L., Fine, B. A., Rosinsky, B. J. and Seely, L. (1990) Patterns of mood states in pregnant women undergoing chorionic villus sampling or amniocentesis. *American Journal of Medical Genetics*, **37**, 191–9.

Tymstra, T. (1989) The imperative character of medical technology and the meaning of anticipated decision regret. *International Journal of Technology Assessment in Health Care*, **5**, 207–13.

Verjaal, M., Leschot, N. J. and Treffers, P. E. (1982) Women's experiences with second trimester prenatal diagnosis. *Prenatal Diagnosis*, **2**, 195–209.

Verp, M. S., Bombard, A. T., Simpson, J. L. and Elias, S. (1988) Parental decision following prenatal diagnosis of fetal chromosome anomalies. *American Journal of Medical Genetics*, **29**, 613–22.

Wald, N. J. (ed.) (1984) *Antenatal and Neonatal Screening*. Oxford: Oxford University Press.

Wald, N. J. and Cuckle, H. S. (1984) Open neural-tube defects. In N. J. Wald (ed.), *Antenatal and Neonatal Screening*. Oxford: Oxford University Press.

Watkins, D. (1989) An alternative to termination of pregnancy. *The Practitioner*, **233**, 990–2.

Watson, E. K., Mayall, E., Chapple, J., Dalziel, M., Harrington, K., Williams, C. and Williamson, R. (1991) Screening for carriers of cystic fibrosis through primary health care services. *British Medical Journal*, **303**, 504–7.

Weatherall, D. J. (1992) Foreword. In D. J. Brock, C. H. Rodeck and M. A. Ferguson-Smith (eds.) *Prenatal Diagnosis and Screening*. London: Churchill Livingstone.

Wertz, D. C., Rosenfield, J. M., Janes, S. R. and Erbe, R. W. (1991) Attitudes toward abortion among parents of children with cystic fibrosis. *American Journal of Public Health*, **81**(8), 992–6.

White-van Mourik, M. C. A., Connor, J. M. and Ferguson-Smith, M. A. (1992) The psychosocial sequelae of a second-trimester termination of pregnancy for fetal abnormality. *Prenatal Diagnosis*, **12**, 189–204.

7

The genetic testing of children: a clinical perspective

ANGUS CLARKE and FRANCES FLINTER

7.1 Introduction

Testing healthy children to identify genetic conditions has been possible for many years by clinical examination, blood tests and other investigations that recognise the relevant phenotype. Examples include the recognition of children with type I neurofibromatosis by examination of the skin, the biochemical recognition of infant boys with Duchenne muscular dystrophy, the identification of unaffected carriers of haemoglobin disorders (e.g. sickle cell disease) by haematological tests, and the identification of some asymptomatic children or adolescents with autosomal dominant polycystic kidney disease by ultrasound examination of the kidneys.

The recent development of molecular genetic technologies has transformed the situation by greatly extending the possibilities for genetic testing. Molecular genetic methods test directly for the relevant gene. These tests can be carried out at any stage of life from conception onwards, using any nucleated tissue, for example white blood cells or a mouthwash sample of oral epithelial cells. There is usually no need to test a tissue affected by the disease process. Many genetic tests are now available for inherited diseases for which there had previously been no diagnostic test. Such tests can identify children who are likely to develop genetic disorders in adult life, and can also identify those carrying recessive disease genes, which have no effect on the health of carriers but may have implications for the health of carriers' future children.

This chapter addresses controversial issues raised by the possibility of testing children that are (apparently) healthy (Harper and Clarke, 1990). None of the issues is completely novel – there are just many more tests available now, but our response to the increasing range of tests available may also lead us to modify some established practices. Before proceeding further, however, we shall briefly set out two sets of less contentious ethical issues concerning molecular genetic testing in childhood. First, when a child has presented with symptoms and signs of a clinical disorder, we regard the use of a molecular genetic diagnostic test as being no different in principle from the use of any other diagnostic test. There may be ethical implications arising from the diagnosis of certain specific conditions, or from attaching a potentially stigmatising diagnostic label to a child; and practical problems may arise if the child's parents are unable or unready to accept the possibility of their child's having a particular condition. These problems, however, are of a general nature, are not specific to molecular genetic testing, and will not be addressed further in this chapter.

The second topic that we shall not pursue at length is the predictive testing of healthy children for disorders that are likely to manifest during childhood, or for which useful medical interventions may be available in childhood. Such testing can raise ethical issues, as with newborn screening for alpha-1-antitrypsin deficiency (Thelin *et al.*, 1985), for the severe and essentially incurable condition of Duchenne muscular dystrophy (Bradley *et al.*, 1993; Parsons and Bradley, 1994), and with predictive testing for adrenoleucodystrophy (Costakos *et al.*, 1991). One potential cause for concern about such predictive tests in childhood applies particularly to newborn screening programmes at the population level. In such programmes, families with no prior history of a particular genetic disease may be informed that their infants have a serious disorder for which no effective treatment is available. These families gain information that they may find useful (e.g. for reproductive decisions or for practical planning), and may be spared months or years of anxiety when their child clearly has a problem but has not yet been given a diagnosis. There is a cost, however, in that their inevitable sadness and distress is brought forward and years of potential happiness ('ignorance is bliss') may be spoiled. There may also be concern about damage to the development of emotional relationships within the family. The balance between the advantages and disadvantages of early diagnosis needs to be studied carefully in practice, and these issues will not be discussed further in this chapter.

The issues that we shall address here are:

(1) predictive, presymptomatic testing of children for disorders that do not usually manifest until adult life, and for which useful medical interventions (including surveillance for complications) are not available in childhood;
(2) testing children to identify unaffected carriers of genetic disease, including autosomal or sex-linked recessive diseases and carriers of balanced, familial chromosome rearrangements.

In both cases, testing generates information that is unlikely to be of practical use to the subject (the tested child) until adolescence or adulthood, when life plans may be made, and we do not know how children and their families will react to receiving such information. The Clinical Genetics Society in the UK has recently decided to adopt a position of caution and to advise against predictive or carrier tests in children, while more research is performed in this area.

7.2 Predictive tests and carrier tests in children: the grounds for concern

Why is there concern about performing genetic tests on children in some circumstances? Such tests are often requested by parents. What reasons could there be for deciding not to carry out the tests?

One of the first concerns to be raised about genetic testing in childhood was in the context of predictive testing for Huntington's disease (HD) (Craufurd and Harris,

1986). In The Netherlands, 41% of 70 individuals at risk for Huntington's disease thought that children under 18 should be allowed to have the test if they so chose (Tibben, 1993), and numerous requests to test minors presymptomatically have been made (Morris *et al.*, 1988, 1989). Discussion among clinicians and scientists involved in developing a molecular genetic predictive test for HD led to a consensus view that children should not be tested (World Federation of Neurology, 1989, 1990; Bloch and Hayden, 1990). It was feared that such children could be stigmatised, even within the family, and knowledge of their genetic status could distort their emotional development, their educational attainments and their future relationships. The test result could effectively blight their lives much earlier and more thoroughly than the disease would do eventually. For these reasons, it was felt advisable to adopt a position of caution, although it is possible that for some individuals the opportunity to grow up from childhood with a definite test result would enable them to come to terms with the result more successfully than they would if they postponed testing until adulthood.

This cautious view has since been reinforced by experience gained in predictive testing of adults for HD. Two findings from predictive testing programmes have been particularly relevant. First, although questionnaire surveys had suggested that a substantial majority of at-risk adults would request testing when it became available, in fact only 10–15% have done so, suggesting that, as adults, only a minority wish to know their HD status (Craufurd *et al.*, 1989; Bloch *et al.*, 1992; Tyler *et al.*, 1992). Secondly, the knowledge of HD gene status can be burdensome, even when the result is low risk (Huggins *et al.*, 1992; Tibben *et al.*, 1992), although in some cases this relates to regrets for missed opportunities and time wasted while living with the burden of risk.

Concerns about testing during childhood arose in the very special case of HD, a particularly devastating condition, but the same principles apply to predictive tests in childhood for other adult onset disorders, and perhaps also to carrier status tests. The arguments against testing in childhood can be summarised as follows.

1. Testing in childhood removes the individual's future right to make their own decision about testing as an autonomous adult. The majority of eligible adults have decided not to take such tests.
2. Testing in childhood removes the confidentiality that would be expected and provided for any adult undergoing the same test, both for the fact of having the test performed and for the test result.
3. A knowledge of the child's genetic status may alter his or her upbringing and the pattern of relationships within the family and also with peers, which could lead to stigmatisation and discrimination. Altered expectations of the child's intellectual abilities, future health and future relationships could affect the deepest levels of self-esteem and have devastating social, emotional, psychological and educational consequences. In addition, there could be serious consequences for life insurance and health insurance (Harper, 1992, 1993) and for employment (Billings *et al.*, 1992).

Breaches of autonomy and confidentiality will inevitably occur in predictive or carrier childhood testing for any disease where the knowledge will not be of any practical use until the child is older. The possible harm inflicted on the child's future development will depend both upon the nature of the disease and upon numerous family-specific circumstances: each family's experience of 'their' disease, their particular set of interpersonal relationships, and the professional support that they receive. Currently, concern about the possible harm that could be caused by testing in childhood is believed to outweigh any potential benefits. There are many unresolved questions, however, that need to be addressed by careful research.

7.3 Arguments for genetic testing in childhood, and our response to them

There are grounds for challenging the perspective on childhood genetic testing presented above. These are set out below, with counter-arguments.

(a) Parents have a right to know the genetic status of their child.

This has been argued from a legal perspective in North America (Pelias, 1991; Sharpe, 1993) but in Britain parents are deemed to have duties towards their children rather than rights over them (Montgomery, 1993, 1994). Weaker versions of the same argument could be proposed: that it would usually be in the best interests of the child to accede to any request for genetic information made by its parents, or that the parents would virtually always be the best judges of the child's long-term best interests. (Other examples of the potential for such a conflict include the possibility of sex selection in pregnancies, and the use of genetic tests to establish paternity in disputed cases).

In general paediatric practice, the only situations in which parental wishes are regularly challenged are where child abuse is suspected or where a parental decision is made that puts the life of a child at risk, e.g. if a life-saving treatment is refused on religious grounds, or if the parents press for the withdrawal of care against medical advice. In genetic testing, however, the conflict of interests between parent and child is not so clear; however, there may be a potential conflict between the parents' wish (for whatever reason) to know about the genetic status of their child, and the child's long-term best interests.

There will be occasions where childhood genetic testing is thought likely to harm the child, and where professionals will want to defer testing until the child is older. It may be harder to identify any potential benefits of testing children. The potential areas of disagreement among professionals will relate to their views about the likely magnitude of this problem: should there be a general presumption against such testing because of the ethical issues (1) or (2) above, or (3) because the potential harm is likely to occur in practice and to cause real damage, or will these anxieties really only be of importance in a small minority of cases?

Parents may experience a strong desire to know about their child's genetic status: a perceived 'need to know'. Even if the geneticist feels that testing would be unhelpful, it would be a mistake to confront such an expressed desire with a blanket refusal to consider testing. Each request should be considered individually. During genetic counselling, time can be taken to explore the reasons and feelings underlying such a request and the likely consequences for the child and the family of the various different outcomes of testing if it were performed. Many families find this process helpful, and careful consideration of the possible consequences of testing often removes the 'need to know'. Counselling can be very difficult if families have previously been told that they definitely can, or cannot, have a certain test performed on a child, and genetic counselling in these circumstances is likely to be particularly time-consuming if the geneticist believes that postponing testing would be preferable.

(b) Testing in childhood may help a child to adjust psychologically to an unfavourable result more readily than if the information were presented to them during adolescence.

Apart from a few anecdotes, there is no evidence to support or to refute this argument. The contrary view, that adolescents may react better to such results if they have been able to determine whether and when to be tested, is equally plausible and equally unsupported by evidence in this specific field of genetic testing, although there is more general evidence that children need to exercise choice in order to acquire self-respect and a sense of responsibility (Clarke and Clarke, 1976).

It may be true that some pre-adolescent children react impassively, or are content to accept the unfavourable results of genetic tests, but this tells us nothing about the reaction of these children to their situation and to the test results when they are some years older. It should be noted that, in the longer term, children do not necessarily adjust any better to medical diagnoses of which they become aware in middle childhood than to diagnoses made during adolescence. For example, many adolescents with diabetes have behavioural problems, and they often disrupt attempts to establish good control of their condition. Perhaps what is crucial here is that the adolescents have no choice about being diabetic. Granting choice and control with respect to the genetic testing to the adolescent, instead of pre-empting it by testing in childhood, may be of value to the adolescent's self-esteem and hence his or her coping strategies, although there is no evidence to prove this. The abilities of children to consider issues of consent should not be underestimated (Alderson, 1993), but there is little experience of obtaining consent from children for genetic tests. No strategies have been developed to assess a child's understanding of the implications of genetic testing, and clearly individual comprehension varies considerably at any age.

(c) A substantial proportion of the tested children (often 33% or 50%) – or their parents – will be reassured by favourable results from the genetic testing.

The counter-argument to this is that the reassurance to (perhaps) half of the tested children is bought at too high a price: the unfavourable results given to the other children. It must also be remembered that a number of adults given low-risk results on predictive testing for HD suffered subsequent psychological distress and disturbance ('survivor guilt'), although we do not know whether this group might have benefited from earlier tests. Furthermore, family psychodynamics could be disrupted if a child is preselected by an irrational family consensus as being affected or as being a gene carrier, or as being unaffected or not a gene carrier, and the test result conflicts with the family preselection (Kessler, 1988; see also Chapter 12). Different problems may arise if the genetic testing reinforces a pre-existing family consensus, but also if a child is prese-lected and remains untested (a problem that can be addressed by careful counselling).

In the longer term any test result, whether positive or negative, may lead to a reduc-tion in the psychological distress experienced by those in some families who know only that they are 'at risk' of carrying a certain gene (Wiggins *et al.*, 1992). Particular prob-lems could arise in a family where one child is identified as destined to develop a partic-ular disease in adult life, or as being a carrier whose children may be at risk of a dreaded disease, and another child is found to be unaffected or not a carrier. Such findings could well lead to difficulties in relationships within the family. This reinforces the concern that favourable test results will not always lead to favourable family outcomes.

(d) Carrier testing in childhood, for a condition that has occurred in the family, ensures that the testing is performed. Professionals can fulfil their responsibilities both to the family and to the tested child in this way, and resolve their own uncomfortable feelings of uncertainty and their concern about the possibility of losing contact with the family in the future.

This argument assumes that the testing of a child – e.g. for a recessive gene carrier state – is of paramount importance. Testing a young child and giving the results to the family do not necessarily ensure that the correct information is imparted to the child in an appropriate manner and at an appropriate age. Many parents find sex education a difficult topic to discuss with their children; how confident can we be that they will be able to discuss genetic issues effectively? This is of particular concern in a society such as ours, with a high rate of family breakdown; it is simply not possible to rely upon the effectiveness and sensitivity of intergenerational communication.

Even if carrier tests are carried out in childhood, the family and the medical profes-sion together still have a certain responsibility to ensure that the young adult is offered genetic counselling. This allows the information given to the family to be updated, and it allows the tested individuals to discuss the issues in confidence from their own per-sonal perspective. Tests may have to be repeated if laboratory methods have improved since the first carrier tests were performed, and occasionally completely different results will be obtained. Some individuals may prefer to defer testing until they have a partner with whom they can discuss all the implications.

(e) How is contact to be maintained with all these families in whom there are children at risk of carrying a genetic disease, but who have not been tested?

To maintain contact with families in this way will, in practice, require a clear transfer of responsibility from the instigating clinician (often a paediatrician) to the family or to another medical practitioner, if the first clinician is not going to maintain contact with the families. If a clear transfer of this responsibility can be achieved, that *ad hoc* arrangement may be perfectly satisfactory. Very often, however, the most practical solution will be the maintenance of an active-contact genetic register (Harper *et al.*, 1982; Read *et al.*, 1986; Burn *et al.*, 1989; Norman *et al.*, 1989); and if the family has not already been referred to a clinical genetics unit this would be a valid indication for referral. The genetics unit would then maintain occasional contact with each family, often by letter or telephone, and only with the willing consent of the family.

(f) Predictive testing in childhood resolves parental uncertainty about the genetic status of their apparently healthy children, and hence relieves anxiety.

The first response to this argument has to be that this only works if the children have favourable results, and that parental anxiety will be increased when the child has an unfavourable result. There are also other factors to be considered. (For disorders that are likely to present in childhood, the concern about removing the child's future adult autonomy is not so relevant, and there may well be practical matters that can benefit from advance planning. In addition, if the testing is not performed, the family may focus on possible early signs of the disorder, for example over-interpreting perfectly normal fatigue or stumbling as evidence of muscle weakness in Becker muscular dystrophy, even when the child does not in fact have the condition. It may therefore sometimes be appropriate to test a child for some 'adult' conditions that can manifest in childhood.)

For other disorders, however, that usually do not manifest until adult life, the resolution of parental uncertainty may exact too great a cost in terms of lost autonomy and confidentiality, and in terms of possible discrimination and stigmatisation, both within the family (distorting child-rearing and family relationships) and in relation to society at large. In the absence of systematic evidence, opinions as to good practice vary, and it may be difficult, or even inappropriate, to draw up rigid criteria for predictive testing in childhood. The natural reluctance of most professionals to cause harm to their clients, however, is likely to lead towards a professional consensus that generally presumes that such testing is inappropriate, but recognises that individual circumstances may justify testing, so that each family situation has to be assessed individually and with care. This may necessitate parents meeting genetic counsellors on more than one occasion to enable the requisite detailed discussions before informed decision-making can occur.

(g) There is no evidence that carrier testing in childhood causes any harm.

While this may be true, it is also true that the results of carrier testing in childhood have never been systematically studied in long-term, prospective studies, and research is urgently required in this area. The same set of considerations as discussed in (f), above, will lead to a similar general presumption that testing children for genetic carrier status flouts their future autonomy and privacy, and is unnecessary and potentially damaging, so that it should usually be deferred at least until adolescence except in very particular circumstances. Such a cautious policy is supported by reports that knowledge of carrier status for a disease gene may generate adverse psychological and social effects in adults, but there is no information about how people might react if they had grown up knowing their carrier status from childhood. There may be stigmatisation and a fear of discrimination, even in an area where the carrier frequency is high (Stamatoyannopoulos, 1974). Adult carriers of Tay–Sachs disease have been shown to view their own future health less positively than non-carriers (Marteau *et al.*, 1992), and an earlier study reported that 19% of Tay–Sachs carriers are still worried by this information several years after testing (Zeesman *et al.*, 1984). In a prenatal carrier screening programme for the haemoglobinopathies in North America, 10% of the carriers were unable to state that being a carrier would not damage their health (Loader *et al.*, 1991).

It is possible that carriers identified in childhood will respond differently from adults to this information, but there are sufficient grounds for concern to justify caution. Carrier testing of the healthy adult sibs of individuals affected by cystic fibrosis can cause considerable distress (Fanos and Johnson, 1993), and the limited British experience with population screening for cystic fibrosis suggests that few people are actively interested in having the test even if it is made available. Most adults will accept an offer of testing if it is offered in person by an enthusiastic research worker able to collect the sample without delay, but the spontaneous levels of motivation and of interest in the test appear low (Watson *et al.*, 1991; Bekker *et al.*, 1993). The fact that many adults who are offered carrier screening for recessive diseases decline to accept the offer is similar to the low uptake in practice of predictive testing for HD. It is interesting that only 20% of the staff in a medical genetics department in London chose to accept the offer of cystic fibrosis carrier screening, indicating that low uptake rates may be caused by factors other than a lack of knowledge or understanding (Flinter *et al.*, 1992).

Although it may be more convenient for health professionals to test all potential carriers in a family on a single occasion, e.g. for a balanced chromosomal translocation or for an autosomal recessive gene, this may not be best for the family. Relatively few adults come forward for predictive or carrier screening genetic tests, and it has been shown that such tests can cause considerable anxiety and distress to adults, particularly if proper counselling is not available, so there are strong grounds for adopting a very cautious policy towards predictive or carrier status tests for children.

(h) Some children in a family may have been tested already, so deciding not to test a subsequent child may lead to an anomalous family situation.

For example, after amniocentesis performed to exclude Down syndrome because of raised maternal age, the fetus may be found to carry an apparently balanced chromosomal translocation (i.e. rearrangement). If this had arisen *de novo* then there would be a significant risk of mental and physical handicap, but if it had been inherited then the risks of associated problems would be negligible. The usual procedure in such circumstances would be to examine the parents' chromosomes, and in most cases one parent is found to carry the identical translocation, which is very reassuring. Chromosomal analysis would be offered in any subsequent pregnancies because of the risk that a future fetus could have an unbalanced translocation, which would almost inevitably be associated with serious problems and could be non-viable. Obviously the parents would need to know the result of the amniocentesis, and although it could be agreed in advance only to state whether or not the result has adverse clinical implications, some parents might object to the withholding of more explicit information, particularly if they already knew the details of the karyotype of their previous child.

Prenatal testing for autosomal recessive conditions can lead to similar information being generated: for example, a couple with a child who has cystic fibrosis (CF) may request prenatal diagnosis. There is a 50% chance that the fetus will be a CF carrier, and the information would be available to laboratory staff. Careful discussion with the parents before the prenatal test being performed is appropriate so that decisions can be made then about the handling of the test result.

When DNA linkage studies are performed before prenatal diagnosis, it is helpful to obtain blood samples for DNA analysis from the affected child, and it may also be helpful occasionally to have samples from unaffected sibs because the latter can assist with the interpretation of results (by helping to establish phase and identify any crossovers). It can be argued that it may be in the unaffected child's best interest not to have an affected younger brother/sister, and also in the family's best interests to minimise the risks of an incorrect prenatal diagnostic result. It is an inevitable consequence of such studies that carrier information may be generated about healthy sibs, and parents may request disclosure of these results. It can be very difficult to decline such a request unless these issues have been discussed in advance and agreement reached about which items of information should be released to the parents and child. If, after discussion of the issues raised above, the parents still demand access to the full result, does the medical profession have authority to withhold it on the grounds that disclosure could be harmful to the child? These issues have not been tested legally, but the likely outcome would be a decision to reveal the full test results. This discussion underlines the importance of only taking samples that are likely to be significant in the interpretation of the overall results, and of considering all the possible implications of any results that may be generated beforehand.

7.4 Decision-making in children

In general, it is assumed that anyone over the age of 18 years (or 16 years) is capable of giving informed consent to a medical procedure or test unless he or she is clearly psychiatrically ill or has significant cognitive impairment. Consent for children is given by their parents or legal representatives. Clearly arbitrary cut-offs based on age alone will exclude some under-age, but perfectly competent, decision-makers from acting as autonomous individuals and include some adults who are incompetent decision-makers. There are no simple ways to assess anyone's ability to give truly informed consent (either as an adult or a child), but the ages at which children reach different levels of cognitive development are being lowered as a result of research in this area (Bryant, 1974; Donaldson, 1978). It would be particularly valuable to establish methods of assessing children's competence to understand the implications of genetic testing – and any tests developed in this area could be equally applicable to adults (Gaylin, 1982; King and Cross, 1989). These issues are considered in more detail in a review paper by Michie and Marteau (1995).

7.5 Professional attitudes

Because of the concerns raised about the ethical implications and the potentially damaging consequences of childhood genetic testing, the Clinical Genetics Society in the UK convened a Working Party to examine the issues. For this report, a questionnaire study of attitudes and practices within Britain was carried out. Clinical geneticists, genetic co-workers, paediatricians, haematologists and other clinicians were asked to give their opinions on these matters and to outline their practice, in so far as these issues arose in practice for them (Clinical Genetics Society, 1994).

It was clear from the findings of this study that professional attitudes and practices vary widely, both within and between professional groups. In very general terms, it was clear that the genetic co-workers (nurses, counsellors) were the most 'protective' of childrens' rights, that many clinical geneticists do (or would) carry out some testing of children but in a selective way, and that most paediatricians and other clinicians would be guided largely by parental wishes. Until now, paediatricians have had little opportunity to arrange genetic testing directly (without involving a clinical geneticist), but this situation is changing. Much of their relative lack of caution may reflect their lack of involvement so far in taking these decisions, but they may well soon be able to arrange molecular genetic testing through commercial channels with a direct interest in performing more tests, rather than through health service channels monitored by clinical geneticists. This commercialisation of genetic testing therefore has some potentially important psychological and ethical implications in this area of childhood testing, as well as elsewhere.

7.6 Conclusion

There are several grounds for recommending caution in carrying out predictive or carrier status genetic tests on children, at least until more information is available about what the effects of testing children could be. Testing during childhood removes the individual's future autonomy and confidentiality, and may cause damage to their self-esteem and future interpersonal relationships. Although there is no direct evidence of harm resulting from such tests, the topic has received little attention in the past because of the restricted range of genetic tests available previously. Certain individuals could benefit from growing up with knowledge about their genetic status, but there could be long-term problems relating to health insurance and life insurance, and to prospects for employment.

Where genetic tests are being or have been carried out on children, there exists a valuable opportunity to evaluate the overall medical, psychological and social effects of testing. Current practice, however, should be to presume against testing children until more research has been done, except in particular, unusual circumstances in individual families. Such requests will have to be considered on their individual merits, and whenever possible the families should be followed up closely afterwards to establish the psychological and social consequences.

7.7 References

Alderson, P. (1993) *Children's Consent to Surgery*. Buckingham, UK, and Pennsylvania, USA: Open University Press.

Bekker, H., Modell, M., Denniss, G., Silver, A., Mathew, C., Bobrow, M. and Marteau, T. (1993) Uptake of cystic fibrosis testing in primary care: supply push or demand pull? *British Medical Journal*, **306**, 1584–6.

Billings, P. R., Kohn, M. A., de Cuevas, M., Beckwith, J., Alper, J. S. and Natowicz, M.R. (1992) Discrimination as a consequence of genetic testing. *American Journal of Human Genetics*, **32**, 476–82.

Bloch, M. and Hayden, M. R. (1990) Opinion: predictive testing for Huntington disease in childhood: challenges and implications. *American Journal of Human Genetics*, **46**, 1–4.

Bloch, M., Adam, S., Wiggins, S., Huggins, M. and Hayden, M. R. (1992) Predictive testing for Huntington disease in Canada: the experience of those receiving an increased risk. *American Journal of Medical Genetics*, **42**, 499–507.

Bradley, D. M., Parsons, E. P. and Clarke, A. (1993) Experience with screening newborns for Duchenne muscular dystrophy in Wales. *British Medical Journal*, **306**, 357–60.

Bryant, P. (1974) *Perception and Understanding in Your Children*. London: Methuen.

Burn, J., Church, W., Chapman, P. D., Gunn, A., Delhanty, J. and Roberts, D. F. (1989) A regional register for familial adenomatous polyposis: congenital hypertrophy of the retinal pigment epithelium as a means of carrier detection. *Journal of Medical Genetics*, **26**, 207.

Clarke, A. M. and Clarke, A. D. B. (eds.) (1976) *Early Experience: Myth and Evidence*. London: Open Books.

Clinical Genetics Society (1994) *Report on the Genetic Testing of Children*. Birmingham: CGS, and *Journal of Medical Genetics*, **31**, 785–97.

Costakos, D., Abramson, R. K., Edwards, J. G., Rizzo, W. B. and Best, R. G. (1991) Attitudes toward presymptomatic testing and prenatal diagnosis for adrenoleukodystrophy among affected families. *American Journal of Medical Genetics*, **41**, 295–300.

Craufurd, D. and Harris, R. (1986) Ethics of predictive testing for Huntington's chorea: the need for more information. *British Medical Journal*, **293**, 249–51.

Craufurd, D., Dodge, A., Kerzin-Storrar, L. and Harris, R. (1989) Uptake of presymptomatic testing for Huntington's disease. *Lancet*, **ii**, 603–5.

Donaldson, M. (1978) *Children's Minds*. London: Fontana.

Fanos, J. H. and Johnson, J. P. (1993) Barriers to carrier testing for CF siblings. *American Society of Human Genetics Meeting, New Orleans, October 1993*, Session 22, Abstract 51.

Flinter, F. A., Silver, A., Mathew, C. G. and Bobrow, M. (1992) Population screening for cystic fibrosis. *Lancet*, **339**, 1539–40.

Gaylin, W. (1982) Competence: no longer all or none. In W. Gaylin and R. Macklin (eds.) *Who Speaks for the Child: The Problems of Proxy Consent*. New York: Plenum.

Harper, P. S. (1992) Insurance and genetic testing. *Lancet*, **341**, 224–7.

Harper, P. S. and Clarke, A. (1990) Should we test children for 'adult' genetic diseases? *Lancet*, **335**, 1205–6.

Harper, P. S., Tyler, A., Smith, S., Jones, P., Newcombe, R. G. and McBroom, V. (1982) A genetic register for Huntington's Chorea in South Wales. *Journal of Medical Genetics*, **19**, 241–5.

Huggins, M., Bloch, M., Wiggins, S., Adam, S., Suchowersky, O., Trew, M., Klimek, M. L., Greenberg, C. R., Eleff, M., Thompson, L. P., Knight, J., MacLeod, P., Girard, K., Theilmann, J., Hedrick, A. and Hayden, M. R. (1992) Predictive testing for Huntington Disease in Canada: adverse effects and unexpected results in those receiving a decreased risk. *American Journal of Medical Genetics*, **42**, 508–15.

Kessler, S. (1988) Invited essay on the psychological aspects of genetic counseling. V. Preselection: a family coping strategy in Huntington Disease. *American Journal of Medical Genetics*, **31**, 617–21.

King, N. M. and Cross, A. W. (1989) Children as decision makers: guidelines for pediatricians. *Journal of Pediatrics*, **115**, 1–16.

Loader, S., Sutera, C. J., Segelman, S. G., Kozyra, A. and Rowley, P. T. (1991) Prenatal hemoglobinopathy screening. IV. Follow-up of women at risk for a child with a clinically significant hemoglobinopathy. *American Journal of Human Genetics*, **49**, 1292–9.

Marteau, T. M., van Duijn, M. and Ellis, I. (1992) Effects of genetic screening on perceptions of health: a pilot study. *Journal of Medical Genetics*, **29**, 24–6.

Michie, S. and Marteau, T. M. (1995) Predictive genetic testing in children: the need for psychological research. *British Journal of Clinical Psychology*, in the press.

Montgomery, J. (1993) Consent to health care for children. *Journal of Child Law*, **5**, 117–24.

Montgomery, J. (1994) Rights and interests of children and those with mental handicap. In A. Clarke (ed.), *Genetic Counselling*. London: Routledge.

Morris, M., Tyler, A. and Harper, P. S. (1988) Adoption and genetic prediction for Huntington's disease. *Lancet*, **ii**, 1069–70.

Morris, M., Tyler, A., Lazarou, L., Meredith, L. and Harper, P. S. (1989) Problems in genetic prediction for Huntington's disease. *Lancet*, **ii**, 601–3.

Norman, A. M., Rogers, C., Sibert, J. and Harper, P. S. (1989) Duchenne muscular dystrophy in Wales: a 15 year study, 1971 to 1986. *Journal of Medical Genetics*, **26**, 560–4.

Parsons, E. P. and Bradley, D. M. (1994) Ethical issues in newborn screening for Duchenne muscular dystrophy: the question of informed consent. In A. Clarke (ed.), *Genetic Counselling*. London: Routledge.

Pelias, M. Z. (1991) Duty to disclose in medical genetics: a legal perspective. *American Journal of Medical Genetics*, **39**, 347–54.

Read, A. P., Kerzin-Storrar, L., Mountford, R. C., Elles, R. G. and Harris, R. (1986) A register-based system for gene tracking in Duchenne muscular dystrophy. *Journal of Medical Genetics*, **24**, 84–7.

Sharpe, N. F. (1993) Presymptomatic testing for Huntington Disease: is there a duty to test those under the age of eighteen years? *American Journal of Medical Genetics*, **46**, 250–3.

Stamatoyannopoulos, G. (1974) Problems of screening and counselling in the haemoglobinopathies. In A. G. Motulsky and F. J. B. Ebling (eds.), *Birth Defects: Proceedings of the Fourth International Conference.* Amsterdam: Excerpta Medica.

Thelin, T., McNeil, T. F., Aspegren-Jansson, E. and Sveger, T. (1985) Psychological consequences of neonatal screening for alpha-1-antitrypsin deficiency. *Acta Paediatrica Scandinavica*, **74**, 787–93.

Tibben, A. (1993) On psychological effects of presymptomatic DNA-testing for Huntington's disease. Ph.D. thesis, Rotterdam University, The Netherlands.

Tibben, A., Vegter-van der Vlis, M., Skraastad, M. I., Frets, P. G., van der Kamp, J. J. P., Niermeijen, M. F., van Ommen, G. J. B., Roos, A. C., Rooijmans, H. G. M., Stronks, D. and Verhage, F. (1992) DNA-testing for Huntington's disease in The Netherlands: a retrospective study on psychosocial effects. *American Journal of Medical Genetics*, **44**, 94–9.

Tyler, A., Morris, M., Lazarou, L., Meredith, L., Myring, J. and Harper, P. S. (1992) Presymptomatic testing for Huntington's Disease in Wales 1987–1990. *British Journal of Psychiatry*, **161**, 481–9.

Watson, E. K., Mayall, E., Chapple, J., Dalziel, M., Harrington, K., Williams, C. and
   Williamson, R. (1991) Screening for carriers of cystic fibrosis through primary health care
   services. *British Medical Journal*, **303**, 504–7.
Wiggins, S., Whyte, P., Huggins, M., Adam, S., Theilmann, J., Sheps, S. B., Schechter, M. T. and
   Hayden, M. R. (1992) The psychological consequences of predictive testing for
   Huntington's disease. *New England Journal of Medicine*, **327**, 1401–5.
World Federation of Neurology (1989) Research Committee Research Group: ethical issues
   policy statement on Huntington's disease molecular genetics predictive test. *Journal of
   Neurological Science*, **94**, 327–32.
World Federation of Neurology (1990) Research Committee Research Group: ethical issues
   policy statement on Huntington's disease molecular genetics predictive test. *Journal of
   Medical Genetics*, **27**, 34–8.
Zeesman, S., Clow, C. L., Cartier, L. and Scriver, C. R. (1984) A private view of heterozygosity:
   eight-year follow-up study on carriers of the Tay–Sachs gene detected by high school
   screening in Montreal. *American Journal of Medical Genetics*, **18**, 769–78.

8

Predictive genetic testing in children: paternalism or empiricism?

SUSAN MICHIE

There is always a problem when new medical technologies lead to clinical services for which we have no or little evidence of their impact. The desirable approach is to carry out clinical trials to determine their effectiveness and safety. Predictive genetic testing is a technological development with possible psychological and social effects that have been much discussed. Because predictive testing can be offered to children, as well as to adults, the discussion has been wide-ranging, encompassing issues such as competence to give informed consent, the rights of the child and autonomy.

The complexity of these issues does not mean that policies should be formulated without the backing of relevant research. Indeed, it could be argued that such testing should only be offered as part of a research protocol determining its effects and the circumstances under which it is most effective and least harmful.

There are three general approaches to the clinical introduction of genetic developments that can be identified in past practice. One is to herald caution, and keep the debate within professional circles. The second is to consult widely, seeking the opinions of potential users of new services and the general public. The third is to promote rapidly the research that will address the questions of psychological and social impact. In Chapter 7, Clarke and Flinter emphasise the first approach; I shall discuss the second two approaches.

The current view of clinical geneticists is summarised by Clarke and Flinter as follows:

> Discussion among clinicians and scientists involved in developing a molecular genetic predictive test for [Huntington's disease] led to a consensus view that children should not be tested. . . . It was feared that children could be stigmatised, even within the family, and knowledge of their genetic status could distort their emotional development, their educational attainments and their future relationships. The test result could effectively blight their lives much earlier and more thoroughly than the disease would do eventually. For these reasons, it was felt advisable to adopt a position of caution. . . .
> *(p. 166)*

8.1 The evidence

There are two pieces of evidence put forward to support the above statements. The first is that a majority of adults at risk for HD said in a questionnaire survey that they would request testing when it became available, yet only 9–15% underwent testing when it did become available (Craufurd *et al.*, 1989; Tyler *et al.*, 1992). These results say nothing about the possible harmful effects of testing. Rather, they confirm a well documented finding that what people say they will do is not always a good predictor of what they will do (for review, see Eiser, 1986).

Flinter and Clarke also argue, from the studies of Huggins *et al.* (1992) and Tibben *et al.* (1992), that 'knowledge of HD gene status can be burdensome, even when the result is low risk'. Neither of these were controlled studies, so we have no information as to whether a lack of knowledge of HD gene status is more, or less, burdensome. Clarke and Flinter mention that in some cases the adverse psychological reaction relates to regrets for missed opportunities and time wasted while living with the burden of risk, which is an argument *for* earlier testing.

There is a possibility of comparing within a testing programme those that have received a result and those who, because of technical difficulties, have not. One such study by Wiggins *et al.* (1992) looked at those choosing to be tested and found that receiving a result, whether positive or negative, was psychologically preferable to not having a result. Those choosing to be tested are, however, a small sub-group of about 10% of those at risk for HD.

8.2 The arguments

Three arguments are put forward by Flinter and Clarke against testing children:

1. It removes the individual's future right to make their own decisions about testing as an autonomous adult.

   Of course this is true, but untested children lose their right to be tested; and adults lose their right to have been tested in childhood.

2. It removes confidentiality that would be expected and provided for any adult undergoing the same test.

   The issue of confidentiality when the testing of one individual has implications for other family members is a complex one for all individuals, adults as well as children.

3. It may alter the child's upbringing and pattern of relationships within the family and also with their peers, which could lead to stigmatisation and discrimination.

   This *may* be true, but it may not be: we lack the evidence. There may also be benefits, such as giving more opportunity to prepare psychologically and practically for the future. Until we know what the actual, rather than the possible, effects are, we should avoid basing policy on speculation.

8.3 Professionals' versus parents' attitudes

The absence of evidence is acknowledged, and varying opinions about practice have been found. Yet, the professional consensus is 'not to test'. Clarke and Flinter write:

In the absence of systematic evidence, opinions as to good practice vary, and it may be difficult, or even inappropriate, to draw up rigid criteria for predictive testing in childhood.

But in the very next sentence, they say:

The natural reluctance of most professions to cause harm to their clients, however, is likely to lead towards a professional consensus that generally presumes that such testing is inappropriate. . . .

If there is a lack of evidence and a variability of practice, why not ask parents and the public what they think about the issue? One of the reasons that parents' views are not given greater weight may be a negative attitude towards parents as responsible agents. This is suggested in the context of communicating about genetics by the following quote from Chapter 7:

Many parents find sex education a difficult topic to discuss with their children; how confident can we be that they will be able to discuss genetic issues effectively? This is of particular concern in a society such as ours, with a high rate of family breakdown; it is simply not possible to rely upon the effectiveness and sensitivity of intergenerational communication. *(p. 169)*

Even if we could always rely upon the effectiveness and sensitivity of professionals' communication, surely the role of professionals in this situation is to facilitate good communication within the family by imparting relevant knowledge and skills, rather than to replace parents as communicators with their children about these issues.

A non-paternalistic view of testing in children has been put forward by Sharpe (1993), from Canada. He suggests that the policy of not testing children is the result of the geneticist imposing her or his own values and beliefs about what is 'best' for the child and the family, and argues that the parents are in a better position to weigh the social, familial, emotional and economic factors:

The issue is whether the decision not to test effectively abrogates what it seeks to protect, the child's personal rights and dignity subordinated to, if not replaced by, the objectives, values and rationality of the geneticist.

When parents have been asked, they are more favourable to testing than are professionals. In The Netherlands, 41% of 70 individuals at risk for Huntington's disease thought that children under 18 should be allowed to have the test if they so chose (Tibben, 1993). This is consistent with other reports (Markel *et al.*, 1987; Meissen and Berchek, 1987; Bloch *et al.*, 1989; Tibben *et al.*, 1992). A UK study of parents from families carrying genetic translocations found that parents were generally in favour of testing being offered (Barnes, 1994). Of 211 children, 75 were untested. Parents of only 8 of these children were against testing in childhood, with 16 undecided and 51 in favour.

180   *Susan Michie*

8.4 Who is involved in the debate?

In Britain, the debate about predictive testing in children has been very much kept within professional circles. A working party was convened by the Clinical Genetics Society to consider the ethical and psychological implications of childhood genetic testing. It sought opinions from many groups of health professional, but not directly from parents or the public. It did not include any behavioural or social scientists, despite these being the professionals with expertise in the key areas of children's psychological development and family functioning.

Developmental psychology, for example, would not suggest that parental anxiety is only relieved if children receive favourable test results, nor that parental anxiety is increased with unfavourable results, as Clarke and Flinter claim.

In interpreting the results of the working party survey, Clarke and Flinter say that the genetic co-workers were the most 'protective' of children's rights. Readers are left to assume that this means that the genetic co-workers were most against testing. This raises two questions, that of protection, and that of children's rights. Since, as Clarke and Flinter admit, there is no evidence of testing harming children, we cannot say that taking one or another course of action is more protective. And on the question of rights, do children not have a right to be consulted about issues that effect them and to be involved in decision-making?

The social climate of the last three decades has resulted in an increased interest in allowing children to participate in decisions about their own health care. The presumption of the child's incapacity to make decisions is weakening and the possibility of children giving valid consent is being entertained (King and Cross, 1989). The patients' rights movement has fostered the growing belief that fairness requires that children assume an increased role in decisions that affect them, including healthcare decisions.

One reason that professionals in the UK have not readily included children in the decision-making process about genetic testing may be that children are not deemed competent to take an informed decision. However, following the legal ruling by the House of Lords in 1986,[1] children under the age of 16 are considered competent to give or withhold consent without their parents' knowledge or agreement if they have sufficient understanding to grasp the implications of the decision they are being asked to make. The issues of informed consent and decision-making – who is competent to make a decision, and who is competent to judge that competence – are complex ones. They, too, should be informed by empirical research. For example, we need answers to questions such as:

> What information is necessary for different types of decision?
> How is that information analysed, interpreted and integrated with values and beliefs?
> How are decisions taken?

There is little research about these processes and their variation, in adults or in children. These questions are discussed in more detail elsewhere (Michie and Marteau, 1995). My

---

[1] *Gillick* v *West Norfolk and Wisbech Area Health Authority* [1986]1 FLR 224

main point here is that we should not use assumptions about children's inabilities to exclude them and their parents from making decisions about genetic testing.

8.5 Who should decide?

Clarke and Flinter conclude:

> Current practice . . . should be to presume against testing children until more research has been done, except in particular, unusual circumstances in individual families. Such requests will have to be considered on their individual merits, and whenever possible the families should be followed up closely afterwards to establish the psychological and social consequences. *(p. 173)*

What are these 'particular, unusual circumstances', and which 'individual families'? Unless we have evidence about effects, why should different things be offered to different people, and who should make this decision? We are told the decision will be made on 'individual merits'. What are these 'merits', and who is competent to make the judgement?

The question of who should make the decision about the desired scope of genetic testing is one that everyone should be involved in answering and one that should be the subject of research. A recent survey asked the general public and three groups of professionals about their attitudes towards decision-making about genetic testing (Michie *et al.*, 1995). Of the 973 members of the public surveyed, none were in favour of individual doctors making the decision and only 6% of the 159 professionals thought that individual doctors should decide. The great majority (78%) of professionals thought than an advisory group of experts and the public should decide, whereas only 25% of the public thought this. The majority of the public thought that professionals should not be involved at all, with 36% favouring the decision being made by individual parents and 18% favouring a public referendum.

8.6 Future research

The suggestion in Clarke and Flinter's last sentence, that whenever possible the families should be followed up to establish the psychological and social consequences, will not give us the research information we need. Unless this is done as part of a well-designed research protocol, such clinical follow-up will tell us only about the individual families.

To ensure that clinical practice is based on empirical evidence, rather than on opinion or clinical impression, research is needed in three main areas. First, we need to know more about the attitudes of children, their parents, the general public and health professionals; where the differences lie; and the reasons for these differences.

The second area is that of informed consent: what should be the criteria for competence, how should they be measured, and who should judge? We know little about informed decision-making in adults, let alone children. For example, how is genetic information analysed and interpreted, how are values and beliefs integrated and what processes are used in decision-making?

The third area for research concerns the psychological consequences of predictive testing in children. Several approaches are needed. First, studies are needed to determine the psychological effects upon children being brought up in families at risk for adult onset diseases. This is a necessary background against which to judge the effects of testing children in these families. Crucially, we need controlled studies of the short-term and long-term effects of testing and of the mediators of these effects.

Research and public consultation needs resources. If this allows the responsible application of the new genetic technological developments, it will be money well spent. Caution in the absence of evidence is 'second best'.

8.7 Acknowledgement

The author of this chapter is supported by a grant from The Wellcome Trust.

8.8 References

Barnes, C. (1994) Genetic testing of children. Presented at Fourth European Conference of Psychosocial Aspects of Genetics, Heidelberg, 12–14 September.

Bloch, M., Fahy, M., Fox, S. and Hayden, M. R. (1989) Predictive testing for Huntington disease. II. Demographic characteristics, life-style patterns, attitudes, and psychosocial assessments of the first fifty-one test candidates. *American Journal of Medical Genetics*, **32**, 217–24.

Craufurd, D., Kerzin-Storrar, L., Dodge, A. and Harris, R. (1989) Uptake of presymptomatic predictive testing for Huntington's disease. *Lancet*, **ii**, 603–5.

Eiser, J. R. (1986) *Social Psychology: Attitudes, Cognition and Social Behaviour*. Cambridge: Cambridge University Press.

Huggins, M., Bloch, M., Wiggins, S., Adam, S., Suchowersky, O., Trew, M., Klimek, M., Greenberg, C. R., Eleff, M., Thompson, L. P., Knight, J., Macleod, P., Girard, K., Theilmann, J., Hedrick, A. and Hayden, M. R. (1992) Predictive testing for Huntington's disease in Canada: adverse effects and unexpected results in those receiving a decreased risk. *American Journal of Medical Genetics*, **42**, 508–15.

King, N. M. and Cross, A. W. (1989) Children as decision makers: guidelines for pediatricians. *Journal of Pediatrics*, **115**, 1–16.

Markel, D. S., Young, A. B. and Penney, J. B. (1987) At risk persons' attitude toward presymptomatic and prenatal testing of Huntington disease in Michigan. *American Journal of Medical Genetics*, **26**, 95–305.

Meissen, G. J. and Berchek, R. L. (1987) Intended use of predictive testing by those at risk for Huntington disease. *American Journal of Medical Genetics*, **26**, 283–93.

Michie, S. and Marteau, T. (1995) Predictive genetic testing in children: the need for psychological research. *British Journal of Clinical Psychology*, in the press.

Michie, S., Drake, H., Bobrow, M. and Marteau, T. (1995) A comparison of public and professionals' attitudes towards genetic developments. *Public Understanding of Science*, **4**, 243–53.

Sharpe, N. F. (1993) Presymptomatic testing for Huntington disease: is there a duty to test those under the age of eighteen years? *American Journal of Medical Genetics*, **46**, 250–3.

Tibben, A. (1993) On psychological effects of presymptomatic DNA-testing for Huntington's disease. Ph.D. thesis, Rotterdam University, The Netherlands.

Tibben, A., Frets, P. G., van der Kamp, J. J. P., Vegter-van der Vlis, M., Roos, R. A. C., van Ommen, G. J. B., Duivenvoorden, H. J. and Verhage, F. (1992) Presymptomatic DNA-testing for Huntington disease: pretest attitudes and expectations of applicants and their partners in the Dutch program. *American Journal of Medical Genetics*, **44**, 94–9.

Tyler, A., Morris, M., Lazarou, L., Meredith, L., Myring, J. and Harper, P. (1992) Presymptomatic testing for Huntington's disease in Wales 1987–90. *British Journal of Psychiatry*, **161**, 481–8.

Wiggins, S., Whyte, P., Huggins, M., Adam, S., Theilman, J., Bloch, M., Sheps, S. B., Schechter, M. and Hayden, M. (1992) The psychological consequences of predictive testing for Huntington's disease. *New England Journal of Medicine*, **327**, 1401–5.

Part III
Social context

9

The troubled helix: legal aspects of the new genetics

DEREK MORGAN

> Sorcerers are too common; cunning men, wizards, and white witches as they call them,
> in every village, which, if they be sought unto, will help almost all infirmities of body
> and mind.
> *(Richard Burton, 1621, cited by Thomas (1973))*

9.1 Epistemology, ethics and genetics

Developments in genetics 'pose challenging questions for the application of traditional legal principles' (Kennedy and Grubb, 1993). This much was recognised, for example, at the Asilomar conference in 1975, where a group of molecular biologists recommended a moratorium on genetic manipulation while arrangements were made to regulate recombinant DNA techniques (Maddox, 1993). Without precedent, the scientific research community was inviting regulation from the legislature for its own activities. Indeed, writing in 1971 James Watson (the elucidator, with Francis Crick, of the structure of DNA) had suggested, of the possible developments in human reproductive research, that techniques for the manipulation of human eggs *in vitro* were likely to be in general medical practice, capable of routine performance in many major nations, within some 10 to 20 years, and that international agreement was a preferred method of control. On some matters there might even be 'a sufficient international consensus . . . to make possible some forms of international agreement before the cat is totally out of the bag' (Watson, 1971).

The particular difficulties that lawyers and ethicists will want to address are the functional analogue of the difficulties that the genome project discloses in general:

> Physically, printing the names of the three billion base pairs would require the number
> of pages in at least thirteen sets of the Encyclopaedia Britannica, and this does not take
> into account the heterogeneity of human beings. The epistemological consequence of
> this huge amount of information is unforeseeable. *(Rix, 1991)*

As it is for society, so it is for individuals; the consequences of knowing so much about oneself, about others and about all humans are unforeseeable. One apparent danger is that because genetic conditions are often regarded as immutable hereditary traits, 'overly deterministic interpretations of genetic information can readily distort genetic risk and become enshrined in institutional policies of social isolation and discrimination' (Jecker, 1993).

One of the difficulties in assessing the human genome project and the associated but independent industry of genetic screening is that we are limited in our assessment because, as Bryan Wynne has written of a cognate area, we 'cannot in any significant sense assess the technology itself for its full "factual" impact' and thus 'we have to assess the *institutions* which appear to control the technology' (Wynne, 1982).

Nevertheless, it has been argued that the development of the human genome project itself raises no new ethical difficulties and perhaps few legal problems (Clothier, 1992), and that such problems as there are arise from the use and development of genetic testing and screening as a therapeutic opportunity rather than the mapping and sequencing of the genome (Maddox, 1993). Even here John Maddox (the editor of *Nature*) has counselled caution in creating 'new' ethical dilemmas: 'the availability of gene sequences, and ultimately of the sequence of the whole genome, will not create ethical problems that are intrinsically novel, but will simply make it easier, cheaper and more certain to pursue certain well-established objectives in the breeding of plants, animals and even people.' It might be objected that it is in this elision between plants, animals and people, in the apparent ease of the assumption that (however well established the fantasy of breeding people scientifically) we are dealing with nothing more than linear progressions of scientific vectors, that the real challenges lie. But Maddox's argument belies the transparency that he has himself claimed for it. He argues that while the molecular causes of conditions such as sickle cell disease, various thalassaemias, Huntington's disease, fragile-X syndrome and cystic fibrosis have all been determined in the past few years, 'this new knowledge has not created novel ethical problems, only ethical simplifications.'

Eschewing such formulations, George Annas has suggested that the uniqueness of the human genome project is not its quest for knowledge. The history of science is filled with little else. What is unique, he claims, is an understanding at the outset that serious policy and ethical issues are raised by the research, and that pre-emptive steps ought to be taken to try to assure that 'the benefits of the project are maximized and the potential dark side is minimized' (Annas, 1993). Annas has suggested that there are three levels of issues that the Human Genome Project raises: (i) individual/family, (ii) society, and (iii) species. Most attention on genetics to date has been at the individual/family level, where questions of genetic screening and counselling predominate. Thus, negligence in failing to offer or to perform properly these tests has already resulted in lawsuits for wrongful birth and wrongful life (see note (1), and Lee and Morgan, 1989), and standards for genetic screening and counselling have indeed been discussed (Chadwick and Ngwena, 1992).

Issues at the second level implicate society more directly. For Annas, the Human Genome Project gives rise to three major societal issues: population screening, resource allocation and commercialization, and eugenics. More specifically, he asks, 'to what uses should the fruits of the project be put in screening groups of people, such as applicants for the military, government workers, immigrants and others?' (Annas, 1993).

How, if at all, should intellectual property laws such as patents be invoked? What funding and resource allocation decisions are there to be made (McLean and Giesen, 1994)?

The results of the EC Human Genome analysis may lead to enhanced possibilities in the prevention and treatment of disease, through new methods of genetic testing and screening. Access to genetic information may also improve the quality and efficiency of public authorities outwith health care; in particular those in the fields of criminal justice, social security and public health and in immigration cases. Each of these areas raises discrete difficulties in terms of protection of individual privacy and integrity, and it may be here that the quantitative difficulties for law will arise (Nielsen and Nespor, 1993).

Third-level issues are more speculative, and involve how 'a genetic view of ourselves could change the way in which we think about ourselves' (Annas, 1993). And, importantly, others; it may affect the way in which we come to view relatedness, otherness and difference. Marilyn Strathern and her colleagues have in a cognate area argued that the deployment of reproductive technologies is affecting assumptions we bring to understandings not only of family life but to the very understanding of family itself and cultural practice (Foucault, 1970; Thomas, 1983; Strathern, 1992; Strathern *et al.*, 1993): 'The way in which the choices that assisted conception affords are formulated, will affect thinking about kinship. And the way people think about kinship will affect other ideas about relatedness between human beings' (Strathern, 1992). And, I would add, the way in which we think about relatedness between human beings will affect the way in which we think about the relationship between individuals, groups and the state (Beck, 1992).

Let me test quickly the hypothesis that underlies my thesis by returning briefly to Maddox's argument. Discussing the possibility that a 'gene' for schizophrenia may confer advantages not yet recognised on those in whom the overt disease does not manifest, he suggests that not only will it be a long time before the genetics of psychiatric conditions is understood, but, more importantly 'geneticists themselves are likely to be the first to recognize the dangers of interfering with the natural flow of genes within a population before the social implications are understood' (Maddox, 1993). Thus, he appears to be suggesting an essentially biological or genetic subset of 'dangers' that are divorced, or separate, from the social implications of what may transpire. To this I want to take exception. Secondly, he avers that 'only geneticists can recognize the dangers.' Herein, it seems to me, lie the seeds of the first new, or radically transformed, ethical dilemma; not simply asserting a professionally proprietorial attitude to knowledge (which would not be new), but additionally to *understanding and application*. This, taken with changes within ethical debate itself, has radical potential. Let me briefly explain and explore this point.

Moral decision-making in medicine (*as in other professional and public organisational settings*) is becoming increasingly institutionalised and subject to formalised procedures and constraints. As Bruce Jennings has suggested, across a broad range in the landscape of contemporary medicine (including research on human subjects, organ procurement and transplantation, assisted reproduction, the rationing of health care and the forgoing of life-sustaining treatment), 'ethical choice and agency are now embedded as never before in a network of explicit rules and formal procedures and processes for making decisions' (Jennings, 1991). These rules stipulate (within certain limits) what types of decision may be made, how they may be made, by whom, and with the assistance of what resources.

This, for Jennings, means that they are embedded in the organisational form of statutes, court opinions, administrative mandates, and institutional protocols. This 'embedment' has important relationships with the kinds of ethical concerns and the way in which they are expressed. There has been an important recent shift away from epistemological questions about the relationship between a rational, knowing subject and a rationally knowable, objective morality as the primary focus of ethical theory towards an approach that aims to understand morality 'as a socially embedded practice'. The 'classical' approaches (exemplified in the different theoretical practices of Plato and Kant) invest far more moral authority in the practice and the results of philosophical reflection than they are able to sustain and seem to require a neutral and insulated space for such reflection, cut off from the space of public discourse and deliberation (Jennings, 1991). The shift thus identified is part of a rethinking of the very nature of ethical theory itself: its relationship to human subjectivity and the cultural context that produces it, the kind of knowledge it can be expected to provide and the force and authority of its claims and its relationship to practice are part of the reconstruction that is under way. This kind of 'post-modern' philosophical reorientation of moral philosophy so fundamentally affects our grasp of the relationship between theory and practice that, Jennings proposes, it cannot but have a profound effect on applied ethics. It exposes the extent to which classical ethical theories 'rest on assumptions about the transcendent character of reason and a "philosophy of the subject" . . . that are no longer tenable.' In other words, it is being claimed that ethical conclusions are produced and constructed rather than found from contemplation. The older questions are being displaced by a newer – post-modern – approach that aims to understand the crucial questions of morality which 'have to do with the way in which the meanings and legitimacy of moral notions are established, reinterpreted or transformed over time.' Again, these transformations have important consequences for the ways in which we conceptualise *and even describe* the setting of a legal framework and the establishment of ethical standards for regulating scientific and technical developments, including genetics. Thus, at the time when it is claimed that genetics is capable of eroding ignorance and delivering to us enlightenment, ethical theory is moving away from a rationally knowable subject and into the realm of constructions. I shall return to these observations in my concluding comments.

First, however, I want to focus on three specific manifestations of the 'new' genetics: (i) the regulation of the use of information derived from genetic testing, taking merely as an example the use by the insurance market; (ii) questions raised by information derived from an individual's genetic makeup, including in this context consent especially where children are concerned; (iii) the particular application of that knowledge in the context of abortion based on fetal anomalies.

9.2 Genetic information and privacy

Genetic information

The use of information about genetic variability in relation to insurance – life, health or social security – raises fundamental questions of discrimination against those with genetic risks and of confidentiality of personal health information. These issues need urgent attention.   *(European Commission, 1992)*

'Genetic privacy' what Erving Goffman once identified as 'the right to reticence' (Goffman, 1964), is essential, if we are to avoid creating a zone in which privacy is earned 'only by having nothing to hide'. We live increasingly in the surveillance society and the global village – panoptic planetary people – in which we are watched over in the bank, the store, the petrol station, the telephone booth; day and night surveillance is facilitated by the possibilities of chip technology. The video camera that can enthral and entertain our children can imprison and impoverish them; information about our movements, our tastes, pleasures and purchases can be transmitted almost instantaneously from one side of the globe to the other.

Genetic knowledge is a form of information technology, and as such it poses three types of disquiet: (i) in facilitating intrusions on personal privacy; (ii) in providing the means for institutions to exercise particular forms of control; (iii) in encouraging practices that threaten certain values. Allied with the biotechnological imperative, which has come to replace nuclear power as the symbol of 'technology out of control,' information technologies can directly affect particular economic interests; they may be a source of risk, and for some, they are a moral threat. Indeed, Dorothy Nelkin has suggested that biotechnology raises many of the same problems as nuclear power; the hazards are invisible and there remains uncertainty about the health effects of low-level long-term exposure. Like nuclear power, biotechnology evokes images of warfare and fantasies of monsters and mutations, demons and chimeras (Nelkin, 1993). The challenge that recent developments in genetic science present is to obtain all the benefits of the knowledge while minimising or eliminating the risks.

The question of genetic information about one person which may identify or suggest a genetic disorder or trait in another poses a particular dilemma in the physician–patient relationship. M. A. M. de Watcher has suggested that

Genetic medicine . . . is greatly expanding . . . views [of privacy and bodily integrity] into a wider concept of corporate ownership of familial and ethnic autonomy. It now seems that the totality of a person's physical existence exceeds the limits of a single person's body. Some already say that genetic information is the common property of the family as 'corporate personality.' Are we then entering a new era of medicine . . . an era where information is governed not only by rules of individual confidentiality but also by duties of common solidarity?   *(in Bankowski and Bryant, 1989)*

Four particular pitfalls are evident: mere biological links may be insufficient to promote the intrusion into the psychosocial components of privacy; it is difficult to draw the line between information relevant to genetic counselling and that which is not;

as more diseases appear to contain hereditary components, possible compromises of confidentiality appear unlimited; removing all control of data from people who are screened may be counterproductive and dissuade them from entering family screening programmes. As genetic tests become simpler to administer and their use expands, a growing number of individuals may be 'labelled' on the basis of predictive genetic information. The use of predictive genetic diagnoses creates a new category of individuals who are not ill, but have reason to suspect they may develop a specific disease sometime in the future: 'the healthy ill' (Bankowski and Bryant, 1989). The loss of autonomy and privacy, which fears of genetic testing have foreshadowed, can be 'the genesis of a life-long psychological prison – the prison of one's perceived genetic "programming"' (Privacy Commissioner of Canada, 1992).

Crude genetic screening has been used for many years by taking into account family history, such that the current health state of an individual is thought to be a useful (if not completely reliable) indicator of that individual's life expectancy and the extent to which he or she is likely to consume health and other resources in the future. Whereas in the past one might have said that an individual's state of health was a combination or amalgam of heredity, environment, behaviour and luck, one of the radical changes that genetic testing introduces is the question of responsibility for health. Over 4000 Mendelian inherited disorders have been identified: some determined wholly by single-gene mutations and inheritances; some causing disease before birth, or shortly after, others observed only in adulthood. Understanding of polygenic conditions (those determined by sets of genes) and predispositions to diseases (where disease onset is determined by a combination of genetic and environmental factors) is growing. Mapping the genome may increase the scientific and medical ability to predict, understand and eventually to prevent or to cure human diseases. This also gives rise to questions as to how this information should be used, and the necessary background laws and norms that will either be applied or that will become more clearly perceived to direct and control, to regulate the use of this information, either by individuals or by societies generally. With the advances in the scope and reliability of genetic testing, information relating to an individual's genetic predisposition will be of considerable use to a potential insurer in that such information is indicative of the risk to the insurer that that individual represents. More than that, the influence that this knowledge, directly and indirectly, could exert over lifestyle and life choices is manifest and major. The knowledge that a woman is at risk of giving birth to a genetically damaged child may ensure that she is unable to obtain insurance cover for the child and hence lead to tacit pressure to have an abortion.

In the European Community Programme for 'Predictive Medicine' (Commission of the European Community, 1989) it was acknowledged that for a large number of common western diseases, such as cancer, coronary heart disease, diabetes, autoimmune diseases and major psychoses, there is a strong environmental component. Given that we are unlikely to be able to remove those environmental factors entirely, the development of information about genetically determined predisposing factors is critical if we are to be able to identify high-risk individuals. The controversial nature of some of the genome research is evident in the reception and reworking of the European 'Predictive Medicine' project: the initial aim of going directly from the discovery of genetic abnormalities to the use of prenatal diagnosis and abortion to prevent their inheritance has been dropped from the programme, although it remains as a clear if controversial backdrop to its development and implementation.

One response in the European Parliament has been to urge that the requirement to submit to genetic screening as a condition for obtaining insurance should be prohibited. In an amendment to the Pompidou Report from the Energy, Research and Technology Committee (Pompidou, 1993) (which had recommended that disclosure of genetic information should be strictly regulated), it has been proposed that 'insurers should not have the right to require genetic testing or to enquire about the results of previously performed tests as a precondition for the conclusion or modification of an insurance contract.' The prohibition on enquiring about previous tests could have a dramatic effect on the insurance industry, described by Brett and Fischer (1993) as 'dangerous.'

The National Heritage Committee has proposed the introduction of a Privacy Bill, and the Third Report of the House of Commons Science and Technology Committee, *Human Genetics: The Science and Its Consequences* (July 1995, H.C. 41–I) calls for the misuse of genetic information to be both a criminal and civil offence (para 225).

Attempts to legislate in the United States have taken the form of Bill HR 5612 first entered the United States Congress in September 1990 as the Human Genome Privacy Act, to be succeeded later by a modified Bill in April 1991. More recently, Bill HR 2045, virtually identical to Bill HR 5612, has been presented to the House of Representatives. Salient sections of the Bill specify that:

> 2(b) The purpose of this Act is to provide an individual with certain safeguards against the invasion of personal genetic privacy by requiring agencies, except as otherwise provided by law, to –
> 1. permit an individual to determine what records pertaining to him or her are collected, maintained, used, or disseminated by such agencies;
> 2. permit an individual to prevent records pertaining to him or her obtained by such agencies for a particular purpose from being used or made available for another purpose without his or her consent;
> 3. permit an individual to gain access to records, to have a copy made of any or all portion thereof, and to correct or amend such records;
> 4. collect, maintain, use, or disseminate any record of identifiable personal genetic information in a manner that assures that the information is current and accurate for its intended use, and that adequate safeguards are provided to prevent any misuse of such information;
> 5. permit exceptions from the requirements with respect to genetic records maintained anonymously for research purposes only; and
> 6. be subject to civil suit and criminal penalties for any damages which occur as a result of negligent, wilful, or intentional action which violates any individual's rights under this Act.

The Bill defines genetic information (at s. 101(2)) as 'any information that describes, analyses, or identifies all or any part of a genome identifiable to a specific individual'.

Privacy protection

Unlike the US law, English law knows no general concept of privacy. In the sense that it gives limited access and ability to be able to control the divulgence of facts or information about oneself, English law may be said at best to afford some tangential recognition of privacy interests, but it admits as yet of nothing so strong as a right of privacy. It is evident, however, that the use of genetic information by insurance companies for the purpose of deciding whether, or at what level, to grant insurance should be debated and resolved before widespread piecemeal applications of genetic testing take place (Keeley, 1981; Beauchamp and Bowie, 1983; Velasquez, 1983; Jecker, 1993; Wells, 1993, ch. 5) to forestall the creation of what Nancy Jecker has identified as an 'underclass of medically uninsurable people' (Jecker, 1993).

Widespread genetic discrimination in the private insurance market would furnish a compelling argument for instituting a public insurance programme or instituting across-the-board regulation of the private insurance market. Those who are dismayed by the reaction of insurance companies to the availability of genetic information might be relatively indifferent to which alternative avenue of public or private provision was preferred, but one ever-present fear is of the creation of at least a two-track insurance system. As Jecker has suggested, any society's attraction to retaining a private health insurance market depends upon private health insurers affirming and meeting responsibilities to the wider society. She argues that how a society and the insurance industry respond in the face of new genetic testing capabilities will be a moral guidepost indicating how we as a society should devise and implement healthcare reform (Jecker, 1993).

If an individual in the United Kingdom currently has had a genetic test, the results will be included in that person's medical records. Insurers can already demand access to those reports as a condition of offering cover. The fear that insurers voice is that as genetic tests become more readily available, cheaper, and thought to be of greater predictive accuracy, more people will have them. Individuals may then use this knowledge adversely to select against insurers, who in turn may respond by demanding, first, access to the results of those tests, as was the early experience with testing methods of seropositivity of HIV. Secondly, and in the more troublesome projections, insurers may then oblige proposers to undergo genetic testing as a condition precedent to insurance cover being underwritten, with premiums then set according to the specific risk that each individual presents. At this point the principle of risk-pooling will begin to evaporate and most people would be unable to obtain many forms of insurance or mortgage cover at standard rates, which are currently applied to some 95% of the insured population (Brett and Fischer, 1993).

Clarification of the position of British insurers has recently come in a letter from Mark Boleat, Director General of the Association of British Insurers, in response to an article in *The Independent* where A. Perutz had argued that confidentiality in respect of genetic matters was of paramount importance (Perutz, 1993). Boleat wrote that 'if an individual wishes to obtain life insurance the insurance company must know as much about the risk it is taking on as the individual. . . . This is fair both to the individual and to other policy holders in the insurance fund.' Although the principle of general fairness to other policy holders may be implicitly acknowledged by reference to general Aristotelian principles, the fairness in this arrangement to the individual concerned (who in the extreme case may discover that he or she is uninsurable in the prevailing market) is left unstated.

Boleat concluded his letter by arguing that genetic information is no different from other medical information that may be held in an individual's medical record, and observed that existing records probably contain references to family history and illnesses. Such information may already be used in making differential assessments of risk, and to allow withholding of genetic information 'distorts such equitable treatment and . . . is akin to offering motor insurance while letting drivers withhold information on their accident record.' This assertion contains a number of assumptions that would need to be explored elsewhere. Importantly, the analysis appears to ignore the possibility that a number of key issues confront the family, the geneticist and society in connection with the genetic counselling process which may ensue following the initiation of what we may call the 'genetic viewing' process. The point has been well put by Ian Pullen: 'What at first sight may seem to be a simple enough transaction, is beset with ethical and legal problems' (Pullen, 1990). These issues include: (i) What is genetic information? Does it apply to family history information and to blood tests? (ii) How *are* insurers to be protected against adverse selection? (iii) Should an underwriter be permitted to request genetic screening when it has not previously been performed? Pullen concludes that there are more *ethical* problems surrounding genetic tests than present-day medicals, from which he argues that 'it is likely' that regulations will be introduced to restrict an insurer's liability to request a genetic test (Kennedy, 1993; Morgan, 1994). Aside from this jurisprudential observation, however, Pullen's point is central; present medical tests can reveal a disorder that a person already has, whereas genetic tests can disclose disorders that a person may get in the future. What liabilities is the underwriter to assume in respect of counselling and follow-up services?

Although there is no extant British legislation, there have been calls for legislation to be introduced to regulate this market in medical futures. Urging continued vigilance in respect of those who would make use of genetic knowledge for personal, political or economic interests and leverage, David Suzuki and Peter Knudtson recall the 'endlessly shifting balances of power that are the inevitable consequence of scientific knowledge and its application' (Suzuki and Knudtson, 1988). Suggestions have included general welfare provision through the establishment of a common insurance fund to cover genetic high-risk individuals, a quota system forcing all insurers to underwrite a fixed number of high-risk cases, or a state-funded insurance pool to prevent the emergence of uninsured individuals. In Canada, the recent report of the Privacy Commissioner has argued for there to be *no* right to inspect an individual's genetic information without consent. It would allow, however, that while there should continue to be a general air of restraint in collecting personal medical information, where a case for collection can be made, 'genetic testing (but *only* with the consent of the subject) may be an appropriate means of acquiring the information.' This type of testing should be subject to strict conditions (Bankowski and Bryant, 1989). Thus, the Commissioner suggests that

   (i)   a person should have the option to be tested by any means that will provide reliable information, including genetic information;

   (ii) the type of information obtained should be strictly controlled, such that they caution strongly public and private sector institutions 'against acquiring more personal information through genetic tests than they would have acquired using other methods';

   (iii) only the information needed to tell whether the person meets the required standard should be collected.

As the Privacy Commissioner concludes, 'the very availability of intrusive technology seems to whet mankind's appetite for its use' (Privacy Commissioner of Canada, 1992). In the process, privacy becomes a casualty. Genetic technology has appeared alongside other biotechnological developments and threatens to surpass them all in its ability to intrude.

> Benevolence can be vulnerable to fear, prejudice, irrationality and the blind drive for efficiency. Taken individually, decisions by employers and insurers to employ biotechnology to their advantage may appear logical. On a societal level, however, they are not. Nor are they necessarily humane. *(Jecker, 1993)*

It is the demystification of health futures through genetic testing that is imposing tremendous costs in the insurability market. Because ignorance meant that luck (or superstition) was ever-present, even in families where there was already genetic risk, insurance could function satisfactorily. It is as though insurers have themselves become frightened of the power that genetic tarot readings offer and have begun to react against adverse selection in the way which they have. This may be described as a radical form of discounting: the chromosomal 'commodification' of fetal futures.

Medical confidentiality and the public interest

Of course, once having obtained information about an individual in respect of his or her genetic make-up, doctors and other health care workers have an established duty not to disclose this information learnt in the course of their practice.[2] This extends to any information about a patient that the doctor has learnt directly or indirectly in the course of the professional relationship or capacity (General Medical Council, 1989). The obligation arises out of the relationship, although it may be reinforced by the nature of the information.[3] The public interest in obtaining information in order to secure public health indicates that certain kinds of information, such as may be obtained by certain kinds of testing, give rise to an obligation of confidence on all concerned (see note (4), and Grubb and Pearl, 1990). This might extend the duty of confidence not only to the healthcare professionals primarily concerned, but also to laboratories and other paramedicals involved in the analytical process. Patients may have a concomitant duty to their doctors in certain circumstances.[5]

The further question arises of whether a healthcare professional may be entitled to or become *obliged to reveal* genetic knowledge to relatives of a primary patient, where that discloses information whose absence would be harmful to those other individuals. And they may have an obligation to do this *despite the objections of the primary patient*, about or from whom they have gathered the initial data. Clearly, once a duty to pass on medical information – or, as here, genetic information – becomes established, the health carer becomes implicated through having that knowledge. A number of circumstances may arise in which information may have to be disclosed to a third party.

The first is where in the course of legal proceedings a breach of confidence is required by a competent court (Grubb and Pearl, 1990) or where disclosure is otherwise authorised by statute – as under the Abortion Act and its regulations, or the Public Health (Infectious Diseases) Act 1984. Secondly, information may, in its nature, come to lose the necessary elements of confidentiality, such that further disclosure would no longer represent a harm to the person to whom it related.[6] Thirdly, there may be circumstances in which the disclosure is expressly or impliedly authorised by the person to whom it relates – in a way this is an example of the information having lost its necessary element of confidence. Finally, there may be circumstances in which, through the behaviour or some other act or omission of the relevant party, it is in the public interest that otherwise confidential information be disclosed.[6]

The case of *W* v. *Egdell* discloses 'in an unusually stark form' the question of 'the nature and quality of the duty of confidence owed to a patient'. Having forwarded a copy of his psychiatric report on W to the Home Secretary without W's permission, Edgell was the subject of an action by W for breach of confidence. The High Court and the Court of Appeal refused W's claim against Egdell. In the Court of Appeal it was said, again, that there were competing public interests, W's in seeking advice and assistance from an independent doctor and a countervailing interest in public safety. In view of the nature and number of killings for which W had been detained, it was vital for those responsible for W's treatment to be provided with full, relevant information concerning his condition.

The Court of Appeal accepted that to justify disclosure there must be

> a real risk of danger to the public,
> a risk of serious or substantial harm,
> and that disclosure must be made to and confined to the proper authorities.

The circumstances in which these conditions will be fulfilled in respect of genetic information are likely to be very rare and to constitute at most narrow exceptions to a general duty of confidentiality.[7]

A final question here is whether there may be circumstances in which a doctor may be held liable to a person for failure to disclose to them risks materialising from information that they have gained in their professional capacity as clinician involved in the care of another. This has arisen in the American case of *Tarasoff* v. *Regents of California*,[8] which confronts the question of whether a doctor, who in treating a psychiatric patient becomes aware of a serious risk to others, is under a duty to warn those others of that risk. The Supreme Court of California held that a psychologist owed a duty of care to a woman murdered by a psychologist's patient. The patient had expressed an intention to kill the woman. The Court accepted that there was a balance to be drawn between the public interest in effective treatment of mental illness and the consequent requirement of protecting confidentiality, and the public interest in safety from violent assault. Duties to warn of contagious diseases have also been recognised by American courts and state legislatures; this has been particularly the case involving fatal conditions such as HIV. The argument has been, and has been accepted in some states of the USA, that (at least) the sexual partners of A should be informed of A's seropositivity by the doctor who has care of A. It has been argued by many commentators in the UK that disclosure would be justified to known sexual partners and needle sharers if the patient himself or herself would refuse to make the disclosure (Grubb and Pearl, 1990).

If the limiting conditions of this advice are extracted, it is clear that it is only in the most limited circumstances – a serious and identifiable risk to a specific and existing individual and not a potential victim in the future – that a duty to disclose might conceivably be imposed on a doctor or other genetic counsellor. This limited scope is in line with developments in the *Tarasoff* doctrine itself, where the duty is anyway narrowly drawn. First, the therapist should not be encouraged to reveal such threats routinely. Secondly, the Californian Supreme Court has itself distinguished *Tarasoff* in a case where a patient made *general* threats against children. In *Thompson* v. *County of Alabama*[9] the court said that *Tarasoff* was limited in that that case involved a specific threat to an identifiable, specifically foreseeable, victim.

There is doubt whether the general principle in *Tarasoff* would be applied in the UK: in *Holgate* v. *Lancashire Mental Hospitals Board*[10] a hospital was held liable for negligently releasing on licence a dangerous patient who had been compulsorily detained. The patient entered the plaintiff's home and assaulted her. The trial judge seemed to assume that there would be a duty of care, which may be justified by the degree of control exercised by the hospital over the patient, analogous to that exercised by a prison over a prisoner. Genetic counsellors and other medical professionals do not exercise this degree of control over their patients and the bases of liability now applied by the common law courts – of proximity, foreseeability and justness in imposing liability on the defendant – do not militate in favour of such liability.[11]

9.3 Genetics and existence: the abortion section

Ruth Chadwick has reminded us that the practice of rejecting 'defective' babies (as she calls them) has a long human pedigree. The Spartans exposed to the elements those neonates deemed unsuitable and Plato recommended that those babies that did not 'fit' established guidelines should be hidden in a dark and secret place (Chadwick, 1987). Contemporary practice with handicapped neonates has sometimes led to prosecution of the medical staff involved, although a line of jurisprudence has emerged according to which it may be permissible to treat some babies for dying, rather than to intervene to keep them alive.[12] And the Abortion Act 1967, since its inception, has allowed for termination of pregnancy on the ground of fetal abnormality. As Chadwick writes, 'Although, of course, abortion is itself by no means uncontroversial, the fact of handicap is generally regarded as one of the best reasons for having one' (Chadwick, 1987).

That this *need not* imply a eugenic policy or preference is clearly articulated by Anne Maclean, who reminds us that there is an important distinction between abortion suggesting that handicapped men and women's lives are not worth living and suggesting that an individual person cannot cope with a handicapped child. 'It is not the case ... that the decision to abort a handicapped fetus implies or presupposes any view about what makes life valuable in the metaphysical sense' (Maclean, 1993). It is important to examine the approach of English law on this point, because the fetal abnormality ground, which is the subject of this part of the chapter, is frequently viewed in precisely this way.

Section 37 of the Human Fertilisation and Embryology Act 1990 amended the Abortion Act 1967 in a number of significant ways. It introduced into the legislation an explicit time limit (24 weeks) after which an abortion would not be lawful, unless special grounds were shown. Those grounds were reformulated and, in a move that surprised many observers, were enacted without time limit. There are now three specific grounds on which an abortion may lawfully be performed up until term; one of these is the fetal abnormality ground. The Abortion Act 1967 (as amended) in section 1 now provides that

> (1) ... a person shall not be guilty of an offence under the law relating to abortion when a pregnancy is terminated by a registered medical practitioner if two registered medical practitioners are of the opinion, formed in good faith – ...
> (d) that there is a substantial risk that if the child were born it would suffer from such physical or mental abnormalities as to be seriously handicapped.

This provision is not without its difficulties, as I shall discuss, but let me dismiss one potential apprehension first. Suppose that a doctor terminates a pregnancy believing, in all good faith, that if the fetus were born it would suffer from 'such physical or mental abnormalities as to be seriously handicapped'. It later transpires that she has made a mistake, and that the fetus was, in fact, in perfect health (there is, of course, a nice ambiguity in describing a fetus in this way). The doctor does not lose the protection of section 1 for an honest mistake, although she may still be open to an action in negligence at the suit of the woman or would-have-been parents.

The fetal handicap provision, now in the reformulated section, had been examined by the Select Committee of the House of Lords on the Infant Life (Preservation) Act 1929, chaired by Lord Brightman (House of Lords Papers, 1987–88). That Act, which had introduced the so-called 28-week time limit for abortions (Morgan and Lee, 1990), had been subject to much debate, and the Committee had taken the opportunity of its review to make comments on wider issues in relation to termination. Of the fetal handicap ground, they observed that if a fetus was diagnosed as

> grossly abnormal and unable to lead any meaningful life, there is in the opinion of the Committee no logic in requiring the mother to carry her unborn child to full term

because the diagnosis was too late to enable an operation for abortion to be carried out before the 28th completed week.   *(House of Lords Papers, 1987–88)*

Leaving aside the philosophical or ethical inquiries to which this might give rise, for present purposes we can abstract from this observation the concept of a fetus being so 'grossly abnormal and unable to lead any meaningful life' as being the closest that we come to an interpretation of the fetal handicap subsection. For, it remains the case that despite being the 'abortion ground' that is likely to command the most widespread sympathy, it remains the most difficult to interpret and, as I have elsewhere called it, 'the unexamined ground.' (Morgan, 1990*a,b*).

This explanation by the House of Lords Select Committee comes close to those guidelines that have been suggested in the common law courts when judicial sanction has been sought for medical decisions affecting the dissolution of neonatal life. In *Re J* (see note (12)), the Court of Appeal was concerned with a preterm baby born at 27 weeks gestation weighing 1.1 kg at birth. Placed immediately on a ventilator, given antibiotics and put on a drip, he was removed a month later from the ventilator. He suffered recurrent convulsions and apnoenic episodes. Four attempts to wean him from the ventilator successively failed and over the following 12–13 weeks, although his condition stabilised, any improvement was from a baseline which was in the words of Lord Donaldson, 'abysmally low.' He had suffered severe brain damage, and the most optimistic assessment of his prognosis was that he would develop serious spastic quadraplegia, be unable to sit up or hold his head upright, and that he was likely to be blind, to all intents mute, and unlikely that he would develop even limited intellectual abilities.

> Most unfortunately of all, there is a likelihood that he will be able to feel pain to the same extent as a normal baby. . . . It is possible that he may achieve the ability to smile and cry. Finally, as one might expect, his life expectancy has been considerably reduced at most into his late teens, but even Dr W would expect him to die even before then. *(Re J)*

This case differed from those that the court had considered earlier, in which guidance had been given about the appropriate approach to the medical treatment of children who are imminently dying and whose deaths can only be postponed for a short while.[13] In a case such as *Re J* the question was whether there could be anything to be balanced against the principle of the sanctity of life, and if so, what that might consist in. Lord Donaldson said that

> What doctors and the court have to decide is whether, in the best interests of the child patient, a particular decision as to medical treatment should be taken which *as a side effect* will render death more or less likely. . . . *Re B* [see note (14)] seems to me to come very near to being a binding authority for the proposition that there is a balancing exercise to be performed in assessing the course that is to be adopted in the best interests of the child. Even if it is not, I have no doubt that this should be and is the law.

The critical question that then arose, of course, was whether anything could displace or question the 'very strong presumption in favour of a course of action which will prolong life.' Lord Donaldson concluded that it was manifest and appropriate that account has to be taken of the pain and suffering and quality of life which the child will experience if life is prolonged. Account also has to be taken of the pain and suffering involved in the proposed treatment itself.[13]

If, in interpreting section 1(1)(d) of the Abortion Act we suggest that *account has to be taken of the pain and suffering and quality of life which the child will experience if life* **after birth** *is prolonged*, then we may be able to suggest a narrow but consistent reading of the abortion ground (which has still not been adequately explored (Mason, 1990)). And, importantly, it is one that avoids *stipulative* conditions about the sorts of people that we are prepared to admit that there may be (a primary concern of those who are worried about the potential for eugenic uses that prenatal diagnosis and other fetal screening programmes offer). It is pre-eminently one that, if given the restricted meaning for which I have sketched one interpretative mechanism, reflects compassion for the child that would be born and not, which is a typically eugenic framework, its failure 'to pass some test which would have secured it the prize of entitlement to life' (Maclean, 1993). In other words, an appeal to the interests of the child, in the way I have argued, is not an argument about the fetus's 'entitlement to life' but one that considers the *kind* of life that the fetus would have as a child *if it were to be born*. This argument sees section 1(1)(d) as primarily a fetal interests provision. Others have suggested, on the contrary, that the fetal ground really relates to the welfare of the parents, or to the public purse (Williams, 1987).

Suppose that this were the preferred reading of the section, it still does not follow that the primary importance of the fetal ground is eugenic. It is true that the care of handicapped people is costly, and it is in part a desire to cut costs that lies behind programmes that screen for certain handicaps. But whether for the individual woman or for the National Health Service itself, offering the screening and the termination services *does not necessarily imply* that the individual woman or the Health Service planners believe people with handicaps to be less 'entitled to life' than people without. While we may be morally censorious (of course, we may not be; that is a different argument), we may none the less accept that the individual decision, like the institutional decision, is based, and based alone, on a *preference* not to raise or care for a handicapped child. In other words, this would be a decision with which the individual may prefer not to have to *cope*, 'without thereby presupposing any view about what makes life valuable in the metaphysical sense, let alone that some people are less entitled than others to life' (Maclean, 1993).

An alternative reading of this story, of course, suggests that what we are witnessing (or rather are failing to see or are wilfully blind to) is a shifting rationality, a moving gradient of responsibility, which may have an explicitly eugenicist slide to it. It is an interesting question for the common lawyer whether this is a new development or merely a continuation of policy that the common law has long adopted (McVeigh, 1994). As technology expands and develops, the very nature of the definitional aspects of questions change. Lene Koch, in a related area (of 'infertility'), has shown how the nature of what constitutes a rational decision may also be metamorphosed (Koch, 1990). Thus the danger might be, with the audacious cocktail of reproductive technologies in societies in which the cult of beauty is tyrannical (Richards, 1991, page 132), that the more sophisticated the instruments of prenatal diagnosis, the more sophisticated the possibilities of diagnosis, the greater the possibility that a defect may be discovered, and the greater the range of possible anomalies that might be disclosed. This, it has been averred, may even have a *preventive* aspect, in which the moral obloquy of abortion is undercut by resort to genetic screening and perhaps replaced by assisted conception. Responsible parents of the future may ask themselves whether 'their own "hereditary material" meets contemporary requirements, or whether it would be preferable for them to resort to ovum or sperm donors, who would of course be carefully selected' (Beck-Gernsheim).

Elizabeth Beck-Gernsheim has argued that there are two powerful forces underpinning the trend towards '"quality control" of progeny'. The first is that in a socially mobile society, parents are exposed to heavy pressure and do all in their power to give the best opportunities to their children from the outset. In other words, a corresponding need for 'genetic engineering' already exists (Beck-Gernsheim, 1990). Secondly, the history of technology teaches that a new technique often itself contributes to the creation of a further need. The promise is followed by growing desire, or as Hans Jonas has put it, 'appetite is aroused by the prospect' (Jonas, 1985). These possibilities have led Beck-Gernsheim to suggest a new and increasing pressure for prospective, responsible parents. Taken together, techniques of assisted conception such as test-tube fertilisation with embryo transfer and techniques of prenatal diagnosis and increasingly sensitive genetic testing may lead to responsible parents of the future being unprepared to accept that their children may have a handicap; 'must they not rather do all in their power to make sure that no impairment exists?' (Beck-Gernsheim, 1990).

9.4 Children and consent

In this final section I shall turn to exploring whether existing children can lawfully be used in a predictive way to gain information about the genetic health of their present and putative siblings and other genetic relations. English law generally admits no principle of altruism, and in respect of young children, who may well be the subject of predictive testing for themselves or others, the courts have evolved and guarded jealously the legislatively adopted principle of the welfare of the individual child. In protecting and advancing the best interests of the child, the courts have worked with a conservative standard which, on occasions, healthcare and other professionals have found to be frustrating and occasionally obstructive.

Generally, the law is jealous of healthcare decisions for those unable, through age, to see through the consequences of what they might themselves otherwise decide, or, *in extremis*, be unable to decide. It has time and again been castigated as paternalistic and out of step with modern notions of young people's needs (let alone demands) for autonomy (Alderson, 1990). I do not wish generally to enter those arguments, but I shall briefly summarise the principles upon which the courts have proceeded, and then apply them in the context of genetics. But one caveat: this has become a motile area of legal development. Commentators have on more than one occasion in the recent past expressed surprise, if not dismay, at the direction of the currents let alone strength in the jurisprudential tide. Attempts to crest along this should, then, be treated with some temporal caution.

The general position in the law of England and Wales stems from the Family Law Reform Act 1969, section 8, and the landmark decision of the House of Lords in the celebrated case of *Gillick* v. *West Norfolk and Wisbech Area Health Authority*.[15] In relation to a person who has reached the age of 16, health professionals may presume that consent to treatment will be as valid as if the person was 18. This presumption is vulnerable if the person is not, as a matter of fact, competent to consent. Below the age of 16 (the 'Gillick competent' test) healthcare professionals must assess the capacity of each person to consent in relation to the proposed treatment or intervention. If the person under 16 is capable of understanding the consequences of their decision, then their consent will be sufficient to render the treatment lawful. If the person cannot understand the question, or if they *refuse* treatment, then in the first case parental consent will be usually required and in the second case either a parent or the court may give the appropriate approval, which again will have the effect of rendering the treatment or intervention lawful. In addition, the Children Act 1989, sections 43(7) and 44(7), sought to provide that a person under 16 may refuse to submit to a medical or psychiatric assessment if 'of sufficient understanding' to make an informed decision.[16] Nothing in the Family Law Reform Act 1969 (which introduced a number of important measures in respect of capacity and age, including lowering the so-called age of majority from 21 to 18), affected common law powers of parental consent (Family Law Reform Act 1969, section 8(3)). The effect of this is that a parent may provide consent to medical treatment for their child until the latter is 18. It was at first thought that this meant that they could consent where the person was *incapable* of doing so, a line of reasoning that appeared to be confirmed in *Gillick*. But in a series of controversial decisions, the Court of Appeal, in panels headed by the then Master of the Rolls, Lord Donaldson, the courts decided that the effect of section 8(3) of the 1969 Act was that parents could consent to treatment of their children under the age of 18 *even when the children themselves were capable of doing so*, and in some cases even where the 'child' was clearly refusing treatment. So, where consent is needed but cannot (or occasionally will not) be given by the child, the consent of one of the parents must be sought. But, as Jonathon Montgomery has expressed it, 'this apparently simple requirement gives rise to three areas of difficulty' (Montgomery, 1993). These concern: Who is a parent? What happens when parents disagree? and What limits are there to parental consent?

I shall not review each of these (what turn out to be) complex questions here. For the present purpose, it is sufficient to draw the following conclusions. In the question of consent to genetic testing, the first question that the healthcare professional must ask is whether the child's parents are married. If they are, the consent of either parent will suffice. If they are not, then prima facie the greater authority will lie with the child's mother, unless the father has one of a number of requisite court orders. In the event of a disagreement between the parents, if the woman consents to any genetic testing, then this will effectively protect the healthcare worker for any properly performed intervention. Where she refuses her consent but apparent consent is given by the child's father, then caution will be appropriate where the mother and father are unmarried. This is because under the Children Act 1989 where parents are married they will both have parental responsibility for their child or children, but where they are not, the mother will have automatic parental responsibility but the father will not, unless acquired by court order.

In the case of operations and the giving of consent in respect of very young children, Lord Donaldson again has set out the philosophy of the common law most clearly:

> it is sensible to try to define the relationship between the court, the doctors, the child and its parents. The doctors owe the child a duty to care for it in accordance with good medical practice recognised as appropriate by a competent body of professional opinion.... This duty is, however, subject to the qualification that, if time permits, they must obtain the consent of the parents before undertaking serious invasive treatment. The parents owe the child a duty to give or withhold consent in the best interests of the child and without regard to their own interests. *(Re J)*

Where the 'intervention' (and here we may explicitly include the question of genetic testing) is expected to benefit the child, the parents may consent to most treatments. For the present, it is only sterilisation of a 'minor' that is thought to give rise to such difficult considerations that such an operation should be referred to the Court. And, following from Donaldson's opinion just quoted, Montgomery has proposed that

> parental consent may be valid because even where care is not expected to benefit the child, the parents have the power to consent if it is reasonable and in the child's best interests to do so. *(Montgomery, 1993)*

The most obvious area in which this is likely to arise is testing for genetic orders that may not affect the child, or at least not until much later in adult life, or may be likely only to affect the putative 'family' of siblings. And, in the light of our previous caution about altruism, Montgomery's advice on the question of genetic testing and best interests is salutary. Three considerations are of particular importance:

(1) in all cases there must be a balancing of the risks of the intervention and the expected benefits *to the subject of the intervention* in assessing the reasonableness of the parental consent;

(2) the urgency of the intervention, or the benefits that it is sought to realise should be considered; thus an operation to remove a healthy organ for transplantation into a dying sibling might be more favourably regarded – even in the absence of a duty of altruism – than a genetic test where the 'beneficiary' would be the not-to-be-born sibling and the subject of the intervention to whom additional parental attention benefits might be expected to flow;

(3) even where the child is (legally) incapable of taking the decision, her or his wishes should be accounted, and if the intervention can be postponed until the child can make it, then it would reasonable for that to be done.

From this brief review, it can be concluded only that difficult questions can arise for individual children, their mothers and fathers and the healthcare professionals working with the family group. The best that the law currently offers is a form of flexibility in the advice that it offers. In any case of doubt, the geneticist, whether clinician or counsellor, is best advised to act cautiously and to make use of formal court procedures in the cases of greatest doubt or difficulty. This analysis does, however, underline the importance to be attached to sensitive handling of genetic information; not everyone wants to be co-opted into the fast lane of the genetic superhighway, and the wishes of those whose demand for ignorance represents the bedrock of their caution should be respected.

9.5 Conclusion

Ulrich Beck has suggested the dangers of uncontrolled or unregulated uses and developments of science and technology. He has suggested that while the latest research results constantly open up possible new applications, because this happens at such a rapid, exponential rate, the process of implementation is practically uncontrolled. A variant on this which I would add is that where it is controlled, countervailing arguments are more easily marshalled on the basis of benign experience or supposed individuation of consequences. Accordingly, although medicine supposedly serves health, it has in fact 'created entirely new situations, has changed the relationship of humankind to itself, to disease, illness and death, indeed, it has changed the world' (Beck, 1992).

Beck is concerned to describe the process of 'reflexive modernisation' within medicine. As this 'institutionalised . . . noiseless social and cultural revolution' is amplified, I shall suggest that law becomes necessarily the voice of the electorate where medicine as a professional power has secured and expanded for itself a fundamental advantage against political and public attempts at consultation and intervention.

> In its fields of practice, clinical diagnosis and therapy, it not only controls the innovative power of science, but it is at the same time its own parliament and its own government in matters of 'medical progress'. When it has to decide on 'malpractice', even the 'third force' of jurisprudence has to take recourse to medically produced and controlled norms and circumstances, which according to the social construction of rationality can ultimately be decided only by medical people and by no one else. *(Beck, 1992)*

There are few better illustrations of this than in the effects of science and technological developments on the beginnings of life.

These examples relate closely to the principles of 'reflexive modernisation' developed and deployed by Ulrich Beck in the sub-politics of medicine – what he calls but 'an extreme case study' of more general questions in and facing scientific and technical societies. The divergence of diagnosis and therapy in the current development of medicine results in a dramatic increase of so-called chronic illness, i.e. '. . . illnesses that can be diagnosed thanks to the more acute medical and technical sensory system, without the presence or even prospect of any effective measures to treat them.' His thesis here is that progress necessarily implies unplanned excess, the harmful effects of which are unintentional. The importance of this observation is this: *if the costs of progress are unintentional, they may well be equally unforeseeable*, while none the less statistically quantifiable as risks. The further impact of this is that the costs become difficult to guard against or anticipate or even calculate, which raises for the lawyer the interesting question of the most appropriate or sensitive response in terms of the regulatory instruments available. As the process of what Beck calls the 'noiseless social and cultural revolution' within medicine ('the institutionalised revolution') is amplified, the role of law, it seems to me, is metamorphosed: whether it keeps pace with the demands of those changes is another question, an empirical or evaluative one to be set alongside a series of conceptual ones.

Thus, an analysis of medical progress as itself institutionalised discloses that there has been 'a revolution of the lay public's social living conditions without its consent' (Beck, 1992). Indeed, in an arresting phrase, Beck describes this process as a 'secret farewell to an epoch of human history' in which the principles of technological feasibility and arrangement encroach on the subjects in such a way that the very foundations of a model of 'progress' that implicates a subject who is supposed ultimately to benefit from the process, are cancelled.

Recalling the basic principles of democratically based societies in which central issues of public policy affecting the future of society are the subject of public debate to shape the political resolve, he fears that the developments of modern technology have set in motion processes that undermine the 'idea of democracy from inside.' Technology, medicine, and reproductive technologies are becoming the instruments of uncontrolled 'sub-politics', where, in the sub-politics of medicine, there is no parliament and no executive in which the consequences of a decision might be examined before it is taken.

> There is even no social forum for decisions. . . . the highly bureaucratised developed Western democracies check every act for its conformity with legal requirements, terms of reference and democratic legitimation; at the same time it is possible to escape all the bureaucratic and democratic controls and to take closed decisions despite the hail of general criticism and scepticism in a world which escapes parliamentary control and in which the very bases of existing life and previous patterns of social control can be completely neutralised. *(Beck, 1992)*

Although there are the beginnings of regulatory regimes now in place in many countries of the European Union, one of the fears is well expressed by Elisabeth Beck-Gernsheim: parliamentary and other official committees find themselves faced with immense and completely innovative issues, which never respect the patterns of scientific disciplines. The struggle to find answers is correspondingly complex. And then a great deal of time is necessary to negotiate, amend and reformulate draft bills, which are passed backwards and forwards in the power struggle between departments. Thus years are wasted. But the genetic engineers do not wait until the material and statutory provisions have been clarified. In their laboratories they have been fertilising and generating life *in vivo* and *in vitro* by homologous and heterologous means (Beck-Gernsheim, 1990).

We must now be engaged with assessing the most appropriate deployment of the law in these massively changed circumstances.

9.6 Notes

1   *Allen* v. *Bloomsbury Health Authority* [1993] 1 All ER 651.
2   *W* v. *Egdell* [1990] All ER 835 (CA).
3   *Langside* v. *Kerr* [1991] 1 All ER 418 (CA).
4   *X* v. *Y* [1988] 2 All ER 648 at 653.
5   *Latham* v. *Stevens* [1913] Macg Cop Cas 83 (1911–1916).
6   *AG* v. *Guardian Newspapers Ltd* [1988] 3 All ER 545 (HL) No. 2.
7   *R* v. *Crozier* [1991] Crim. LR 138.
8   *Tarasoff* v. *Regents of California* [1976] 551 P 2d 334.
9   *Thompson* v. *County of Alabama* [1980] 614 P 2d 728.
10   *Holgate* v. *Lancashire Mental Hospitals Board* [1937] 4 All ER 19.
11   *Caparo Industries* v. *Dickman* [1990] 1 All ER 568.
12   For one of the most publicised (and discussed) prosecutions see *R* v. *Arthur* [1981] (Crown Court) 12 BMLR 1, and for the leading case on 'treatment for dying' see *Re J (a minor) (wardship: medical treatment)* [1990] 3 All ER 930 (it would be lawful to withhold life-saving treatment from a very young child in circumstances where the child's life, if saved, would be one irredeemably racked by pain and agony). For a discussion of some of the earlier legal cases and clinical assessments see Wells (1989), and for an examination of criticism of the most widely announced philosophical arguments brought into play in this area, consider Maclean (1993), especially chapters 2 and 3.
13   *Re C (a minor) (Wardship: medical treatment)* [1989] 2 All ER 930.
14   *Re B* [1981] 1 WLR 1421.
15   *Gillick* v. *West Norfolk and Wisbech Area Health Authority* [1985] 3 All ER 402.
16   This summary may be deduced from a patchwork of (more or less controversial) cases following from Gillick; the more important are *Re R* [1991] 4 All ER 177, *Re E* [1991] 1 FLR 386 and *Re W* [1992] 4 All ER 627.

9.7 References

Alderson, P. (1990) *Choosing for Children: Parents Consent to Surgery*, Oxford: Oxford University Press.

Annas, G. (1993) *Standard of Care*, pp. 149–50. New York: Oxford University Press.

Bankowski, Z. and Bryant, J. (1989) *Health Policy, Ethics and Human Values: European and North American Perspectives*. Geneva: CIOMS.

Beauchamp, T. and Bowie, N. (1983) Corporate social responsibility. In T. Beauchamp and N. Bowie (eds.), *Ethical Theory and Business*, second edition, pp. 52–7. Englewood Cliffs, NJ: Prentice Hall.

Beck, U. (1992) *Risk Society: Towards a New Modernity* (transl. M. Ritter). London: Sage.

Beck-Gernsheim, E. (1990) Changing duties of parents: from education to bio-engineering? *International Social Science Journal*, **42**, 451–63.

Brett, P. and Fischer, E. P. (1993) Effects on life assurance of genetic testing. *The Actuary*, July, 11–12.

Chadwick, R. (ed.) (1987) *Ethics, Reproduction and Genetic Control*, pp. 93–135. London: Croom Helm.

Chadwick, R. and Ngwena, C. (1992) The development of a normative standard in counselling for genetic disease: ethics and law. *Journal of Social Welfare and Family Law*, 276–95.

Clothier, C. (1992) *Report of the [Clothier] Committee on the Ethics of Gene Therapy*, Cm 1788.

Commission of the European Community (1989) *Adopting a Specific Research and Technological Development Programme in the Field of Health: Predictive Medicine*. Brussels: CEC.

European Commission (1992) Working Group on the Ethical Social and Legal Aspects of Human Genome Analysis (WG-ELSA), Final Report. Brussels: EC.

Foucault, M. (1970) *The Order of the Things*. London: Tavistock.

General Medical Council (1989) *Professional Conduct and Discipline: Fitness to Practice*, para. 80. London: GMC.

Goffman, E. (1964) *Stigma: Notes on the Management of Spoiled Identity*. Harmondsworth: Penguin.

Grubb, A. and Pearl, D. (1990) *Blood Testing, Aids and DNA Profiling: Law and Policy*. Bristol: Jordans.

House of Lords Papers (1987–88) HL 50.

Jecker, N. (1993) Genetic testing and social responsibility of private health insurance companies. *Journal of Law, Medicine and Ethics*, **21**, 109–16.

Jennings, B. (1991) Possibilities of consensus: towards democratic moral discourse. *Journal of Medicine and Philosophy*, **16**, 447–63.

Jonas, H. (1985) *Philosophical Essays: From Ancient Creed to Technological Man*. Englewood Cliffs, NJ: Prentice-Hall.

Keeley, M. (1981) Corporations as non-persons. *Journal of Value Enquiry*, **15**, 119–55.

Kennedy, I. (1993) Law and Ethics. In *Proceedings of the Second Symposium of the Council of Europe on Bioethics: Ethics and Human Genetics*, Strasbourg, 30 November–2 December. Oxford: Oxford University Press.

Kennedy, I. and Grubb, A. (1993) *Medical Law: Text with Materials*, second edn. London: Butterworths.

Koch, L. (1990) IVF – a rational choice? *Reproductive and Genetic Engineering: International Journal of Feminist Analysis*, **3**, 235–42.

Lee, R. (1989) To be or not to be: is that the question? The claim of wrongful life. In R. Lee and D. Morgan (eds.), *Birthrights: Law and Ethics at the Beginnings of Life*, pp. 172–94. London: Routledge.

Lee, R. and Morgan, D. (eds.) (1989) *Birthrights: Law and Ethics at the Beginnings of Life*. London: Routledge.

Maclean, A. (1993) *The Elimination of Morality: Reflections on Utilitarianism and Bioethics*. London: Routledge.

Maddox, J. (1993) New genetics means no new ethics. *Nature*, **364**, 97.

Mason, K. (1990) *Medico-Legal Aspects of Pregnancy and Parenthood*, pp. 106–7. Aldershot: Dartmouth.

McLean, S. and Giesen, D. (1994) Legal and ethical considerations of the Human Genome Project. *Medical Law International*, **1**, 159–76.

McVeigh, S. (1994) *Socio Legal Studies Association Annual Conference*, Nottingham, April 1994.

Montgomery, J. (1993) Consent to health care for children. *Journal of Child Law*, **5**, 117–24.

Morgan, D. (1990*a*) Abortion: the unexamined ground. *The Criminal Law Review*, 687–94.

Morgan, D. (1990*b*) Legal and ethical dilemmas of fetal sex identification and gender selection. In A. A. Templeton and D. Cusine (eds.), *Reproductive Medicine and the Law*, pp. 53–7. Edinburgh: Churchill Livingstone.

Morgan, D. (1994) Problems and possibilities: the case of regulating United Kingdom health care. Paper prepared for the AAAS NATO Advanced Workshop, Developing an Infrastructure for Science and Technology in Eastern Europe: the Role of Scientific and Technical Societies, Visegrad, Hungary, 27–31 October 1994.

Morgan, D. and Lee, R. (1990) *Blackstone's Guide to the Human Fertilisation and Embryology Act 1990: Abortion and Embryo Research, The New Law*, pp. 43–7. London: Blackstone.

Nelkin, D. A. (1993) *Against the tide of technology. The Higher*, 13.

Nielsen, L. and Nespor, S. (1993) *Genetic Test, Screening and Use of Genetic Data by Public Authorities in Criminal Justice, Social Security and Alien and Foreigner Acts*. Copenhagen: The Danish Centre of Human Rights.

Perutz, A. (1993) The right to know your own genes. *The Independent*, 21 August 1993, p. 828.

Pompidou, A. (1993) A3-0000/93; Doc. EN/PR/218992, 1 March.

Privacy Commissioner of Canada (1992) *Genetic Testing and Privacy*, p. 30. Ottawa: Privacy Commission of Canada.

Pullen, I. (1990) Family genetics. In E. Sutherland and A. McCall Smith (eds.), *Family Rights; Family Law and Medical Advance*. Edinburgh: Edinburgh University Press.

Richards, S. (1991) *Epics of Everyday Life: Encounters in a Changing Russia*. Harmondsworth: Penguin.

Rix, B. A. (1991) Should ethical concerns regulate science? The European Experience with the Human Genome Project. *Bioethics*, **5**, 250.

Strathern, M. (1992) The meaning of assisted kinship. In M. Stacey (ed.), *Changing Human Reproduction*, pp. 148–69. London: Sage Publications.

Strathern, M. *et al.* (1993) *Reproducing the Future: Anthropology, Kinship and the New Reproductive Technologies*. Manchester: Manchester University Press.

Suzuki, D. and Knudtson, P. (1988) *Genetics: The Ethics of Engineering Life*. London: Unwin Hyman.

Thomas, K. (1973) *Religion and the Decline of Magic: Studies in Popular Beliefs in Sixteenth and Seventeenth-Century England*. Harmondsworth: Penguin.

Thomas, K. (1983) *Man and the Natural World*. Harmondsworth: Penguin.

Velasquez, M. (1983) Why corporations are not morally responsible for anything they do. In T. Beauchamp and N. Bowie (eds.), *Ethical Theory and Business*, second edition, pp. 69–76. Englewood Cliffs, NJ: Prentice Hall.

Watson, J. D. (1971) Potential consequences of experimenting with human eggs. Paper presented at the 12th meeting of the panel on Science and Technology, Committee on Science and Astronautics, US House of Representatives, Washington. Cited in J. Gunning and V. English (1993) *Human In Vitro Fertilization: A Case Study in Medical Innovation*. Aldershot: Dartmouth.

Wells, C. (1989) Otherwise kill me: marginal children and ethics at the edges of existence. In R. Lee and D. Morgan (eds.) (1989) *Birthrights: Law and Ethics at the Beginnings of Life*, pp. 195–217. London: Routledge.

Wells, C. (1993) *Corporations and Criminal Responsibility*. Oxford: Oxford University Press.

Williams, G. (1987) *Textbook of Criminal Law*, second edition. London: Stevens.
Wynne, B. (1982) Technology, risk and participation: on the social treatment of uncertainty. In J. Conrad (ed.), *Society, Technology and Risk Assessment*. London: Academic Press.

10

Human pedigree and the 'best stock': from eugenics to genetics?

DEBORAH THOM and MARY JENNINGS

> Natural selection rests upon excessive production and wholesale destruction; Eugenics
> on bringing no more individuals into the world than can be properly cared for, and
> those only of the best stock.
> *Francis Galton,* Memories of My Life *(1908)*

10.1 Introduction

Can genetics be confused with eugenics? This is one of the questions raised by the
Nuffield Council on Bioethics in their report on the ethical issues of genetic screening,
which suggests that the 'potential for eugenic misuse of genetic testing will clearly
increase' (Nuffield Council on Bioethics, 1993). In the report, eugenics is defined as the
selective breeding of some members of the population and the elimination of undesirable individuals, which found its worst expression in the racial hygiene policies of Nazi
Germany. Britain, however, has a long history of a eugenic movement involving many
leading members of society, including members of the medical establishment, a movement that continued through the 1950s and 1960s. This chapter investigates the British
eugenic movement as one way to help address the difficult questions emerging today
from the 'new genetics'.

We examine the origins of eugenics in the early twentieth century in the work of
Francis Galton, who was a cousin of Charles Darwin. The middle-class concern with
social problems of the 1920s and 1930s, unemployment, pauperism, alcoholism and the
decline of the nation, found expression in eugenic policies on population control.
Scientists, too, participated actively in debates on the hereditary aspects of social problems. These debates continued after World War II but in modified ways. We shall investigate traces of eugenic thought in the construction of the British welfare state and in the
development of medical genetics. Our chapter will provide a framework for the emergence of the new genetics and contributes a historical background to current debates.

Katherine Hepburn's first starring role in cinema in the film *A Bill of Divorcement*,
which came out in 1932, demonstrates the potency of the idea of a hereditary taint. The
film concerned an impetuous and hot-tempered young woman whose father has been a
stranger to her but reappears, bringing slow dawning understanding that he is mad,
and creates so strong a fear of the madness she believes to be in her that she breaks her

own engagement and decides to devote herself to her father's care. How did such a scenario, with all its unstated common understandings of the hereditary nature of madness, become believable? It is here, in commonsense common knowledge, that we need to look at the impact of hereditarian thought in history as well as in the present day. Lay or popular understandings of inheritance have historical consequences as significant as those of the professional groups, particularly doctors, who systematised a theory of heredity and generated demands for policy and practice to be reformed to follow theory.

Galton systematised his account of the hereditary transmission of human qualities, both physical and mental, into both a science and a process of reform, using one word 'eugenics', defined in 1907:

> The term National Eugenics is here defined as the study of the agencies under social control that may improve or impair the racial qualities of future generations either physically or mentally.   *(cited by Blacker, 1952)*

What was eugenics then depended very much on the question of whether it was positive eugenics as emphasised by Galton in his 1901 essay, 'On improving the human breed . . .',

> The possibility of improving the race of a nation depends on the power of increasing the best stock. This is far more important than that of repressing the productivity of the worst.   *(Galton, 1909)*

or negative eugenics that Galton was to emphasise only seven years later:

> namely the hindrance of marriages and the production of offspring by the exceptionally unfit. The latter is unquestionably the more pressing subject of the two.   *(Galton, 1908)*

Galton created the Eugenic Education Society in 1907 to provide propaganda for eugenic thought, and a research laboratory at University College, London, in 1909 to further the study of eugenics. At the key moment when this occurred it is possible to investigate what the context for this new science was and how far it represented a substantial innovation in thought. By this means we can evaluate whether eugenics did actually influence social policy, popular beliefs and social and medical practices.

10.2 Galton and the invention of eugenics

The invention of eugenics was very much part of debates over what historians call the 'Condition of England' question. Degeneration, meaning physical degeneration, mental deficiency and moral degeneracy, had been described as part of the analysis of a declining Empire in which Imperial black subjects had, in the persons of the Mahdi's forces in the Sudan and the Zulu in Southern Africa, defeated British soldiers. In England the working class was reproducing much faster than the middle class, whose birth rate had been declining since the first extensive modern census of 1851. Daniel Pick and Elaine Showalter have identified the literary working of these fears, which were so much inflected by public anxiety about the *fin de siècle* (Pick, 1989; Showalter, 1991). This was not the mere accident of dates. It represented a sense of political and moral crisis that unemployed rioters in Trafalgar Square in 1888, socialist agitators challenging policemen for the right to speak on street corners, and growing enthusiasm for universal enfranchisement did nothing to allay. There was also increased concern with sexuality and reproduction, which resulted in legislation on the age of consent to sex for girls, the criminalisation of homosexuality and a continuing concern for the extent of infection with both syphilis and gonorrhoea, usually expressed by discussion on prostitution (Walkowitz, 1980; Weeks, 1981; Mort, 1987; Walkowitz, 1992). Galton also reflected older debates that followed from his cousin Charles Darwin's theories of natural selection and fears of the direct social effects of a morally undesirable group becoming the fittest whose survival seemed assured.

Galton's writings were influential, but as a rhetorical trope rather than the direct inspiration of policy in his lifetime. He set up and ran an anthropometric laboratory, which is described in full by his biographer and disciple, Karl Pearson. Galton envisaged this as the basis of marriage guidance in his Utopia Kantsaywhere (Pearson, 1914–30, vol. 2, p. 346; vol. 3, pp. 416–418). The laboratory measured physiological characteristics, and mental powers were registered by using reaction times. Some 9000 people passed through the laboratory, each paying a threepence fee. They continued to come when it was located in the Museum in South Kensington (now the Science and Natural History Museums) so that there were 13000 records. Galton wanted all school-children to be measured and followed through into adult life. Pearson claimed this was novel in two ways: first, what Blacker calls 'the impulse to count and measure, to classify, grade and assess', but secondly the significance of the numbers counted in relation to society. This Pearson also lays at Galton's door but he could, less modestly, have claimed this for himself:

> the novel statistical methods, which . . . lead him to the correlational calculus, the *fons et origo* of that far-reaching ramification – the modern mathematical theory of statistics. *(Pearson, 1914–30, vol. 2, p. 357)*

Donald MacKenzie has written about the origins of population statistics in the investigations and techniques of both Galton and Pearson, but the eugenic impulse should also be credited with some of the purchase that such statistics begin to have on more widespread non-technical understanding (MacKenzie, 1981). This process takes longer. Charles Booth, who wrote a survey of London as a scientific contribution to the question of whether poverty was a danger and a growing phenomenon, concluded that the myth of hordes of barbarians waiting to sweep out of the slums and destroy civilisation was untrue. They were, he wrote, 'a disgrace but not a danger', and concluded that the two lowest classes of society, A and B, were a minute fraction of the poor (Booth, 1902). These, however, were to be seen as a risk to society through their **reproduction**, which should be discouraged. Galton quoted him as writing about them thus,

> Their life is the life of savages, with vicissitudes of extreme hardship and occasional excess. . . . They degrade whatever they touch.

Galton's solution to the problem Booth had described was to eliminate this class by limiting their reproduction:

> It would be an economy and a great benefit to the country if all habitual criminals were resolutely segregated under merciful surveillance and peremptorily denied opportunity of producing offspring.   *(Galton, 1909)*

Measures such as the registration of midwives would, it was hoped, improve the general standard of nurture of children and encourage more to be born. Galton recommended several moves to encourage a eugenic frame of mind and social acceptance of eugenics, which he summed up most succinctly in 'The possible improvement of the human breed' as 'dowries . . . assured help in emergencies during the early years of married life, healthy homes, the pressure of public opinion, honours and above all the introduction of motives of a religious or quasi-religious character' (Galton, 1909).

From the accounts of meetings of the Sociological Society, which discussed Galton's views on marriage and reproduction, it is clear that he believed that people could be encouraged to behave in a socially responsible way in their attitudes to marriage, and that this could be made into a part of the everyday public opinion; that was most contentious in his programme, whereas the fundamental basis of eugenics in statistical observation of existing populations seems to have been easily acceptable. Galton wrote, 'social opinion operates powerfully without us being conscious of its weight' (Galton, 1909, p. 24). How far did he and his heirs succeed in remoulding public opinion?

The Boer Wars and the rejection of one third of recruits as medically unfit gave a direct popular reference to the concept of fitness. Indeed, the language of military processing was to enter the language in the categories of World War I when the whole male population was graded A1 to C3 to allocate them to appropriate activities. The enquiry of the Inter-Departmental Committee into Physical Deterioration, which reported in 1907, also publicised the ideas of fitness and the remedies for unfitness that were to provide the twin poles around which eugenic thought was to spin. They concluded that much of the unfitness of the child population could be attributed to environmental factors that could be remedied, a conclusion that was also reached by Sir John Gorst, in his survey *The Children of the Nation* published in 1906. This book was dedicated to the new Labour MPs who would, he believed, help to bring about some of the necessary changes. It provides an interesting check to those who believe that eugenist ideas swept all before them, for its analysis is almost entirely unaffected by hereditarian thought. The chapter headed 'Hereditary diseases' is about syphilis, which can of course be inherited, but is not genetic, and alcoholism, of which there seems to be some evidence of a predisposing gene. (Although some of Gorst's contemporaries and indeed the next generation also believed that alcoholism could be inherited, Gorst attributed it almost entirely to poor conditions and absent education.)

In 1909 William Cecil Dampier Whetham and Catherine Durning Whetham published *The Family and the Nation*, and Galton, *Essays on Eugenics*. They warned of a tendency, accelerated by environmental improvement, to deterioration in the quality of the race:

> If by increased medical and hygienic knowledge, the feeble-minded and weak-bodied stocks be allowed to survive, and if, as seems to be the case at present, they reproduce themselves faster than do the better stocks, the relative numbers of such persons in the country must increase, and the average quality of the race deteriorate. If, by economic and social conditions, children be made too heavy a burden on the more desirable elements of the population, there is danger that the thrifty and far-seeing members of the community will postpone marriage, and when married, restrict the number of their offspring. Thus, while the weak and careless elements grow at an increasing rate, the good stocks of the people check their rate of growth or even diminish in number, and the selective deterioration of the race is hastened in two ways. *(Whetham and Whetham, 1909, p. 3)*

The book recommends a historical study of heredity that has been influential in succeeding periods – that is, the assumption that pedigrees will generate understanding. This tendency to reify social characteristics is evident in Whetham in ways that are quite new. Galton looked to general qualities, as his recognition that heiresses tended to be unhealthy or infertile, because they were the result of parental failure to produce more than one live female child, provides an example of this sort of thinking, but he was talking about statistical probability, whereas the Whethams talked about 'unsoundness' as a real, human quality. In the chapter 'The rise of families' they provided an example in Galton's own family. This information came from Britain's intellectual elite, and with the peculiar parochialism of English public life, a disproportionate amount of the Whethams' book does refer to Cambridge data such as the marital fate of women graduates, performance in the Tripos examinations and the recent arrival of married dons, of which Whetham was himself an example.

This period was one of social turmoil and critical intellectual enquiry in which the role of women came under particular scrutiny, partly because of the rise of feminism and partly because of official complaints about the quality of mothering. These were expressed by Sir George Newman, Chief Medical Officer at both the Local Government Board and at the Board of Education (Davin, 1978; Dwork, 1987), and used a general concept of fitness, including the biological. Two discussions among professionals particularly affected the perception of eugenic theory in the years 1900–1914, but curiously did not prompt medical research, taking for granted as they did the assumption that many qualities were transmitted. The first was the question of mental defect. The Royal Commission on mental deficiency drew on a rich strain of hereditarian psychiatric thought, particularly strong in the work of Henry Maudsley, who had helped to professionalise psychiatry and in the process removed from its central concerns those groups who were effectively deemed incurable – especially the 'idiots, aments and morons' of the discourse of mental deficiency of the day. Tredgold and Potts, who between them had cornered the market in mental deficiency, both believed that a large proportion of the mentally defective were incurable and had inherited their defect. The Royal Commission and the Act of 1913 that followed it supported this view and created a new institution that was to be purpose-built for this group and would replace the mixture of placements in lunatic asylums, idiots' asylums, the Poor

Law Infirmaries and the prisons. The theory was that children would be identified in the compulsory medical inspections that all were to undergo on entering compulsory education at five and be segregated in special schools, if not 'severe', and in special institutions, if 'severe'. The theory of the innate nature of mental defect and its inherited aetiology was thus enshrined in a law that was designed to provide care but also insisted on custody to ensure that they did not reproduce (Simmons, 1978; Barker, 1983).

The second and less controversial area of public debate led to the passing of the Incest Act in 1908. This, however, had less to say than one might imagine, in view of folk beliefs about intermarriage, about the dysgenic results of incestuous sex. Probably more reproduction was prevented in the name of the Poor Law than was in the name of eugenics. These debates were given greater salience by the sense of crisis of Empire, coming war and upheaval in society. Local government was much preoccupied with the state of paupers, idiots and inebriates, and many of its activists looked nationally to eugenic connections to get rid of some of their local problems. Women active in the strategy towards mental deficiency, like Mary Dendy in Manchester and Mrs Hume Pinsent in Birmingham, regarded eugenic theory as the basic principle of much of their work, but, in practice, prewar eugenic thought confined them to a belief in hereditary determinants of socially undesirable conditions but no particular mechanism for detecting precisely how transmission worked, nor how it could be prevented (Hollis, 1987). The data collected on pedigrees were almost the only information that was solicited at the Galton laboratory, and although local school medical officers often attempted to create their own datasets they were again generally along the lines popularised by the infamous comparison between the Kallikaks and others. Hence the Victorian theme of institutionalisation was merely given an added twist because incarceration also made 'undesirables' incapable of reproduction.

Prewar popular support for eugenic thought came down to a belief in the need for more data in the form of both anthropometric data and social surveys as well as the belief in the family history as an essential component in any case study. About the only outcome of much of the substantial bequest that Galton made in his will to the establishing of the Galton Laboratory was a collection of data, much of which Edgar Schuster, the first Galton research fellow, had garnered from *Who's Who* and the *Dictionary of National Biography*. Ethel Elderton, who succeeded Schuster, spent most of her time computing on this basis. Karl Pearson, for whose benefit Galton had endowed the chair in biometrics, and indeed supported the magazine *Biometrika*, continued to develop the correlation coefficient as the basis of most population investigation. Thus academic institutions were firmly in place when war broke out, although academics were divided because most British eugenists believed that Mendelian explanations could explain hereditary variation whereas Pearson, who remained at odds with the scientific and social elite for all of his working life, supported his own theory of evolution, which was based more on a notion of variation (Kevles, 1985). Members of the Eugenics Education Society, who were mostly, as historians have argued at great length, members of middle class professional groups who found some notion of innate talent more comforting than assumptions that the aristocracy deserved to retain a leading social position (Sutherland, 1984), were divided too on the question of birth control, or Malthusianism as it was known. Some thought that its effects had been dysgenic because the prudent, thrifty and intelligent were using birth control whereas the improvident, feckless and stupid were reproducing fast; others recognised it as an attractive way of encouraging infertility among the unfit and controlling it for the fit (Soloway, 1982). The spread of welfare provision for members of the infant National Insurance scheme and all children could be said to reflect the emphasis given by eugenics to children as the focus for welfare improvement while it underlined the different political and personal responses to the theories that lay behind it. Both Francis Galton himself and Major Leonard Darwin, General Secretary of the Eugenics Education Society, were childless, Darwin because he felt there were strong eugenic arguments against his reproducing because of ill-health in the family. Yet the Whethams used the Darwin family pedigree to explain the most desirable capacities in a model of enhanced reproduction.

10.3 War and birth control

The outbreak of war in August 1914 did not inhibit eugenic thought: in fact, it encouraged it. Thoughts of the next generation became even more pressing faced with the deaths of so many young men on European battlefields. The National Birth Rate Commission, which began its deliberations in 1916, enjoyed a semi-official status, which meant that it was reported in major newspapers and journals. Other semi-official initiatives also encouraged greater public attention to reproduction. National Baby Week and the National Jewel Fund both urged the wealthy to support the nation's infants and to view them as a national resource. Despite the protests of negative eugenists like Saleeby, these efforts were directed at all children, regardless of 'quality'. Even illegitimate children were now to be subsumed under the umbrella of infant life preservation as recorded in McLeary's accounts, and various philanthropic organisations, which did not consign all mothers of illegitimate babies to the workhouse, began to function. (Indeed, the Ministry of Munitions provided nurseries for children of all mothers in the same institution.) In a welter of wartime pronatalism, specifically eugenic considerations were rather hidden; in fact, Saleeby protested that too much welfare provision was damaging the future of the nation. However, Marie Stopes, a botanist, had begun her long career as a commentator on the matrimonial and sexual life of the nation as a member of the national Birth Rate Commission, and published the most popular text ever produced by a eugenist in 1918. *Married Love* was reprinted three times in its first year of publication and continued to sell well throughout the 1920s and 1930s, as to a lesser extent did its successors, *Radiant Motherhood* and *Wise Parenthood*. In these books Stopes outlined a set of proposals focusing on birth control as the means of achieving 'racial hygiene' and social progress. The technique of birth control or even, perhaps, abortion was about the only direct eugenic intervention possible. All other eugenic proposals were, like

Galton's, designed to provide a fruitful environment for eugenically desirable qualities to flourish (Stopes, 1918; Hall, 1991).

In 1922 the Eugenic Society produced a film that demonstrates the main purpose of their activities, which was to get people to see their reproductive behaviour as a part of their civic responsibilities. The film's viewers were encouraged to look at their pedigrees and calculate their own family's fitness for marriage. Oral history indicates very little effect of these prescriptions upon ordinary people's choice of partner or expectations of marriage. Stopes propagated knowledge of sex and birth control techniques (Humphries, 1988; Gittins, 1986; MacLaren, 1977). The impact of eugenics was simply to enlist its support in pursuit of what people wanted to do for a variety of other reasons: limit family size. Another factor was the effectiveness of feminism, which had moved from the issue of the vote to a concentration on maternal health. When Dora Russell and other members of the Workers' Birth Control group argued that 'it was four times more dangerous to bear a child than go down a coal mine', they were making the same claim based on the centrality of reproduction to women's lives that eugenics was also making (Russell, 1975). The Eugenics Society sponsored a group in the mid-1930s, the Population Investigation Committee, which attempted to identify the cause of population decline, because of the declining European birth rate, and to assess the implications. Demographers created the notion of the net reproduction ratio as a way of predicting future population trends by relating outcomes to the number of fertile, sexually active women in the population as whole. At a stroke, women became the sole reproductive sex statistically, as they had always been ideologically, eliminating men from the calculations entirely. Hogben's Social Biology Unit at the London School of Economics began the task of population projections, which was to cause such anxiety among legislators over the decline in the birth rate. Debate led to greater public discussion about projections into the future, because the health being agonised over was a general phenomenon, of all the nation's children, not just eugenically desirable groups. Anxiety over population size and balance was to be vital to the ideologues of an egalitarian, redistributive reform of postwar society, ably summed up by Richard and Kay Titmuss in the 1938 pamphlet *Parents Revolt*, and gave weight to demands for improvement in housing, health care and free secondary education, but it did not lead to much research of a specifically genetic, rather than sociological, kind.

10.4 Fit or unfit for marriage

Popular marriage guides reflect another route through which ideas about the technical possibilities of intervening in human reproduction receive their most accessible and influential translation into everyday speech. Thomas Van de Velde, who had written in *Ideal Marriage* a very popular manual, almost as influential as Stopes's work, wrote, in a book published in English in 1934, a book that attempted to explain the variety of ways in which marriage was appropriate for different people. In *Fit or Unfit for Marriage*, Van de Velde (1934) was judicious in his account of the nature of the human constitution:

The personal qualities of an individual are therefore the product of the interaction of all the factors of his environment on the hereditary elements with which he has been endowed.

He attributed the hereditary degeneration that makes up an unhealthy constitution as due to the deterioration of the germ cells or to inbreeding, but went on to list a variety of environmental influences as most potent in affecting the inheritance of a child from its parents, alcohol, bacterial poisons and malnutrition. The rest of the chapter is almost entirely composed of an analysis of the sexual compatibility of different physical types: the asthenic, the pycnic, the athletic and the dysplastic (the misshapen). These types are discussed in terms of their capacity for sexual love rather than reproduction. Epilepsy is directly described as causing

> injurious consequences, mental diseases of the most varied kinds, also dementia, which may go to the most extreme degrees, often comes into existence. Likewise, we often find among the descendants of epileptics, degenerate, asocial or criminal individuals. *(Van de Velde, 1934)*

Yet he did not recommend a close personal following of eugenic precept because, as he pointed out, 'we are still far from being able to make definite practical demands which might guarantee a higher psychic development of the child.' He cited Beethoven and Goethe as outstanding children of problem fathers, although later he supported Mendelian theory to explain why inbreeding, which promotes many undesirable recessive qualities, is to be deplored and denies the likelihood of breeding genius. He passed with some obscurity over the question of race, arguing that though Jews are a race, Aryans are only a group of languages, and that mixed marriages are probably not a good idea because of the problems that the members of the marriage face. He wrote approvingly of marriage advisers, generally envisaging that this function would be undertaken by doctors. In the 1930s the Eugenics Society had tried to encourage doctors to recommend a thorough investigation into family history before setting a marriage in motion. However, the profile also posed the problem of what to do if the pair turned out to be unsuitable – in that it was usually too late to delay marriage, only possible to prevent conception. The same strategy was tried at Peckham, in the clinic associated with the early Peckham experiment in socialised medicine, where families were treated as a whole at a health centre, and had the same problems concerning when genetic counselling should be best applied.

Eugenists had thus penetrated public views in so far as popular writings about marriage and parenthood suggested that heredity should be considered, but provided very little by way of concrete indicators against procreation except for a general assumption that madness and epilepsy could cause grave problems for future children. The eugenists who were to have the most significance in British public life came to the forefront in the 1930s: Carlos Blacker and Richard Titmuss. The effect of their investigations and agitation was, however, primarily in the areas of social policy, which attempted to remove traditional inhibitions on the flowering of talent. Blacker, a psychiatrist, became General Secretary of the Eugenics Society in 1936, and Titmuss was to succeed him in 1939. A wheat farmer's bequest meant that the society had resources to sponsor research and research fellows independently of the academic structures already in place at University College, London. However, their success in keeping eugenic theory in the public eye also owes something to the decline of negative eugenics as a major activity of the British movement. Eugenics had a very different public profile in the USA, where it had also influenced public debate though it produced little by way of scientific research. The assumptions of the hereditary transmission of madness, mental defect, alcoholism and promiscuity led directly to enforced sterilisation programmes as an alternative to the large, expensive segregative institutions that had also developed to 'control' the 'problem of mental defect' in more than half the states of the Union. This was an assumption, and no studies were done to investigate the therapeutic effect on this procedure, nor, shockingly, did any physician investigate the long-term sequelae for the sterilised 'patient' or the patient's family or community. The discussion of race as an entity was also based upon a notion of heritability that had affected discussions in Britain very little (Barkan, 1992).

The success of the mental deficiency legislation in winning the battle for public belief in institutional segregation had led to failure when eugenists tried to introduce to Britain legislation like that in the USA, Denmark and Fascist Germany: to sterilise 'defectives'. The attempt was made after the 1934 Report of the Departmental Committee on Sterilisation, which recommended that sterilisation should be made legal,

> in the case of:–
> (a)   A person who is mentally defective or who has suffered from mental
> disorder. . . ;
> (b)   A person who suffers from, or is believed to be a carrier of, a grave physical
> disorder which has been shown to be transmissible. . . ;
> (c)   A person who is believed to be likely to transmit mental disorder or defect.

The committee reported that out of 60 witnesses, only 3 had been opposed to sterilisation. Their reasoning was that such people should be able to marry and to undertake care in the community without what they call the 'harassing uncertainty of contraceptives'. Thus, on the face of it, eugenic thought appeared to have become the dominant form of thought for reformers and legisators, particularly those related to medicine.

This impression is accentuated by two other publications in the same year. Blacker edited a medical text *The Chances of Morbid Inheritance* in 1934, 'to supply to the general practitioner the means of dealing with requests for a eugenic prognosis', a term used to mean 'an estimate of the probability of a given disease or defect being inherited'. The book opens with an account of Mendelian inheritance that carefully distinguishes dominant from recessive genes and their associated pedigrees, sex-linked characters and lethal factors and concludes that 'it is clear that advice against marriage or procreation should only be given when the perpetuation of serious defects is involved.' There follows a substantial list of diseases of the nervous system. In a chapter on epilepsy, W. Russell Brain stated, despite contrary claims, that there is 'no doubt that heredity plays an important part in the aetiology of epilepsy'. Yet Aubrey

Lewis, in the longest chapter in the book, was much less certain about the mode of transmission of mental disorders of which, he argued, there was clearly no single predisposing genetic factor and that consequently some notion of polymorphism (different versions of symptoms deriving from one single genetic cause – inherited mental illness) was implausible. Some disorders, he argued, were clearly partly hereditary; the important thing was to encourage and reassure the healthy as well as the chance to 'save from being born many people destined to misery and disease'.

Lewis supported the argument that undesirable social qualities were being transmitted by descent to a group identified by many at the time as the 'Social Problem Group'. Lewis described them thus:

> Many who do not now fall within the range of certifiable insanity are themselves socially undesirable because of abortive mental disorders or psychopathy – many tramps, criminals, swindlers, irresponsibles and chronic dependents, agitators, hysterics – a large part of the social problem group. These people may transmit the predisposition to maladaptation of various kinds.

Lewis thus veered between his major argument, which is that a healthy member of a tainted family should marry a healthy member of a family without the taint; and his minor one, that society as a whole then suffers from the spread of such taints more generally as recessive genes enter the population. He suggested that the problem should be discussed with the recognition that these qualities are already widely spread – the concern of the eugenist and the sociologist rather than the clinician.

10.5 Population and equality of opportunity: reform eugenics in the 1930s

Both Blacker and Titmuss dominated the Eugenics Society after 1936 and they argued that the focus of eugenics lay in maximising the welfare of the talented, rather than impeding births of the unfit. Titmuss in 1938 even wrote to Newhouse, another leading eugenist, 'You can do anything you like and call it eugenics', and some of the older Establishment figures deplored these innovations. Blacker and Titmuss were leading figures in the growth in demographic investigations at the London School of Economics, in the discussion over the declining birth rate, advocacy of birth control, attempts to encourage the further reproduction of the fit through the Galton Fund, to encourage mothers of high-quality children to have more, and a general concern for differential population change that led, in 1943, to the setting up of the Royal Commission on Population, which reported in 1949. When the membership of the Commission was announced there was great rejoicing recorded in the minutes of the Council of the Eugenics Society because so many members were eugenic sympathisers, including the economists who were so large a part of the group of witnesses, as well as the psychologists who gave evidence on the decline in the national intelligence, and the doctors who advocated eugenic means of creating a more successful public health polity.

Those members of the professional middle classes, especially doctors, who belonged to the movement were influential in the direction taken by sociology and in the package of welfare measures of the postwar welfare state. How much did they encourage new medical techniques or research in Britain? Some wished to prove eugenic arguments wrong or at least incorrect in emphasis. Lionel Penrose worked in the field of mental deficiency when he started his medical career. His first major publication, *The Biology of Mental Defect*, was his reponse to the Wood and Brock assumptions about sterilisation (Penrose, 1949). He concluded, after looking at the 2000 or so inhabitants of the Royal Eastern Counties Institution, that only 8% of those cases were certainly attributable to inherited conditions of a genetic kind. In other words the assumptions of the mental deficiency legislation were incorrect. He also criticised, although less vehemently, the concept of the Social Problem Group. The idea that a group of social problems could all be described and reified as the Social Problem Group was not of course new. Earlier definitions of a small, genetically created group of people with multiple defects spanning the social, mental and physiological had been earlier sustained as debates about the condition of England had come up with the submerged tenth, the residuum, Outcast London or people of the abyss (Stedman Jones, 1971; Walkowitz, 1980). However, in the 1930s unemployment and government expenditure cuts had meant a revival of hereditarian explanations for the plight of those at the bottom of the social pyramid and hence cheap hereditarian solutions. The majority of the unemployed were increasingly seen as a problem for society whereas the minority, the 'unemployables', were seen as a problem in themselves. Disadvantage, in this argument, was a genetic condition. Moral attitudes could be reproduced, it was being argued, as a part of an individual's hereditary endowment. The group

> would include, as everyone who has extensive practical experience of social service would readily admit, a much larger proportion of insane persons, epileptics, paupers, criminals (especially recidivists), unemployables, habitual slum dwellers, prostitutes, inebriates and other social 'inefficients' than would a group of families not containing mental defectives. The overwhelming majority of the families thus collected will belong to that section of the community, which we propose to term the social problem or subnormal group. This group comprises approximately the lowest 10 per cent in the social scale of most communities.   *(in Blacker, 1937)*

However, the dominance of eugenical thinking was to some extent undermined by the examination of the constituent member of the group then described. Blacker himself pointed out that poverty was what turned a family problem into a social one, and at the end Caradog Jones could sum up the only shared conclusion about this group (apart from the repeated demand for more research, ill met by the pedigrees cited throughout the collection of essays), which was that

> If a significant positive correlation was definitely established between defective or retarded intelligence and other subnormal or abnormal conditions, considerable weight would be added to the view that every effort ought to be made to discourage the fertility of the social problem group, defined as a group of subnormal intelligence. *(in Blacker, 1937)*

The concept of mental deficiency used in this and other contemporary texts seemed, to Penrose at least, at best unproven and, at worst, downright mischievous. Penrose was part of a group described by Gary Werskey (1988) and, more recently, by Dianne Paul (1984) that was both left-wing and hereditarian. The enquiry on which he embarked was to shift social policy on mental deficiency substantially, but not until 1957, when the Royal Commission on Mental Health and Mental Illness reported and recommended diminishing reliance on institutional care for the mentally defective.

World War II and its aftermath, in which was discovered the horrors of medical investigation and extermination in the name of eugenics, are often seen as the decisive factors in the change from eugenics to genetics. However, the shift in naming the investigation of both clinical and policy concerns arising from an understanding of genetic endowment happened earlier and did not represent a major change of research or of motivation for research. In fact, many of the concerns of eugenists continued effectively unaltered until the late 1950s. Throughout, the notions of 'normal' (desirable) heredity and 'abnormal' (undesirable) remained. The mapping was different, needing a microscope rather than pedigrees, but the questions remained the same. The eugenists neither drove social concern as much as has been assumed in past histories nor were as different from other medical sciences as posterity would wish them to be. Penrose disclaimed contemporary eugenics in his analysis of the late 1930s without knowing anything of Dr Mengele, but he did not alter the name of the journal from *Annals of Eugenics* until 1954, and he continued to teach a course called 'Eugenics' at University College, London, until 1951. Josiah Wedgwood talked out the sterilisation legislation in the 1935–6 Parliamentary session before many of these experiments had even begun. Indeed, Blacker was chosen by the administration of the Occupied Zone to investigate medical atrocities in concentration camps in order to assess whether they had had any medical justification or value. His report was not published, but its conclusions were: he condemned, unequivocally, the experimenters and the experiments as unmedical, inhumane and effectively nothing to do with eugenics. The legacy of eugenics in the war years lay less in any widespread assertion of new policies in relation to eugenically desirable marriages, and far more in the belief that eugenically desirable qualities were being hindered by the absence of equality of opportunity. This belief surfaced in debates over educational reform, improvements in health care, housing, and the prevention of unemployment. The popularity of eugenically inspired notions of equality of opportunity is shown by their frequent reproduction in the publications of bodies as diverse as the Army Bureau of Current Affairs, the Fabian Society, the Workers' Educational Association and many more. But negative eugenics, particularly the anxiety about the multiplication of mental deficiency, left the public cold. When the Royal Commission on Population reported, in 1949, they concluded that they could not understand the argument that the tendency of the National Intelligence was to decline along with the birth rate, and the fact was further disproved by the investigation into the intelligence of a representative group of Scottish schoolchildren that concluded that in fact the average intelligence of children was rising, not declining.

The area of research that remained open to eugenical concerns was thus still that of mental defect, because eugenist social reformers had won (for 40 years at least) the argument about letting native endowment flower unfettered by social constraints. Thus, the hunt turned to the question of the inheritance of mental ability, or intelligence. Here eugenists such as Hans Eysenck, funded by the Eugenics Society for some of his early research projects, remained convinced of the heritability of certain qualities of mind and continued psychological investigations to demonstrate the efficacy of this relationship. More contentiously at the time, the question of race as a genetic characteristic continued to surface despite the impossibility of isolating a single gene for race, or any factor more easy to demonstrate than skin, eye and hair colour, or curl in the hair. Which eugenics counts? The postwar history of eugenics shows that it was a clinical analysis based on pedigrees and the more general construction of a notion of parental risk that continued to influence policy and practice in the postwar years.

From 1880 to 1950, research into human heredity was conducted under the auspices of eugenics, no matter what the political allegiances of the participants were. After World War II, with the revelations of what happened in Nazi Germany – the enforced sterilisations and extermination of people under its racial hygiene policies – eugenics began to be associated with Nazi science. Because of our memory of Nazism, it is often assumed that eugenics and eugenic thinking disappeared after the war, yet the British Eugenics Society continued to exist and publish its journal for over 20 more years after 1945. Many of those who were active in the eugenic movement in the 1920s and 1930s continued to call themselves eugenists after the war.

10.6 Carlos Paton Blacker and the activities of the Eugenics Society

Carlos Paton Blacker (1895–1975) was General Secretary of the Society from 1931 to 1952, and again from 1952 to 1961. After the war he became a physician at the Maudsley Hospital, which was a hospital for the mentally ill. He was an adviser to the Ministry of Health, and a member of the Royal Commission on Population that sat during the late 1940s. He ran five miles before breakfast every morning and was described in an obituary notice as being 'every inch the ex-Etonian and ex-Coldstream Guards officer'. Blacker felt that the issue of Nazism had to be faced by the Eugenics Society. Blacker had worked for the British government's War Crimes Committee in 1947, investigating crimes against humanity committed by the Nazis. In a paper given to a meeting of the Eugenics Society in 1951, called 'The eugenic experiments conducted by the Nazis on human subjects', Blacker said:

> Some of you may think that, in view of how irrelevant to what we understand as eugenics are the human experiments I have described, the title of my paper is misleading. You may think that I should have referred to 'so-called' eugenic experiments. Perhaps I should. But the inexorable fact remains that whatever our own views may be, the word eugenics has, through the events I have described, suffered degradation in the eyes of many people and organisations.

How then to continue the eugenic agenda? Why not construct a new history for eugenics by creating a 'hero' of the founder of eugenics, Francis Galton? As an eminent Victorian, and cousin of Darwin, with an interesting life and a wide range of interests, it was a less contentious path to take, which could reformulate eugenics and diminish the connections between eugenics and Nazism. As part of Blacker's policy to regenerate and to rehabilitate eugenics in the postwar years, he wrote *Eugenics – Galton and After* (Blacker, 1952), an account of Galton's life followed by an apologia for eugenics and an explanation of Galton's views on race. Blacker presented eugenics as 'a system of thought, of feeling, and of behaviour; or as science, sentiment and policy'. He argued that eugenics as a system of values was no longer 'a monopoly of experts' and that the 'ideals and beliefs of the common man . . . must determine the tempo of eugenic progress'. He argued that the ordinary person should be allowed to contribute to the discussion of eugenics and human genetics, and complained of the use of 'complex mathematical procedures . . . which intimidate not only the layman but also many geneticists devoid of mathematical bent': 'Few have the temerity to offer challenge' and 'fears of a discrediting scientific castigation may limit the range of creative controversy'. He identified the reasons for the mistrust of eugenics by geneticists as the technical complexity of human genetics, the leftward movement of the scientific world with its rejection of class values, and Nazism. 'The result,' he wrote, 'has been that the word eugenics, like the word race, has come under a cloud.' Blacker was adamant that his book was not a textbook. The Eugenics Society decided, in its policy discussions in November 1949, that a change in emphasis was needed:

> The sphere in which we look for a clear expression of eugenic values is thus the family. This is safe ground which commends itself to everyone: much safer than the 'ground' of class distinctions.

Earlier, in 1946, the Council of the Eugenics Society had welcomed the incorporation of previously private functions into the new National Health Service: 'Birth Control; Advice as to Infecundity; Guidance as to Eugenic Prognosis; Premarital Health Examinations; Marriage Guidance'. It looked therefore as though the concern for the state of population expressed in the Royal Commission on Population, which was to report in favour of expanding the welfare state, and the Royal Commission on Marriage and the Family demonstrated that the eugenic agenda had now become a part of official health provision. However, the practical effects were small, partly because most of these policies were also supported by others, not for specifically eugenic reasons.

10.7 Genetic counselling in Britain after 1945

The first genetic counselling clinic was set up in Britain in 1946, at the Hospital for Sick Children, Great Ormond Street, London, by John Fraser Roberts, author of one of the first textbooks on medical genetics, and a member of the Eugenics Society for 37 years.

In 1957 the Medical Research Council set up the Clinical Genetics Research Unit at the Institute of Child Health, London, with Roberts as its director. When Roberts retired in 1964, the directorship was taken over by Cedric Carter, an active member of the Eugenics Society, secretary of the society and editor of the *Eugenics Review* for many years. Lionel Penrose, one of the world's experts in medical genetics, and head of the Galton Laboratory in London, preferred not to give genetic advice. Roberts has been described by Polani (1992) as 'a mild reform eugenicist', one of the founders of clincial genetics and a 'model genetic counsellor'. Robert's texbook became a standard in the field, and his principles of counselling, including that advice should be given in terms of risk and that it should be non-directive, became widely accepted (Roberts, 1959). For Carter, a major aim of genetic counselling was to prevent an increase in the number of children born with a genetic predisposition to serious disease. He advocated that special efforts be taken to encourage 'the most ignorant' and 'least gifted groups' in society to use family planning (Carter, 1962). Another major centre for genetic coun-selling was set up in 1960 as the Paediatric Research Unit at Guy's Hospital, London, with the aim of preventing congenital disorders through research and genetic advice, with Roberts as one of its geneticists. Genetic counselling was developed and promoted in Britain during the 1950s and 1960s by medical practitioners, some of whom contin-ued to maintain links with the Eugenics Society.

When Roberts retired from the Clinical Genetics Unit, the directorship was taken over by Cedric Carter (1917–1984), again an active member of the Eugenics Society (secretary for 1952–1957, librarian from 1959 until 1965, and editor of the *Review* for some years) who remained more consistently committed to eugenics than Roberts throughout his life, and practised what he preached by having a large family with seven children! It was said of Carter in an obituary that 'If he had a quiet passion then it was for positive eugenics and nothing gave him more pleasure than to note the tendency for Social Class 1 families to have more children.' He was director of the MRC Unit from 1964 to 1982, and founder of the UK Clinical Genetics Society in 1971, which pub-lished several reports on genetic counselling and screening services in the late 1970s and 1980s under the auspices of the Eugenics Society. In contrast to Roberts, Carter does not seem to have been the most sympathetic counsellor: it was said of him that 'his unemotional attitude exasperated many who visited the unit' (Baraitser, 1989, p. 79).

Throughout this period, genetic advice was given in terms of the risk to a couple of having a child with a particular condition, called 'probability counselling', and many who sought advice already had a child with a genetic disease. From the literature, it seems that only married couples were given advice, and both parents were seen. In 1979, researchers reported that probability counselling was still the only option for a substantial proportion of families (Polani *et al.*, 1979).

Initially about 30 couples per year were seen at the Clinical Genetics Unit, increasing to more than 300 annually by 1970. In general, couples were referred by their GPs or paediatricians, although the initiative to seek advice came from the couples themselves. The major conditions for which advice was sought were: mental subnormality, espe-cially Down syndrome; disorders of the nervous system like anencephaly; skeletal malformations; and disorders of the alimentary system like cystic fibrosis. Risks were assessed as 'high' if there was greater than 1 in 10 chance of occurrence, and 'low' if there was a less than 1 in 10 chance. Risks were explained in terms of 'odds'. Roberts particularly liked using the football pools to help parents appreciate risk factors. Carter found that couples in general took 'responsible decisions', i.e. those in the high-risk groups refrained from having children, although some had 'unplanned pregnancies'. Carter remarks that the majority of couples who sought advice were 'self-selected for intelligence'. Possibly because of this, when he found that they were put off by a risk factor which he considered a low risk and no doubt because he wanted to encourage more Social Class 1 families, he did begin to intervene and say, 'In your place I would not take a risk of this kind too seriously', although he did say that the potential parents should take the decision themselves. Before the 1967 Abortion Act, difficulties were experienced by some women in the 'high-risk' category in getting a termination of pregnancy, although terminations were obtained (Carter *et al.*, 1971).

The membership of the Eugenics Society gradually declined during the 1950s. In 1957, Blacker was concerned that the number of Fellows and Members of all categories had decreased from a high-water mark of 768 in 1932 to 456 in 1956. But the society still had prominent individuals and scientists as members like Sir Charles Darwin, President from 1954 to 1959, and Julian Huxley. Huxley was a national figure, regularly appearing in the media, founder of the World Wildlife Fund, Director of Unesco and President of the Eugenics Society from 1959 to 1962. In the early 1960s, as Penrose was finally succeeding in getting his work called 'human genetics', Huxley continued to advocate 'negative eugenics':

> Negative eugenics has become increasingly urgent with the increase in mutations due to atomic fallout, and with the increased survival of genetically defective human beings, brought about by advances in medicine, public health and social welfare. But it must, of course, attempt to reduce the incidence, or the manifestation, of every kind of genetic defect. *(Huxley, 1963)*

Genetic defect for Huxley included myopia, some kinds of sexual deviation (he did not specify which), diabetes, as well as Huntington's chorea (the contemporary term).

> Against the threat of genetic deterioration through nuclear fallout there are only two courses open. One is to ban all nuclear weapons and stop bomb-testing; the other is to take advantage of the fact that deep-frozen mammalian sperm will survive, for a long period of time, and accordingly to build deep shelters for sperm-banks – collections of deep-frozen sperm from a representative sample of healthy and intelligent (of course!) males. *(Huxley, 1963)*

Francis Crick and others who were not inculturated within the eugenics world also believed in sperm banks and licensing of those who should reproduce children. In 1957, Blacker suggested that the society should adopt a less evocative name and that it should devote itself to biosocial science, including the study of conservation, evolution and the progress of mankind. The society organised several symposia during the 1960s on the biological aspects of social problems. As these were successful, the Society decided to stop publication of the *Eugenics Review* and to start publication of a new journal, the *Journal of Biosocial Science*, concerned with 'the common ground between the biological and social sciences', and including the old favourites of education and criminology. The Eugenics Society closed its doors on eugenics in 1968 and opened its doors to the Galton Foundation. Yet this was not the only institutional location for the study of genetics: the direct heir to Galton at the laboratory he had founded was Penrose, who had criticised some eugenic assumptions before the war, and continued a long career into the 1960s.

10.8 Lionel S. Penrose and the Galton Laboratory

One of those from the leftist tradition that constantly attacked the eugenic agenda, both during the 1930s and after the war during the 1950s, 1960s and 1970s, was Lionel S. Penrose. In 1945 he became the only Professor of Eugenics in the country and held the post for 20 years in charge of the Galton Laboratory for National Eugenics. Penrose is renowned for his work on Down syndrome. For Penrose, mental defect was not a disease but merely the expression of normal variation in the intellectual capacities of the human species. One of his first actions on taking up the chair was to quietly delete the phrase National from the letterhead of the laboratory notepaper. Contesting the terrain of human heredity from a particularly advantaged position, Penrose began to implement his criteria for a 'science' of human heredity, his standards for indicating who can speak with authority about human heredity, and his method and strategies for moving the study of human genetics away from eugenics. These strategies included the collection of data, the use of numeric tables, the application of mathematical techniques, the appropriation of eugenic icons and concepts for anti-eugenic ends, direct technical challenges to central eugenic concepts within the realm of scientific debate, and the subdued use of language with occasional interjections of personal comment (often humorous). These Penrose continued to employ in eugenic, scientific and public arenas over the next 20 years in the transformation of research on human heredity to human genetics. Privately he thought eugenics was a mixture of propaganda and pious hopes about the future of the human race, supported by blood-curdling threats about the inexorability of Mendelian inheritance. Publicly he tried to make the subject a little more grammatical and methodical. For many scientists today, Penrose is almost the Galileo of genetics, establishing 'true science' in the face of the religiosity of the eugenicists. Penrose strategically and judiciously encapsulated the central tenet of his science in his textbook *The Biology of Mental Defect* (Penrose, 1949) in the last sentence of his text, where it could be easily found by curious readers and reviewers:

> Subcultural mentality must inevitably result from normal genetical variation and the genes carried by the fertile scholastically retarded may be just as valuable to the human race, in the long run, as those carried by people of high intellectual capacity.

What was it like to work in the field of human heredity after the war? What would the day-to-day work be like? One of the people interviewed by Kevles (1985) describes the working conditions as follows: 'a few deep freezes, some pipettes, and a sort of old microscope that Pasteur would have thrown out. For the most part, people sat at tables and desks working with numbers and papers.' The Galton gained a worldwide reputation for genetics research and the visitors book resembled a *Who's Who* of the key researchers in human genetics of the time (Kevles, 1985). The Galton employed a large number of women, some of whom had children; this was unusual in the 1950s when in the main the cultural climate only encouraged women to be housewives. This Galton tradition dates back to Pearson's days and his interest in social reform. Much of the work consisted in collecting and analysing data, measuring birth weights, and using dermatoglyphics to identify and classify various conditions; connecting the age of the mother and risk of Down syndrome came out of this work. This again was a history going back to Galton's interest in fingerprinting. And it was cheap to do! The main reason that conditions were so spartan at the Galton was that the laboratory was not very well funded by British funding bodies such as the Medical Research Council, but did get funding from the Rockefeller Foundation. In sharp contrast to Blacker and other eugenists, Penrose was not an establishment figure.

A significant strategy that Penrose used was to make a sustained challenge to eugenic science from the powerful platform of the major journal for publishing research on human heredity, the *Annals of Eugenics*. The *Annals* was one of only two journals worldwide that published research exclusively on human heredity in the 1950s. The second journal was the *American Journal of Human Heredity* set up in 1947 by the American Society of Human Genetics. The *Annals* had already acquired a reputation for academic excellence with Fisher as editor, but its main focus then was on the statistics of human populations. One of Penrose's first actions on taking over the editorship in 1946 was to change the sub-title to 'a journal of human genetics'. In 1946, Penrose introduced a bias towards medical genetics, and as 'there was no other journal in English devoted primarily to human genetics, this venture should be very promising'. Penrose quickly used the journal to publish his own research with six papers in the next two years as a means to establish his methodology for future authors. In 1954, Penrose changed the name to the *Annals of Human Genetics*, using as a reason that 107 out of a total of 190 papers published under his editorship were about human genetics, and the circulation had increased in consequence. The journal had originally been set up to provide a scientific basis for eugenics; now that basis for eugenics had been removed. For almost 20 years, until the establishment of the *Journal for Medical Genetics* in 1964, researchers in human heredity had to pass through that gate-keeper of human genetics, Penrose. The journal provided a powerful and strategic tool for continuing the agenda he set in his inaugural lecture and in his textbook, and a method for training researchers in the language and style of science he promoted. Eugenists were less advantaged in contesting the scientific basis of eugenics as they had no reputable academic journal from which to do so. The *Eugenics Review* was established as a journal to popularise eugenics and to educate the public in eugenic ideas, not as a research journal. Of course, researchers in human heredity could publish in medical journals, and many did so including Penrose, but the significance of their work could be missed or could be contested within that world. Penrose's surveillance and policing of research on human heredity through the *Annals* was a significant force in the closure of debate on the political nature of human heredity and in the transformation of eugenics. The process of expunging the word 'eugenics' from the polite world of science was gradually completed for Penrose when the title of his Chair was changed to the Professor of Human Genetics. Although for 20 years Penrose had tried to have the name changed, 'to something descriptive of what was actually studied and taught in the department', the catalyst for the change of name came from a newspaper item about the Galton Laboratory in *The Times*, subtitled 'work on eugenics will be hastened', about the Wolfson Foundation's gift of money to University College, London, for the building of a new laboratory. Penrose was outraged and wrote to the Provost, 'It has been my consistent policy to work on and teach human genetics, not eugenics: human genetics is a science and eugenics is an ideology.' The change of name involved consultation with solicitors over the legal implications of Galton's will. Public opinion was changing and 'a new building for eugenics seemed less practical than one for human genetics'. Eventually, as Penrose came to retirement age, the name of his chair was changed.

Penrose was not alone in his creation of a human genetics devoid of eugenics. Although not an establishment figure, he did not have the physical resources of other research institutions, and was a pacifist in his politics. He had the support of scientists who wanted to work in human heredity freed from accusations of Nazism. With the demise of the Social Problem Group due to the Welfare State, and the creation of 'normality', the social order seemed not to be threatened internally. The threat to social order came from outside: the Russians and radiation fallout. Self-proclaimed eugenic scientists found expression for their work in genetic counselling and the promotion of 'positive eugenics', as the abnormal or defective individuals were less visible in society. Techniques for studying human chromosomes were described in the late 1950s (karyotyping), and led to an explosion in research in medical genetics in the 1960s, particularly on the cytogenetics of the sex chromosomes. The research was reported in a *New York Times* leader in 1959:

> the first fruits of a technical advance which may revolutionise human genetics . . . an enormous territory awaiting exploitation with nothing less than the first real exploration of the human chromosome map as the first prize if the early promise is even half-fulfilled.

So the idea of mapping the human genome is quite old, more than 30 years old. Chromosomal anomalies were found to be associated with conditions such as Down, Kleinfelter and Turner syndromes. In 1969, the Genetic Centre at Guy's Hospital began to offer a prenatal diagnosis service using amniocentesis. The tests were quite labour-intensive, particularly as cells from 'normal' people had to be cultured as controls. The majority of assays (38%) were done because of maternal age to test for chromosomal disorders of the foetus, 35% were done for neural tube defects, the remainder for metabolic disease like Tay–Sachs. Where the outcome of pregnancy was known, about 4% of amniocenteses produced evidence of 'fetal abnormality' and the pregnancies were terminated, and about 90% of pregnancies produced live babies. By the late 1970s, genetic counselling had moved from limiting a risk-giving situation to one where direct information was given on a high-risk pregnancy. The Genetic Centre moved towards a more outgoing approach, to informing the professions and the public of the services available, using circulars, lectures and the provision of library and information services (Polani *et al.*, 1979). The Department of Health and Social Security memorandum on Human Genetics in July 1972 listed 25 clinics giving genetic advice outside London and 10 in Greater London for the benefit of GPs, and advised that it consisted

> essentially in giving as accurate information as possible to the extent that knowledge permits, on the risks of transmission of inherited or partly-inherited conditions. Family doctors are well placed to undertake simple genetic counselling with supporting specialist help. Routine premarital counselling is not practicable but advice could be given where there is a specific problem.

Thus, when new techniques in chromosomal diagnosis came along, there were institutions prepared to offer advice as to how they should be used in which eugenic concerns had played a leading role. However, the mixed motives of eugenists did not always get translated into practice in these clinics because a key factor in the take-up for such practices is public understanding that they exist and the construction of parental risk. In this respect eugenists have not had the influence they thought they should have – over public policy – in that reproduction remains, for the majority, white adult population, a private matter. Eugenists have, however, been influential in the way that population in general remained a central concern of policy, until the rolling back of the state in the early 1980s.

The legacy of the eugenic construction of the concern for human heredity that led directly to the mapping of the human genome has been indirect. We would suggest three main areas of influence. How parents construct risks and understand pedigrees is also in part a product of this history, which has in its turn been re-approached every time new reproductive techniques become available. The pattern of recording data through family pedigrees has been a consistent feature of this history, as has been the assumption of standard family organisation of married parents approached primarily through the mother. The question of who is the patient for clinicians has been overlaid by the problem, in some instances, of identifying the disease. Even if a microscope can add to the information available once the patient at risk has been identified, the populations where the microscope should be used is still derived from the assumptions of population statistics developed out of eugenic concerns. The allocation of research funding is also crucially determined by the political profile of population and heredity at the time. The issue of whether reproduction of life-threatening inherited diseases should be prevented or discouraged in the interests of future generations, the afflicted person or society as a whole is not any better understood or agreed than it was in the debates over the Social Problem Group. Perhaps it is no bad thing that some of the nastier excesses and inhumanities of a notion of a better stock should not be brought to mind from time to time as a way of dealing imaginatively with the power to intervene that genetics can provide, and to understand how weak some of the causal connections on which policy has been based in the past have been. Human genetics owes its origins to eugenics, but as the contrasting careers of Carter and Penrose show, origins do not determine outcomes.

10.9 Bibliography

Primary sources

The records of the Eugenics Society are mostly deposited at the Contemporary Medical Records Centre at the Wellcome Institute of the History of Medicine, the Wellcome Foundation in the Euston Road, London. The papers of Lionel Penrose are deposited in the Library of University College, London. Otherwise the journals of eugenics, *Eugenics Review*, *Annals of Eugenics* and *Biometrika* provide a large amount of relevant material. Kevles, cited below, has a very full bibliography on both the US and UK; we have limited our listing to works directly quoted or from which substantive arguments derive.

Eugenic thought has influenced several Royal Commissions and Parliamentary Committees but the two most important are PP1934, Cmd. 4485, *Report of the Departmental Committee on Sterilisation*, the Brock report; and PP1949, Cmd. 7695, *Report of the Royal Commission on Population*.

Secondary sources

Baraitser, M. (1989) Obituary for Cedric Carter. In G. Wolstenhome (ed.), *Munk's Roll, Lives of the Fellows of the Royal College of Physicians*, vol. 8, pp. 78–80. Oxford: IRL Press.

Barkan, E. (1992) *The Retreat from Scientific Racism*. Cambridge: Cambridge University Press.

Barker, D. (1983) How to curb the fertility of the unfit: the feeble-minded in Edwardian England. *Oxford Review of Education*, **9**, 197–211.

Blacker, C. P. (1934) *The Chances of Morbid Inheritance*. London: Routledge.

Blacker, C. P. (ed.) (1937) *A Social Problem Group?* London: Oxford University Press, Humphrey Milford.

Blacker, C. P. (1952) *Eugenics – Galton and After*. London: Duckworth.

Booth, C. (1902) *Life and Labour of the People of London*, 17 vols. London: Williams & Norgate.

Carter, C. O. (1962) *Human Heredity*. London: Penguin.

Carter, C. O., Evans, K. A., Fraser Roberts, J. and Buck, M. R. (1971) Genetic clinic: a follow-up. *Lancet*, **i**, 281–5.

Davin, A. (1978) Imperialism and the cult of motherhood. *History Workshop Journal*, **5**, 9–65.

Dwork, D. (1987) *War is Good for Babies and Other Young Children*. London: Tavistock.

Farrell, L. A. (1979) The history of eugenics: a bibliographical review. *Annals of Science*, **36**, 111–23.

Galton, F. (1869) *Hereditary Genius: an Inquiry into its Laws and Consequences*. London: Macmillan.

Galton, F. (1908) *Memoirs of My Life*. London: Methuen.

Galton, F. (1909) *Essays in Eugenics*. London: Eugenics Education Society.

Gittins, D. (1986) *Fair Sex*. London: Hutchinson.

Gorst, J. (1906) *The Children of the Nation*. London: Methuen.

Gould, S. J. (1981) *The Mismeasure of Man*. New York: Norton.

Hall, L. (1991) *Hidden Anxieties: Male Sexuality, 1900–1950*. Cambridge: Polity Press.

Hollis, P. (1987) *Ladies Elect*. Oxford: Clarendon Press.

Humphries, S. (1988) *The Secret World of Sex: Forbidden Fruit – the British Experience, 1900–1950*. London: Sidgwick & Jackson.

Huxley, J. (1963) The future of man – evolutionary aspects. In Wolstenholme, G. (ed.), *Man and His Future*, pp. 1–22. London: Churchill.

Kevles, D. J. (1985) *In the Name of Eugenics*. New York: Alfred Knopf.

Love, R. (1979) Alice in eugenics land: feminism and eugenics in the careers of Alice Lee and Ethel Elderton. *Annals of Science*, **36**, 145–58.

MacKenzie, D. (1981) *Statistics in Britain 1865–1930: the Social Construction of Scientific Knowledge*. Edinburgh: Edinburgh University Press.

MacLaren, A. (1977) *Birth Control in 19th Century England*. Devon: Croom Helm.

MacNicol, J. (1989) Eugenics and the campaign for voluntary sterilization between the wars. *Social History of Medicine*, **2**(2), 147–69.

Mazumdar, P. (1992) *Eugenics, Human Genetics and Human Failings: the Eugenics Society and its Critics*. London: Routledge.

Mort, F. (1987) *Dangerous Sexualities: Medico-moral Politics since 1830*. London: Routledge.

Nuffield Council on Bioethics (1993) *Genetic Screening: Ethical Issues*. London: Nuffield Council.

Paul, D. (1984) Eugenics and the Left. *Journal of the History of Ideas*, **45**, 567–89.

Pearson, K. (1914–30) *Life, Letters, and Labours of Francis Galton*, 4 vols. Cambridge: Cambridge University Press.

Penrose, L. S. (1938) A clinical and genetic study of 1280 cases of mental defect (Colchester Survey). *Special Report of the Medical Research Council* no. 229. London: His Majesty's Stationery Office.

Penrose, L. S. (1949) *The Biology of Mental Defect*. London: Sidgwick & Jackson.

Pick, D. (1989) *Faces of Degeneration: a European Disorder 1848–1918*. Cambridge: Cambridge University Press.

Polani, P. (1992) John Alexander Fraser Roberts. *Biographical Memoirs of Fellows of the Royal Society*, **38**, 305–22.

Polani, P. E., Alberman, E., Alexander, B. J., Benson, P. F., Berry, A. C., Blount, S., Daker, M. G., Fenson, A. H., Garrett, D. M., McGuire, V. M., Fraser Roberts, J. A., Sellar, M. J. and Singler, J. D. (1979) Sixteen years of counselling, diagnosis and prenatal detection in one Genetic Center: progress, results and problems. *Journal of Medical Genetics*, **16**, 166–78.

Roberts, J. A. F., (1940) *An Introduction to Human Genetics*. Oxford: Oxford University Press. [Second edition 1959.]

Russell, D. W. (1975) *The Tamarisk Tree*. London: Virago.

Searle, G. (1979) Eugenics and politics in Britain in the 1930s. *Annals of Science*, **36**, 159–69.

Showalter, E. (1991) *Sexual Anarchy*. London: Bloomsbury.

Simmons, H. G. (1978) Explaining social policy: the English Mental Deficiency Act of 1913. *Journal of Social History*, **11**, 387–403.

Soloway, R. A. (1982) *Birth Control and the Population Question in England, 1877–1930*. Chapel Hill: University of North Carolina Press.

Stedman Jones, G. (1971) *Outcast London*: Oxford: Oxford University Press.

Stopes, M. (1918) *Married Love*. London: A. C. Fifield.

Sutherland, G. (1984) *Ability, Merit and Measurement*. Oxford: Oxford University Press.

Van de Velde, T. (1934) *Fit or Unfit for Marriage*. London: Chapman & Hall.

Walkowitz, J. (1980) *Prostitution and Victorian Society*. Cambridge: Cambridge University Press.

Walkowitz, J. (1992) *City of Deadly Delight*. London: Virago.

Weeks, J. (1981) *Sex, Politics and Society: the Regulation of Sexuality since 1800*. London: Longman.

Werskey, G. (1988) *The Visible College: A collective Biography Scientists and Socialists of the 1930s*. London: Free Association Books.

Whetham, W. C. D. and Whetham, C. D. (1909) *The Family and the Nation*. London: Longmans.

11

Public understanding of the new genetics

JOHN DURANT, ANDERS HANSEN and MARTIN BAUER

11.1 Introduction

In recent years there has been growing concern about the public understanding of science throughout the industrialised world. Such concern has fostered a wide range of practical initiatives aimed at promoting greater public understanding of science on the part of governments, scientific institutions, science-based industries and the mass media (for an international review, see Schiele *et al.* (1994)); at the same time it has facilitated the growth of research concerned with the interrelations between science and the public. In the UK, for example, a scientific inquiry into the public understanding of science (Royal Society, 1985) stimulated the Royal Society to join forces with the British Association for the Advancement of Science and the Royal Institution of Great Britain in the establishment of the Committee for the Public Understanding of Science (COPUS), and it provoked the Economic and Social Research Council to establish a substantial research programme in this area (Wynne, 1991; Ziman, 1991).

Within the broad field of the public understanding of science, medical science occupies a very special position. For obvious reasons, research on health and illness is of great relevance to everyone. Research reveals a substantially higher level of public interest in medical science than in other branches of science and technology (Durant *et al.*, 1989), and this is reflected in the fact that the mass media consistently allocate more space and time to medical science than to other sciences (Hansen and Dickinson, 1992). Durant and colleagues have argued that medical science is paradigmatic for the popular representation of science in Britain; in other words, medical science is a principal source of the ideas and images that are commonly associated with science as a whole within popular culture. These ideas include a generally utilitarian view of science and a broadly positive attitude towards both science and scientists (Durant *et al.*, 1992).

The new genetics illustrates well the reasons for medical science's comparative prominence in the public domain. Not only does the new genetics hold out the promise of new genetic diagnoses and therapies, but also it raises a large number of ethical, social and legal questions that are increasingly the subject of public debate. In Britain, a spate of recent popular books by scientists and journalists (e.g. Wilkie, 1993; Bodmer and McKie, 1994) testifies to the high level of public interest in this area.

In this chapter we shall offer an overview of the problem of the public understanding of the new human genetics. First, we shall consider a number of conceptual and theoretical issues in the emergent research field of the public understanding of science. Then we shall review some of the literature on the public understanding of the new human genetics. After this, we shall present in summary form some results from our recent study of British public perceptions and media coverage of the Human Genome Project. Finally, we shall point to a number of key issues raised by all of these studies and attempt to draw some general lessons from them.

Our principal thesis is easily stated at the outset: in general, what may be termed public understandings of the new genetics are not passive reflections of professional, scientific understandings; rather, they are active constructs, the products of multiply-mediated historical and cultural (including mass media) influences, which may be expected to diverge significantly from those professional understandings of science with which they coexist. Public understandings of science are important, not least because they are capable of exercising a powerful influence over both scientific research and clinical practice. For this reason, among others, it is in the interests of healthcare professionals, patients and the general public that greater attention should be paid to the public understanding of the new genetics.

11.2 The public understanding of science

In approaching the public understanding of the new genetics, it is worth taking account of a number of conceptual and theoretical issues that are currently the subject of discussion in the wider field of the public understanding of science. These issues concern the meaning that is attached to the phrase 'public understanding of science'. The phrase is, of course, multiply ambiguous: what is the 'public'? what is 'understanding'? and what is 'science'? There are no uniquely correct answers to these questions. All these terms are open to being characterised in many different ways. Particular characterisations often reflect particular scientific or social interests, and to this extent they tend to be controversial. In this situation, the researcher's goal should be, not uniquely correct definitions, but rather definitions that are appropriate to the particular case.

'Public' is a term that is used in at least three different ways: first, it may refer to things that derive from or depend upon the state (hence, 'the public sector'); secondly, it may refer to a conceptual field in which actions are deemed to be open or accessible to scrutiny (hence, the notion of things being 'in the public domain'); and thirdly, it may refer to outsiders who are taken to lack particular characteristics (hence, 'the general public'). In general, commentators on the public understanding of science have tended to operate implicitly with the last of these usages; but recently the German sociologist Friedhelm Neidhardt has advocated a variant of the second usage, according to which the public is viewed as a particular kind of communication system comprising speakers, media and audiences (Neidhardt, 1993).

The term 'understanding' may be used to suggest intellectual grasp or comprehension; but equally it may be taken to suggest empathy, sympathy or even overt support.

This ambiguity reflects a larger question, namely: are efforts to promote the public understanding of science really professional public relations under another name? (For a historical review of this issue, see Lewenstein (1992); for a conceptual critique, see Wynne (1991).) According to the so-called 'deficit model', science itself is the yardstick by which to measure the relative deficiencies of particular public understandings of science. On this view, members of the public bring nothing significant to their encounter with science except ignorance and (it is to be hoped) the aptitude and the willingness to learn. By contrast, advocates of what might be termed 'contextual' models reject the yardstick of science itself and take much more seriously the informal or lay understandings that may exist far beyond the professional boundaries of the scientific community (see Hilgartner, 1991; Wynne, 1992*a*; Ziman, 1991).

Finally, the term 'science' itself requires clarification. Leaving aside the somewhat pedantic question of what are to count as scientific subjects, it seems important to decide whether the public understanding of science is concerned with scientific facts, scientific methods, or the social and institutional structures of science. The American researcher Jon Miller has advocated a threefold definition of 'scientific literacy' embracing factual knowledge, scientific method(s) of inquiry and the impact of science on society (Miller, 1983); and a number of other commentators have advocated a view of the public understanding of science that places even greater emphasis upon the social and institutional processes as contrasted with the factual and theoretical products of science (see, for example, Shapin, 1992; Collins and Pinch, 1993; Durant, 1993).

These general reflections are relevant to a consideration of the public understanding of the new genetics. Neidhardt's approach to the public as a particular communication system incorporating speakers, media and audiences suggests a view of the public understanding of the new genetics as a conceptual space populated with a great diversity of actors and actor groups: researchers, who are developing new understandings of human genetics; clinicians, who are treating genetic diseases; counsellors, who are advising particular 'at-risk' groups; patients and patients' families, who are receiving treatment or advice; lawyers, who are debating the patenting of novel human DNA sequences; journalists, who are reporting the latest ideas and issues to readers and viewers; special interest groups, who are seeking to influence other actors and actor groups; policy-makers and politicians, who are developing regulatory frameworks for the new human genetics in light of the multiple influence of other actors; and so on. No single approach is likely to do justice to the needs and interests of all these different actors and actor groups.

Recent criticisms of the 'deficit model' suggest that we should think carefully about approaches to the public understanding of the new genetics that focus solely upon the problem of conveying the formal contents of Mendelian and molecular genetics to non-geneticists. Although we may well share the desire to foster what Bodmer and McKie (1994) term 'genetic literacy' among the general populace, this laudable desire should not lead us to neglect the other factors that may militate against the ready compliance of particular audiences with such a project. These factors include the existence of informal or lay understandings of inheritance that may cut across the categories of

formal scientific knowledge, and the existence of public doubt or disquiet about particular clinical applications. Commenting on the limitations of a 'deficit model' approach
to the problems encountered in genetic counselling, for example, Richards (1993) has
recently pointed to the need to take seriously the 'lay beliefs' about inheritance that
patients and patients' families bring with them into the consulting room.

11.3 Public understanding of genetics

Much of the existing research on the public understanding of the new genetics has been
concerned with problems encountered in the clinical context. Here, patients and
patients' families are increasingly likely to be drawn into a direct encounter with
Mendelian and molecular genetics, not only through the clinical processes of diagnosis
and treatment but also through the supportive processes associated with genetic counselling. Research has concentrated on the (mis)understandings that may arise in the
course of treatment and/or counselling, and particularly on the vexed question of risk
perception. Among the commoner topics to have been addressed are: the reception of
relevant scientific information, including probabilistic risk assessments; the uptake of
screening programmes; and attitudes towards new developments in clinical molecular
genetics (see, for example, Lubs and de la Cruz, 1977; Evers-Kiebooms and Van den
Berghe, 1979; Lippman-Hand and Fraser, 1979; Kessler, 1989; Marteau *et al.*, 1989;
Lippman, 1991; Lerman and Croyle, 1993; Richards and Green, 1993).

Useful as these studies undoubtedly are, there are reasons for believing that research
on the public understanding of the new genetics should extend well beyond the professional encounters involved in diagnosis, treatment and counselling. For one thing,
Richards points out that cognitive and cultural processes far removed in space and time
from the clinical setting may help shape patients' conceptions of inheritance; and for
another, whereas in the past human molecular genetics was largely confined to the consideration of single-gene disorders that affect more or less clearly defined risk groups,
today the new genetics is rapidly expanding to embrace multi-factorial illnesses (e.g.
some cancers, and cardiovascular disease) that are of central concern in the field of
public health. Finally, it is important to recall that throughout this century human
genetics has consistently raised larger ethical, legal, social and political questions that
are of widespread public interest and concern. With the Human Genome Project now
well under way, we may expect the new genetics to continue to expand beyond the con-
fines of the clinical context; and as this expansion takes place, so the need to deal with
other and wider publics will be felt more keenly.

At present, the best evidence we have concerning wider public understanding of the
new genetics comes from two sources: random sample social surveys, and quantitative
media (especially newspaper) analysis. In the USA, the General Social Survey of 1990
explored public awareness of and attitudes towards genetic screening (*US General
Social Survey*, 1990): 13% of respondents reported a 'great deal of knowledge' of
screening, whereas 26% reported knowing 'nothing at all' about it; 60% of respondents
expected 'more good than harm' to come from genetic screening, but at the same time

85% of respondents opposed the use of genetic tests in the selection of employees and 75% of respondents supported the principle that sole control over access to the results of genetic tests should rest with the tested individual.

A more recent study of public understanding of biomedical science in the USA reveals relatively low levels of both self-assessed and independently assessed knowledge about molecular genetics (Miller and Pifer, 1993). For example, 24% of respondents stated that they possessed a 'clear understanding' and 35% stated that they possessed a 'general sense' of the meaning of the term DNA; but in response to the follow-up invitation to define the term in their own words, only 20% of respondents were able to provide a minimally correct definition, while a further 21% made more general references to genes and/or chromosomes. On a subsequent, closed response item, 'True or False: DNA regulates inherited characteristics for all plants and animals', 59% opted for the correct answer.

A random-sample survey conducted in Germany in 1993 revealed a pattern of relatively little knowledge about genetic screening coupled with a relatively high level of willingness to undergo genetic testing (Hennen *et al.*, 1993). Before the interview, 42% of respondents expressed a positive attitude towards genetic screening, whereas after the interview, this figure fell to 32%; 73% of respondents opposed the use of genetic tests in the selection of employees, but only 23% favoured prohibiting the use of genetic tests altogether; a slightly larger proportion (26%) thought that people have a duty to have their genetic make-up tested.

A limited amount of international comparative research has been undertaken in this area. For example, in a preliminary comparative analysis of public perceptions of gene therapy in Japan, Europe, the USA and New Zealand, Macer (1992) found a common pattern of growing acceptance of gene therapy coupled with awareness of both benefits and risks. This study provides some evidence that, for example, fears relating to eugenics are more prevalent in some countries (e.g. Germany) than in others (e.g. Japan), but more systematic research is required on this and other possible cultural differences in public perceptions of the new genetics.

The second area of research on wider public understanding of genetics involves media analysis. Kepplinger *et al.* (1991) analysed trends in media coverage of genetics in a wide range of German daily and weekly newspapers and popular science magazines. He found that human genetics accounted for 22% of all genetics coverage, and that much of this coverage featured controversial issues. In general, the use of genetic tests for purposes of employment, eugenic programmes for 'breeding the human race' and germline gene therapy were all portrayed very negatively, whereas prenatal diagnosis and somatic cell gene therapy were portrayed very positively. Ruhrmann *et al.* (1992) have conducted a similar study of coverage of human genome analysis in eight German newspapers.

There are obvious weaknesses in the existing research literature on the public understanding of the new human genetics. First, most of the studies that have been undertaken in the clinical context have been implicitly informed by a deficit model of public understanding (but see Richards (1993) and Chapter 12 for an exception to this rule).

Secondly, relatively few studies have been conducted outside the clinical context. Most work of this kind appears to have been done in the USA and in Germany, but even in those countries the picture is far from complete. Thirdly, as we have already seen there is extremely limited comparative data concerning patterns of cultural (e.g. regional and national) variation in the public understanding of the new genetics. Fourthly, survey studies have generally been conducted in isolation from media studies, and this means that it is difficult to characterise the relationships that may exist between speakers, media and audiences in particular public arenas.

11.4 The Human Genome Project and the British public: a case study

We have recently completed a study of British public understanding of the Human Genome Project in particular, and the new genetics more generally (for a full account of methods and results, see Durant *et al.* (1993)). The study involved parallel analyses of two key elements within a particular public arena: the media (as represented by a range of British daily and Sunday newspapers) and the audience (as represented by a number of focus group interviews conducted in the London area). In tracing parallel public and press discourses, we made no assumptions about causal relationships between the two. Rather, we were concerned to identify similarities and differences between audience and media in one particular place, Britain, at one particular time, 1992.

In undertaking this study, we drew heavily upon social representation theory (Moscovici, 1961; Farr and Moscovici, 1984; Farr, 1987; Jodelet, 1989; Cranach, 1992). This theory emphasises the reconstructive nature of the processes by which new knowledge and/or expert practices become 'common sense' for a particular social group or system. Typically, a (re)presentation is a new form of knowledge that is organised around a limited number of core themes. Within a given theme, the unfamiliar is 'anchored' in the familiar; and at any particular moment, key issues are characterised by means of recurrent images, icons or metaphors. Where a 'lay belief' is an individual phenomenon, a social representation is a phenomenon of the group that reveals itself in recurring patterns of knowledge, belief and expectation.

We explored public discourse about the new human genetics in 12 focus group interviews in the Greater London area, and we investigated press discourse about the same subject by means of traditional quantitative content analysis of articles selected by keywords from six national daily and three national Sunday newspapers (for further methodological details, see Durant *et al.* (1993)). In the focus groups, we found a high level of interest in the new genetics, combined with a low level of awareness of the Human Genome Project itself; similarly, in the broadsheet and the tabloid press we found frequent references to genes, genetics, biotechnology and DNA, but relatively few references to the Project itself. Other similarities (as well as a number of differences) between public and press discourses emerged in the course of analysis, and these are most easily summarised with the help of a simple diagram (Figure 11.1).

Among non-specialist focus group interviewees we found no prior awareness of, or knowledge about, the Human Genome Project at all. However, there was a consider-

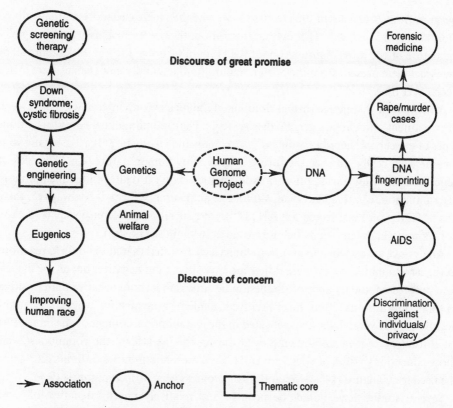

Figure 11.1. The social representation of the Human Genome Project in Britain in 1992. The figure maps the principal associations of the Human Genome Project in Britain, organised in two dimensions so as to illustrate the fundamental tension between the discourses of great promise and concern. The figure maps the core of the social representation of the project around September 1992; however, a more complete map would include a third dimension concerning control/lack of control over the project.

able amount of general background awareness of related subjects, and especially of genetic engineering and DNA fingerprinting, which we regard as core themes for public discourse about the new genetics in Britain at the time the study was undertaken. The core theme of genetic engineering was strongly associated with issues to do with genetic and personal manipulation, and the core theme of DNA fingerprinting was strongly identified with issues to do with genetic and personal identification. Prominent anchors were: (i) for genetic engineering, one or two relatively well-known disorders such as Down syndrome and cystic fibrosis, and eugenics; and (ii) for DNA fingerprinting, rape and murder cases, and (via the issue of discrimination against individuals on grounds of biological/health status) AIDS.

In the newspaper coverage, we found frequent references to the new genetics in general, but far fewer specific references to the Human Genome Project. The suggestion to map the human genome was first mentioned in the British press in 1986, but there were only 10 articles referring to this subject before 1988. The amount of coverage grew from 24 articles in 1988 to 30 in 1990, and then increased more dramatically to 57 in 1991 and 55 in 1992. Throughout this period, there were strikingly few changes in the nature of the press discourse about the Human Genome Project. Essentially all of the issues were present from the outset, and no fundamentally new themes were introduced over time.

A number of distinctive themes dominated the focus group interviews. These were: the treatment and cure of genetic disease (e.g. 'They're finding new cures for various types of cancer and hereditary illness.'); screening and testing (e.g. 'It's like having your fortune told, really; you see the future.'); improving human nature (e.g. 'The idea of genetic engineering started off with Adolf Hitler, didn't it really.'); identification and discrimination (e.g. 'It's like when you fill out an insurance form, it is always on there now: "Have you been tested for AIDS?"'); and social control/conspiracy (e.g. 'You don't know what's happening behind closed doors with scientists.').

Analysis of transcribed focus group interviews revealed that at the outset, reactions to the new human genetics were rather more positive than negative; but as the discussion progressed, the balance of discussion became rather more negative than positive. In total, only around 10% of all evaluative statements were positive, whereas 31% were negative. This overall balance is reflected in the metaphorical content of the interviews. Of a total of 172 metaphors employed during the interviews, the commonest were those concerning issues of social control (22%), moral boundaries or limits (20%) and the need for vigilance (14%). Metaphors of progress were fifth commonest (9%).

Several studies have noted the positive and relatively uncritical nature of news reporting on genetic engineering, genetic fingerprinting and biotechnology as a whole (Nelkin, 1985, 1987; Goodell, 1986; Plein, 1991). However, in our study we found that press reporting of the Human Genome Project in Britain is characterised by a mixture of two discourses: first, enthusiastic accounts of the great promise of this research; and secondly, less sanguine discussions of the many ethical, legal and social concerns that are raised by the project. In Figure 11.1, these contrasting discourses are represented in the vertical dimension. The discourse of great promise dwells on the size, scale and significance of the Human Genome Project, as well as on the potential medical and social benefits that it will bring. From the outset, the Project is described as 'big science', 'pioneering science', or 'frontier science'. It is variously portrayed as the 'Holy Grail of biology', as the biggest science project since the moon shot, and as the biological equivalent of the development of the atom bomb.

The discourse of concern encompasses a cluster of issues, of which the most prominent are: the safeguarding of genetic information against potential misuse by employers, educational establishments, insurers and government agencies; discrimination on the grounds of genetic information in employment, insurance and education – leading to the creation of a 'genetic underclass'; the resurgence of eugenic practices of social and racial 'hygiene'; selection of people with resistance properties that would allow them to work in high-risk industries; 'made-to-order babies' and 'designer children', together with associated genetic vetting of marriage partners; and, much less prominently, concerns about scientists 'playing God' and 'tampering with nature'. A simple indication of the prominence of the 'ethical concerns' discourse is the finding that during the four-year period 1989–92 just over a third of articles mentioning the genome also made reference to 'ethical' or 'moral' issues.

The vertical dimension in Figure 11.1 represents the fact that both public and press display considerable ambivalence about the new genetics. In the focus group interviews, as we have already noted, the discourse of concern tends to prevail, at least in quantitative terms; but in the press coverage, the reverse is the case. Only a small number of press articles mention potential risks, safety questions, doubts over the precision of genetic manipulation, or other issues that point to potential problems within the science itself, and many of the references to moral problems are formulaic and routine. For example, a highly enthusiastic account of the great promise of genome research may be followed by a brief comment to the effect that such research also raises ethical problems such as the use of genetic information to discriminate against individuals in employment and insurance.

To a greater extent than the broadsheet newspapers, the tabloid newspapers report genetics and genetic research within the discourse of great promise. Thus, on 8 May 1992 the *Daily Mirror* reported: 'Farmers hope new VIP will help save their bacon: Superpig! She is the shape of things to come . . . a true aristocrat'. Just a few days later, the *Daily Mail* reported on 'Pure genius: the long life tomato'. Although the discourse of great promise clearly predominates, the discourse of concern is not entirely absent even from the tabloid press. Three of the 96 articles identified in the tabloid press had as their main focus moral and social problems associated with the new human genetics. On 13 May 1992, the *Daily Mail* asked its readers, 'What price will we pay for these medical miracles?'; and a few weeks later, the same newspaper announced, 'Doctors' warning over gene screening at work.' Similarly, on 12 May 1992 the *Daily Mail* reported: 'Your baby is going to be homosexual so we'll abort him: scientists' nightmare vision of DNA revolution.'

There were very few images of 'run-away' science, or of scientists as weird and sinister madmen. Two articles from our sample that presented such images were actually concerned with the television play *The Cloning of Joanna May* by authoress Fay Weldon. However, these were exceptions to the more general rule that tabloid newspaper coverage of genetics or DNA is predominantly a discourse of hope and promise. Where the coverage is concerned with forensic investigations of rape, murder or paternity, genetic fingerprinting is portrayed as a unique, ingenious and infallible technique; and where it is concerned with health and illness, genetics is portrayed predominantly in the context of spectacular breakthroughs, offering hope (a recurring expression) for sufferers from disease and holding out the promise of cures for common afflictions like asthma and dreaded diseases like cancer and AIDS.

Figure 11.1 captures many of the key features of the social representation of the new genetics in Britain in 1992. In other words, the figure portrays the principal themes, associations and anchors that are present in public and press discourse about the new genetics in a particular public arena at a particular point in time. Within this social representation, the single largest and (by most professional accounts) significant scientific endeavour in the field – the Human Genome Project – is only, as it were, 'virtually present': it exists principally through a network of associations with the adjacent and better-known fields of genetic engineering and genetic fingerprinting. The lack of prominence of the Human Genome Project within the social representation appears to reflect a relatively large amount of public and press interest in immediate practical applications, and a relatively small amount of public and press interest in longer-term strategic developments.

Thus far, we have mapped key features of the social representation of the new genetics in two dimensions: manipulation/identification, and promise/concern. Our data suggest that a more complete map or model of the social representation of the new genetics will require a third, orthogonal dimension having to do with control/lack of control. This third dimension is required to take proper account of the substantial concerns that were expressed in both the public and (to a lesser extent) the press discourses about the extent to which the new genetics is adequately regulated in the public interest. Metaphors used to convey such concerns included images of exclusion (e.g. 'they're keeping us in the dark', 'I think it is kept under wraps') and expressions of a process that is running out of control (e.g. 'sort of galloping along out of control', and 'outpacing the monitoring that is done').

Figure 11.2 provides the outlines of a model of the social representation of the new genetics that incorporates these three dimensions: manipulation/identification, promise/concern, and control/lack of control. At first sight, this model may appear overly abstract as a basis for analysing public understanding of the new genetics. Developed more fully, however, we believe that such a model would prove extremely useful, particularly for purposes of comparison, because it provides a theoretical three-dimensional space within which particular social representations – particular public understandings – may be characterised and compared. In our view, something of this sort will be required if shifting historical and cultural patterns of the public understanding of the new genetics are to be the subject of systematic comparative analysis.

11.5 Conclusion

In this chapter we have argued that there is no single, uniquely valid approach to the question of the public understanding of science in general, and the new genetics in particular. There are several significantly different ways of characterising 'the problem' of public understanding, and each characterisation generates different questions and (hence) different agendas for research. It is perfectly valid to ask about the extent to which the general populace is familiar with at least the rudiments of Mendelian and molecular genetics; however, at the same time we have given reasons for supposing that it is equally valid (and at least as important, practically speaking) to enquire about the extent to which professional and popular understandings of these subjects may be systematically different from – as opposed to simply better or worse than – one another.

The deficit model informs the great majority of the research that has been undertaken in this area to date. While this model may reveal significant absences (ignorance)

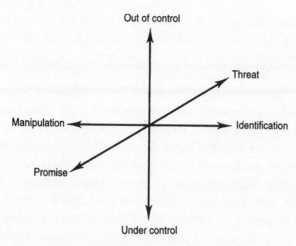

Figure 11.2. An outline three-dimensional model of the social representation of the new human genetics.

in the public domain, it is less good at revealing significant presences (informal or lay beliefs) whose existence may undermine the efforts of clinicians, counsellors and educators to improve public understanding. At the same time, the deficit model neglects the possibility that rich and relatively highly structured discourse about the new genetics may exist in the public domain in the large-scale absence of formal scientific knowledge. We have argued that analysis of social representations of the new genetics offers one way of exploring this discourse more meaningfully; this, in the end, may prove useful for practical as well as research purposes.

Given the increasing scientific and medical importance of the new genetics, the scope for further research on public understanding of this area is very great. In addition to better accounts of the history of genetics – and especially of the history of popular genetics – we need more detailed studies of particular arenas for the public understanding of genetics. Ideally, such studies should incorporate accounts of how speakers, journalists and audiences view one another. We need to know not only what particular publics think about geneticists but also what particular geneticists think about the public; for here, as elsewhere in the world of relationships, the presumptions and expectations of both parties shape the outcome of particular encounters.

There is scope for detailed studies of the way in which particular developments in human genetics are presented to, and received by, particular audiences. Increasingly, research 'breakthroughs' in this field are the subject of high-profile news and features coverage in the media. By following the course of specific research work in the public domain – from forecasts and 'previews', through immediate news reports, to wider discussions and debates – it will be possible to chart the reconstructive processes by which new knowledge becomes incorporated into the 'common sense' representations of particular social groups. In this way, empirical flesh can be put on the theoretical bones of the notions of lay beliefs and social representations.

At present there are very few systematic comparative studies of the public understanding of the new genetics. Richards (1993) has set out an agenda for research that extends well beyond the immediate setting of the doctor–patient or counsellor–patient relation, and embraces the nature of lay beliefs about inheritance, the nature of communication within families about inheritance, and the cultural differences in understanding of, and attitudes towards, inherited disease. Although much may be learnt in this area from detailed studies of particular social groups, much more may be learnt by comparisons between different social groups. Here, key questions concern the nature of particular cultural/historical assumptions about the nature of inheritance, and the influence of such assumptions on the reception of medical information and advice.

Finally, it is worth noting the increasing interest in recent years in the subject of ignorance. On the deficit model, ignorance tends to be viewed as a negative or passive phenomenon: it is the result of neglect on the part of the scientific and the educational communities, and/or lack of interest on the part of the public. However, a number of sociologists have suggested the usefulness of viewing ignorance as a positive or active phenomenon: the result, perhaps, of particular social groups' disinclination to become involved with particular sorts of information, or even their dislike of taking on some of the responsibilities that attach to particular forms of knowledge (see, for example, Smithson, 1985; Wynne, 1992b). Ultimately, investigations of public rejection of scientific knowledge may prove every bit as illuminating as investigations of public acceptance of it.

The main consequence of embedding public understanding of the new genetics within a contextual rather than a deficit model of the public understanding of science is that it points to the need for a far more subtle and textured characterisation of the interrelation between speakers, media and audiences. On their own, simple statistics about the numbers of people who can accurately define the nature and function of DNA fail to do justice to the extent to which DNA has become part of public discourse in recent years – a routine point of reference, not only in newspaper science coverage but also in television plays (e.g. *The Cloning of Joanna May*) and Hollywood films (e.g. *Jurassic Park*). Today, DNA is not merely the name of the genetic material, it is also the name of a perfume. It is only by taking account of the shifts in meaning, in connotation and in significance that are involved in the transformation of molecular genetics into mass culture that we shall do justice to the public understanding of the new genetics.

11.6 References

Bodmer, W. and McKie, R. (1994) *The Book of Man*. London: Little, Brown.
Collins, H. and Pinch, T. (1993) *The Golem. What Everyone Should Know about Science*. Cambridge: Cambridge University Press.
Cranach, M. (1992) The multi-level organisation of knowledge and action – an integration of complexity. In M. Cranach, W. Doise and G. Mugny (eds), *Social Representations and the Social Basis of Knowledge (Swiss Monographs in Psychology)*. Lewiston, New York: Hogrefe & Huber.
Durant, J. (1993) What is scientific literacy? In J. Durant and J. Gregory (eds), *Science and Culture in Europe*. London: Science Museum.

Durant, J., Evans, G. and Thomas, G. (1989) Public understanding of science. *Nature*, **340**, 11–14.

Durant, J., Evans, G. and Thomas, G. (1992) Public understanding of science in Britain: the role of medicine in the popular representation of science. *Public Understanding of Science*, **1**, 161–82.

Durant, J., Hansen, A., Bauer, M. and Gosling, A. (1993) *The Human Genome Project and the British Public: A Report to the European Commission.* London: Science Museum.

Evers-Kiebooms, G. and Van den Berghe, H. (1979) Impact of genetic counselling: a review of published follow-up studies. *Clinical Genetics*, **15**, 465–74.

Farr, R. (1987) Social representations: a French tradition of research. *Journal for the Theory of Social Behaviour*, **17**, 343–70.

Farr, R. and Moscovici, S. (eds) (1984) *Social Representations.* Cambridge: Cambridge University Press.

Goodell, R. (1986) How to kill a controversy: the case of recombinant DNA. In S. Friedman, S. Dunwoody and C. Rogers (eds), *Scientists and Journalists: Reporting Science as News.* New York: The Free Press.

Hennen, L., Petermann, T. and Schmitt, J. (1993) *TA-Projekt 'Genomanalyse' – Chancen und Risiken Genetischer Diagnostik. (TAB-Arbeitsbericht* Nr 18) Bonn: TAB.

Hilgartner, S. (1990) The dominant view of popularisation: conceptual problems, political uses. *Social Studies of Science*, **20**, 519–39.

Jodelet, D. (1989) *Folies et Representations Sociales.* Paris: Presses Universitaires de France.

Kepplinger, H., Ehmig, S. and Ahlheim, C. (1991) *Gentechnik im Widerstreit, Zum Verhaltnis von Wissenschaft und Journalismus.* Frankfurt: Campus.

Kessler, S. (1989) Psychological aspects of genetic counselling. VI. A critical review of the literature dealing with education and reproduction. *American Journal of Medical Genetics*, **34**, 340–53.

Lerman, C. and Croyle, R. (1993) Psychological impact of genetic screening. In R. Croyle (ed.), *Psychosocial Effects of Screening for Disease Prevention and Detection.* New York: Oxford University Press.

Lewenstein, B. (1992) The meaning of 'public understanding of science' in the United States after World War II. *Public Understanding of Science*, **1**, 45–68.

Lippman, A. (1991) Research studies in applied human genetics: a quantitative analysis and critical review of recent literature. *American Journal of Medical Genetics*, **41**, 105–11.

Lippman-Hand, A. and Fraser, F. (1979) Genetic counselling: provision and reception of information. *American Journal of Medical Genetics*, **3**, 113–27.

Lubs, H. and de la Cruz, F. (eds) (1977) *Genetic Counseling.* New York: Raven Press.

Macer, D. (1992) Public acceptance of human gene therapy and perceptions of human genetic manipulation. *Human Gene Therapy*, **3**, 511–18.

Marteau, T. M., Johnston, M., Shaw, R. W. and Slack, J. (1989) Factors influencing the uptake of screening for neural tube defects and amniocentesis to test for Down's Syndrome. *British Journal of Obstetrics and Gynaecology*, **96**, 739–41.

Miller, J. D. (1983) Scientific literacy: a conceptual and empirical review. *Daedalus*, Spring, 29–48.

Miller, J. and Pifer, L. (1993) *Public Understanding of Biomedical Science in the US.* Chicago: Chicago Academy of Sciences.

Moscovici, S. (1961) *La Psychoanalyse: Son Image et Son Public.* Paris: Presses Universitaires de France.

Neidhardt, F. (1993) The public as a communication system. *Public Understanding of Science*, **2**, 339–50.

Nelkin, D. (1985) Managing biomedical news. *Social Research*, **52**, 625–46.

Nelkin, D. (1987) *Selling Science: How the Press Covers Science and Technology*. New York: W. H. Freeman & Co.

Plein, L. (1991) Popularising biotechnology: the influence of issue definition. *Science, Technology, and Human Values*, **16**, 474–90.

Richards, M. P. M. (1993) The new genetics: some issues for social scientists. *Sociology of Health and Illness*, **15**, 567–86.

Richards, M. P. M. and Green, J. (1993) Attitudes towards prenatal screening for fetal abnormality and detection of carriers of genetic disease: a discussion paper. *Journal of Reproductive and Infant Psychology*, **7**, 171–85.

Royal Society (1985) *The Public Understanding of Science*. London: The Royal Society.

Ruhrmann, G., Stoeckle, T., Kraemer, F. and Peter, C. (1992) *Das Bild der Biotechnischen Sicherheit und der Genomanalyse in der Deutschen Tagespresse, 1988–1990. (TAB-Diskussionspapier no. 2.)* Bonn: TAB.

Schiele, B., Amyot, M. and Benoit, C. (1994) *When Science Becomes Culture: World Survey of Scientific Culture*. Boucherville Quebec: University of Ottawa Press.

Shapin, S. (1992) Why the public ought to understand science-in-the-making. *Public Understanding of Science*, **1**, 27–30.

Smithson, M. (1985) Toward a social theory of ignorance. *Journal for the Theory of Social Behaviour*, **15**, 151–72.

US General Social Survey (1990) Chicago: National Opinion Research Centre.

Wilkie, T. (1993) *Perilous Knowledge. The Human Genome Project and its Implications*. London: Faber & Faber.

Wynne, B. (1991) Knowledges in context. *Science, Technology and Human Values*, **16**, 111–21.

Wynne, B. (1992a) Public understanding of science research: new horizons or hall of mirrors? *Public Understanding of Science*, **1**, 37–43.

Wynne, B. (1992b) Misunderstood misunderstanding: social identities and public uptake of science. *Public Understanding of Science*, **1**, 281–304.

Ziman, J. (1991) Public understanding of science. *Science, Technology, and Human Values*, **16**, 99–105.

12
Families, kinship and genetics

MARTIN RICHARDS

I am the family face;
Flesh perishes, I live on,
Projecting trait and trace
Through time to times anon,
And leaping from place to place
Over oblivion. *Thomas Hardy, 1840–1928:* Heredity.

12.1 Introduction

Genetics concerns families and kinship. It is the study of the ways in which heritable characteristics and conditions are passed from parents to children through the generations. As the new genetics develops, the possibilities of describing the gene mutations that an individual may carry and may pass to children are increasing dramatically. This new knowledge and the ways in which it is employed may have profound consequences for family life and relationships.

The public's knowledge and beliefs about inheritance have not arisen *de novo* with the coming of the new genetics, or even with Mendelian genetics at the turn of the century: they have long been part of family culture. Much family talk is about particular characteristics of family members, who these may have been acquired from, and who they may be passed to. Witnessing family members greet a new baby demonstrates the important process of 'placing' the new baby in terms of characteristics shared with forebears. Increasingly, we have photographs and other visual evidence of the appearance of our forebears and we use these to point to similarities and differences. In my own family I am said to get my nose from my mother's family but have a temperament more like some of my father's male relatives. One of my sisters, however, is said to be very similar in appearance to one of the great aunts on my father's side of whom we have a rather faded portrait. Such knowledge about family resemblances at least implies a notion of a process of biological inheritance. I shall argue that such notions may have important consequences because scientific accounts of inheritance that may be provided at school or, for some, in the genetic clinic, will need to be assimilated with whatever pre-existing lay knowledge an individual may hold. Communication about matters of inheritance within families is of particular importance for those in families who carry a recognised genetic disorder. As can be seen from the personal accounts in Chapter 1, some may learn at quite a young age from their parents of risks of developing a serious genetic condition. For others this news may first come during a professional encounter. In the genetic clinic the basic investigative tool, even in the age of the new genetics, is the pedigree – the family tree constructed from a person's knowledge of their family relations and the conditions each may or may not have had.

As yet, very little research has been done on genetic disorders, or the use of genetic services, from a family perspective. There is no very substantial body of work that can be drawn upon for a chapter of this kind. Rather, it is a matter of bringing together evidence and ideas from a diversity of sources. Inevitably at times I shall be speculative, but I hope it will be clear where the speculation begins and ends. My aim is to provide a geography of a relatively newly discovered continent; I shall try to draw an outline map even if details of some of the coastline are still vague.

In this chapter I shall begin by clarifying some terms needed for a discussion of family relationships, genetic disorders and knowledge about inheritance.

Family life varies widely with ethnicity (see Chapter 14, by Hannah Bradby). Unless statements I make are qualified, they should be taken to apply to what has become known as the Euro-American family. The application of this term is not well defined. I take it to apply to English families but not necessarily to families of ethnic minorities living in the UK. Many, but not all, North American families of European descent would fall within the category, as would those in Australia and New Zealand. But to discuss the full range of diversity is beyond the scope of this chapter. However, it is an important and urgent issue in the age of the new genetics to increase our knowledge of the diversity of family life and to be mindful of this when dealing with genetic issues.

12.2 Concepts of the family

In our culture we use the term 'family' in multiple and potentially confusing ways. For the discussion in this chapter we need to be clear about what we mean by the term. Consider the following:

> Will you be seeing your family at Christmas?
> They have recently married and don't yet have a family.
> On which side of the church is the bride's family sitting?
> My family originally came from Wales.
> It's a family car.
> He lodged with my parents for many years and we thought of him as a member of the family.
> The family in Britain is on its last legs.
> He is family really. He has been going out with my sister for years.
> She grew up in a single-parent family.

The list could be continued, further illustrating the diversity of meanings. For our present purpose we need to notice that 'family' can mean those recognised as being connected by 'blood'[1] or affinity, or simply those who are not related in this way but who share the same household. In a slightly different sense it can mean the line of descent (or house). There is an implication that members of a family are regarded in a different way from those who are not family. But the latter may acquire the status of family membership by marriage, through a relationship or simply by long co-residence. Family membership implies rights and obligations (see Finch, 1989).

It might seem that the 'blood' part of the definition of family is most easily specified. Indeed, many geneticists (and others) speak as if shared DNA sequences were the definition of these family relationships – parent and child, grandparent and so on. However, in the lived-in social world, the definition is a social one that usually presumes a biological connection but is not determined by this. In some situations, including a genetic clinic, the discovery that a blood relation, such as a father, is not biologically connected can cause profound changes in family relationships. But for others the lack of biological connectedness is known and accepted and does not determine the relationsip. This, for example, is acknowledged in law where a child of a marriage is not defined in solely biological terms. Conversely, the existence of a biological link does not always ensure a family relationship: a child born to a parent outside a relationship needs to be socially 'recognised' before he or she becomes part of a family.

Another implied connotation of 'family' in the statements I quoted above is one about values. A family car is one thought appropriate for parents and children to use, as opposed to sports cars, for example. Similarly, a family butcher is one you might go to for the Sunday joint rather than a shop aimed at caterers. These kinds of usage warn us of the emotional and ideological baggage that the term carries.

For present purposes it is useful to use three terms: kinship, close family and household. Kinship may be defined as a set of beliefs, values and categories that structure relationships and social perceptions and actions (Harris, 1990). As I have noted already, kinship is not a map of reproductive relationships, rather it reflects popular (and often legal) concepts of reproduction and family relationships. Our kin are the wider family that we recognise: grandparents, uncles and aunts and cousins. English kinship is described as bilateral, meaning that we recognise similar relatives on both sides of the family and the two sides have equal status. Most kin terms are the same for each side of the family. However, because it is most often women who retain and pass on information about kin, the knowledge of kin may be more extensive, and the relationship with them closer, where the link is via a female relative. A bilateral kinship system parallels a pedigree in genetic terms where there is an equal contribution of genetic material from each parent. A simple prediction that follows from this is that in societies with bilateral kinship it should be easier to accept and understand a scientific account of inheritance (see Strathern, 1992) than in societies where the kinship system awards different roles and importance to the two sides of a family. This is a point we shall return to. On the whole, the kin that are generally known and recognised are those for whom we have kin

[1] The use of the term 'blood' for the genetic connection can be traced back at least to Aristotelian biology. Despite the point that few believe that in a literal sense blood is exchanged between generations, the term remains in wide use.

names. Many people will be able to name their grandparents and will include them as kin even if they have never met them. But typically they know a lot less about brothers and sisters of grandparents and often very little about their descendants.

We should note that the English system is fuzzy around the edges. There are distant cousins who may or may not be considered members of the family – in American terminology, are they 'kissing cousins'? An aunt may divorce and, particularly if her ex-partner has not developed strong relationships with other family members, he may effectively cease to be a family member after the divorce. But children will often make the difference. A son-in-law or daughter-in-law is less likely to be written out of the family after a divorce if the marriage has produced children.

Divorce, and the cohabitation that often follows it, have certainly added to the complexities of defining a kinship (see Finch, 1994). If the ex-son-in-law remarries and has further children, those children may develop a relationship with their half-siblings. This may draw their mother (their father's new partner) into the network.

In recent years, partly as the result of concerns related to high divorce rates, some research has been undertaken asking children to define their families (see, for example, Gilby and Pederson, 1982; Funder, 1991). Such work suggests that children may be less constrained by cultural notions of kinship and more influenced by the nature of their social relationships, so that they may include family friends and household pets among family members. A similar shift has been indicated in a recent Dutch study of inheritance, which shows that those in close relationships are increasingly benefiting in wills at the expense of kin members (de Regt, 1994). However, there are very few recent studies of inheritance (of property) in England (but see Finch and Hayes, 1994).

We may contrast the notion of kin relations with that of the close family. This may be defined as kin with whom it is usual to co-reside, at least for part of the life cycle: your own parents, your siblings and your children and, to a degree, the cohabiting partners of these people. Whereas the kin system is permissive – you can decide whether or not to activate a relationship, to meet your cousins, invite them to your wedding and so on – close family are those with whom you will have emotionally charged relationships and shared experience. Close family relationships are usually enduring whether or not you remain in close contact, and distance need not be an impediment to their maintenance. For kin it is rather different; face-to-face contact is required to establish a social relationship, though not subsequently to maintain it (see Firth *et al.* (1970), Wilson and Pahl (1988), Finch (1989), Finch and Mason (1992) and Strathern (1992) for studies of English kinship, and Schneider (1968) for those in the USA of European descent).

The final concept we need to discuss is the household. Usually household members will be a subset of the close family: parents with non-adult children, a couple with an aged parent and so on. But given relatively high rates of divorce, remarriage, marriage and cohabitation, household structures and the relationships between them and within the close family can be very complex. Sometimes there is regular movement of family members between households and so the household should not necessarily be seen as a fixed entity. Our household system may be termed nuclear. This refers not to household composition but to household formation (Harris, 1990). At marriage or cohabitation a couple will aspire to setting up a new household, if it is economically feasible.

12.3 Kin and biological relationships

Technological developments in reproductive medicine allow both new disjunctions
between biological and kin relations and new possibilities for defining biological relationships. Now a woman can give birth to a child with whom she shares no genetic
material, or one who is her grandchild, or her niece or nephew. A child may have more
than one mother, as defined in traditional terms. At the same time the new genetics has
provided simpler and more certain techiques for determining the genetic relationships
of individuals. These are already being deployed to settle disputes about paternity, as is
attested by the advertisements for DNA fingerprinting that appear in legal journals (see
Figure 12.1). Although the identity of a child's mother is usually obvious at birth, for
the first time the new genetics provides a straightforward means of determining paternity. Some social theorists have long held that many cultural and family arrangements
are the results of fathers' needs to reduce uncertainty about paternity. If these interpretations are correct, a simple means for determining paternity may lead to changes in
such arrangements.

In drawing up a pedigree in a genetic clinic, the geneticist infers biological relationships
from kin relations. Even in the pre-DNA days, geneticists became familiar with the situation where the genetic data did not fit with the social relationship. Non-paternity, as geneticists term it, may now be revealed more directly in situations where DNA from family
members is being analysed for linkage testing. Clearly some tact and sensitivity may be
needed when such situations arise in genetic counselling. Usually the geneticist shelters
behind a supposed technical problem in the DNA techniques as a way of not disclosing the
lack of a biological relationship. Interestingly, as Macintyre and Sooman (1991) have discussed, there is a widespread tendency to suppose that non-paternity is much more
common than the evidence from genetic clinics suggests. Indeed, there are stories of supposed scientific studies that are said to reveal very high rates of non-paternity that circulate
in the scientific community in the same manner as urban myths of headless hitch-hikers
and driverless cars. In fact, there is little evidence about rates of non-paternity from genetic
clinics (usually assumed to be about 5–10%). It may well be that those who know that for
some of their family relationships the social and the biological do not fit will avoid consulting geneticists or may refuse to take part in linkage testing.[2]

12.4 Collecting family histories

As already mentioned, drawing up a family tree remains the basic technique of the clinical geneticist. The pedigree, as it is usually known, sets out a family tree, usually for at
least three generations. The convention in drawing these has been taken from the
writers of school textbooks and family historians with a series of lines moving across
and down the page representing the links between parents and children and between
siblings (see Figure 12.2). Although such conventions are widely followed in the com-

---

[2] For an imaginative clinical response to non-paternity in a family with Huntington's disease see Klawans
(1988).

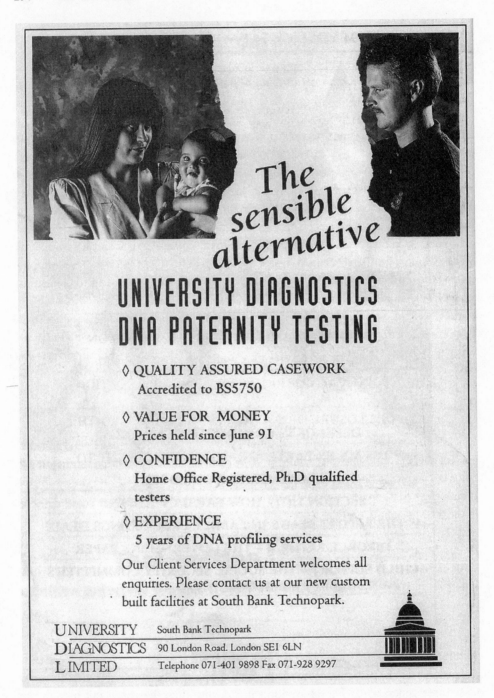

Figure 12.1. Examples of advertisements from the legal press for commercial services offering DNA profiling to aid in the establishment of family relationships. Reproduced courtesy of University Diagnostics Limited and Cellmark Diagnostics.

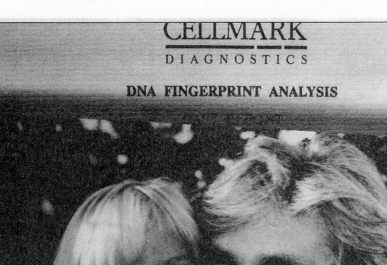

CELLMARK
DIAGNOSTICS

DNA FINGERPRINT ANALYSIS

The DNA fingerprint of Mr John Brook contains all 13 of ......

It is therefore one thousand million times more likely that Mr John Brook is the true father of Sarah Brook rather than being unrelated.

That settles it.

Definitive paternity testing from Cellmark Diagnostics

Cellmark Diagnostics, Blacklands Way, Abingdon, Oxfordshire OX14 1DY.
Telephone: Abingdon (0235) 528609. Fax: (0235) 528141

The above names are fictitious.

Family ref no. | Name of propositus | Diagnosis | Primary community worker, health visitor or social worker

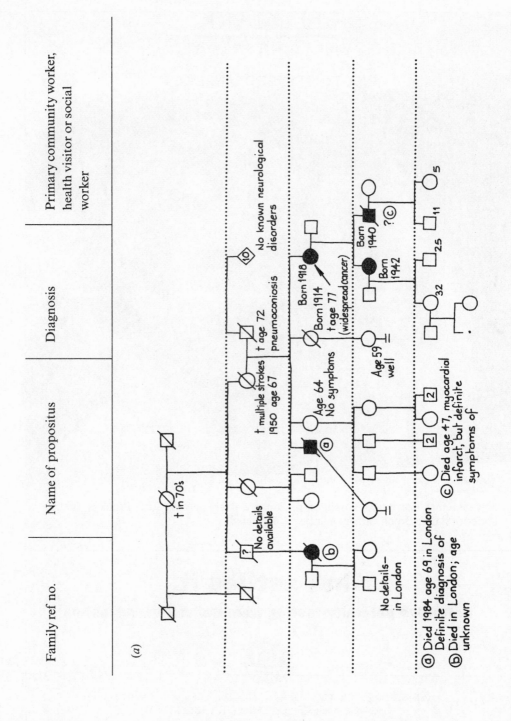

(a)

(a) Died 1984 age 69 in London
    Definite diagnosis of
(b) Died in London; age
    unknown

(c) Died age 47, myocardial
    infarct, but definite
    symptoms of

No known neurological
disorders

† age 72
pneumoconiosis

Born 1918
Born 1914
† age 77
(widespread cancer)

Age 59
well

Born 1940
?(c)

Born 1942

5

11

25

32

† multiple strokes
1950 age 67

Age 64
No symptoms

No details
available

† in 70's

No details—
in London

?

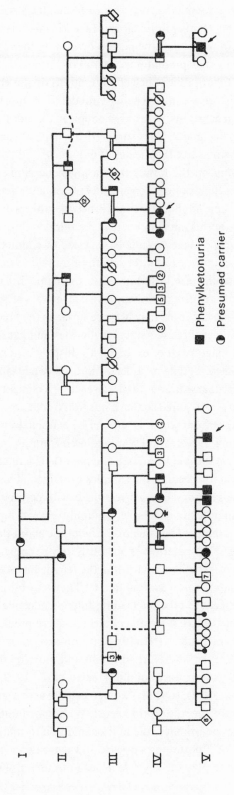

Figure 12.2. Examples of family trees drawn up by a clinical geneticist. (*a*) A 'working pedigree' showing how information on a family is recorded in a genetic counselling session. (*b*) Working pedigrees may be redrawn, often with the help of a computer program that will draw it out with conventional symbols. This figure shows a large 'inbred Welsh gypsy kindred' who carry phenylketonuria. The filled symbols indicate affected individuals, and half-filled symbols carriers of this recessively inherited disorder. Marriages indicated by a double rather than a single line are between related individuals. The individuals marked with an arrow are the 'proband', 'propositus' or 'propositi', i.e. the individual who has attended a genetic clinic and has been the primary source of information about the family. Both figures are from *Practical Genetic Counselling*, 4th edn., by Peter Harper (published by Butterworth/Heinemann), with permission; (*a*) redrawn.

munity and, indeed, clinic attenders may arrive with such trees sketched out on paper, not everyone presents their family in this form. Like kinship terms themselves which define family members in relation to the person describing them, some people, when asked to draw their family tree, present it as a series of lines radiating out from themselves at the centre (see Figure 12.3). Such configurations do not emphasise descent as do the conventional historian's and geneticist model. The most commonly drawn family tree covers three generations, with only a couple in the oldest of these (siblings of grandparents are seldom included). The characteristic shape of such a pedigree has led to its being called a Christmas tree (Schneider, 1968).

The process of drawing up the family tree in a genetic clinic can have profound effects on family relationships, quite independently of whatever genetic interpretations may follow. Sometimes clinic attenders are forwarned that they will need to bring information about the family, or this information may be collected through an initial visit or telephone call from a genetic nurse. An intensive period of contacting members of the family may follow with enquiries about the health of individuals, new births or about branches of the kinship who are little known to the clinic attender. There may be renewed contact with 'lost' relatives, or it may revive memories of severed relationships and family rifts. In the clinic, people may become upset as they recount details of the illness and death of relatives. Precise details of illnesses and causes of death may be required for the genetic diagnosis. For example, if a history of a potentially inherited cancer is being investigated, details of a dead mother's condition may be needed to establish the site of the primary tumour in the absence of a clinical record; or the place of death of a relative may be required so that a hospital can be contacted to see if a histology specimen still exists from which DNA can be extracted for linkage testing. In order to do linkage testing, DNA samples are required from several kinship members who have and do not have the condition (see Chapter 2). The relatives will have to be contacted and the reason for needing the samples explained. This may be done by a family member or by the geneticist or a genetic nurse, with permission from the family member. The process can raise not only issues of family relationships but also ethics, as it can result in a kinship member learning for the first time about their own risk of carrying a genetic condition. Non-paternity is obviously another potential problem.

In almost all aspects of genetics activity, it is the female members of a family who most often take the initiating role (see Chapter 16). They may initiate the contact with the clinic and they are likely to take the leading role in collecting information about their kinship and that of their partner. In conditions where predictive testing for late onset conditions is possible, such as Huntington's disease, more women seek counselling (see, for example, Tyler and Harper, 1983), and more of them opt for testing than male family members (see, for example, Tibbens *et al.*, 1990; Evers-Kieboom *et al.*, 1991; Holloway *et al.*, 1994). In short, women usually act as genetic housekeepers for the kinship. It is part of the role of kin-keeper. Within the family itself, women are often the ones who take the primary role in exchanging information about a genetic condition. In one study of Huntington's disease, it was women who had married into the families who seemed to play the key role in such discussions (Shakespeare, 1992).

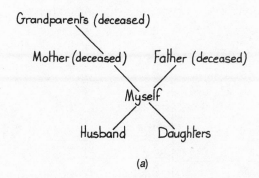

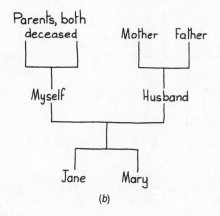

(a)

(b)

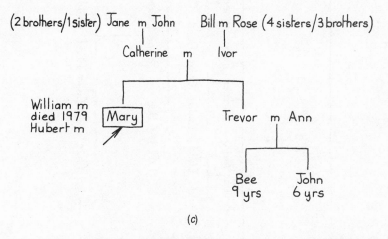

(c)

Figure 12.3. (*a–c*) Family trees drawn as part of a research study by family members asked to 'draw a family tree which includes everybody who is a member of your family'. Names have been changed to preserve anonymity. From an unpublished study of members of families that carry genetic disorders, Centre for Family Research, Cambridge. (*d*) A typical 'Christmas tree' drawn by a respondent asked to draw her family tree. Note which family members are designated by name (these have been changed to preserve confidentiality) and which simply by relationship. In this family tree the respondent, Jenny, chose not to portray her partner's family. (Overleaf.)

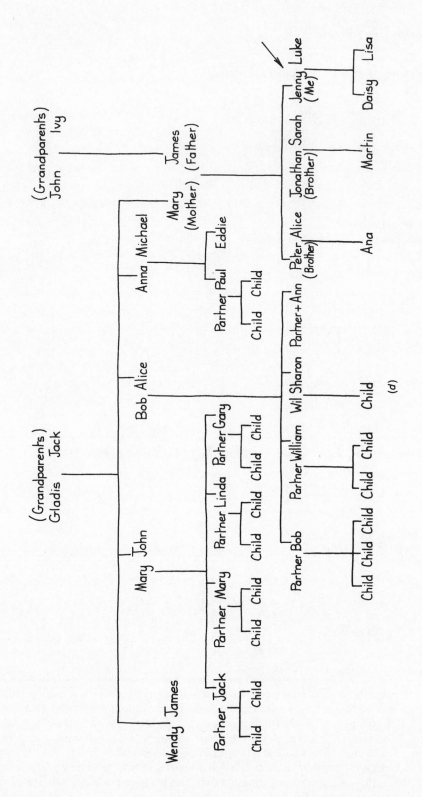

Figure 12.3 (cont.)

Even in diseases that affect only men, such as prostate or testicular cancer, there are anecdotal reports of women retaining this role of genetic housekeepers. Where a disease is female and inherited, such as some cases of ovarian and breast cancer, the emphasis on female members of the kinship is such that both the professionals involved and lay family members may fail to consider kin members who might have inherited the gene mutation via a father. Similarly, the family histories of clinic attenders from families with these cancers suggest a bias towards over-representation of those who inherit via the female line.

It is widely assumed that it is men rather than women who have the greatest concern about a genetic link with their children. However, the little evidence we have about this issue suggests that this may not always be true. In a study of couples who were both carriers of a recessively inherited disorder and who in most cases had produced an affected child, when asked about the possible use of egg or sperm donation to avoid further births of affected children, it was the women who were more likely to give the lack of a genetic connection as a disadvantage of the technique (Snowdon and Green, 1994). However, care must be exercised in the interpretation of findings such as these. Here women may be adopting the role of emotional housekeeper. It may be their concern that the lack of a genetic connection would worry their partner and so have negative effects on their relationship that leads them to see a disadvantage. This same study also indicates that the notion of a genetic balance may be important to some couples. Overall, couples preferred an option where neither spouse had a genetic connection with a child – adoption – to those such as egg or sperm donation where only one of the parents shared genetic material with the child.

The extent to which family members are aware of conditions that run in the family (the phrase they most often use) depends not only on communication within the kinship and the nature of the condition but also the pattern of inheritance. In recessively inherited conditions where sibships are relatively small, as has been true in recent generations in England, there may be too few cases to make up a pattern that family members may recognise. This is even more likely to be the case if the condition is variable in its symptoms and has been poorly diagnosed in the past. Thus many couples are quite unaware that they or their families carry cystic fibrosis mutations until a child is born with the condition. Dominantly inherited conditions are much more obvious in a kinship, and family members are frequently all too well aware that members of the kinship are 'prone' to conditions like the inherited forms of breast and ovarian cancer (see Chapter 1 for examples of this). Although it is only over the last couple of decades that clinicians have recognised that dominantly inherited gene mutations are associated with some cases of these cancers, some family members report that past members of their families had realised for themselves that there was an inherited risk but had failed to find any clinicians who would take this idea seriously. Perhaps the fact that these cancers almost exclusively affect women and that women tend to be the kin-keepers has meant that they are likely to recognise patterns of inheritance so that these may become part of a family culture without any professional knowledge or advice necessarily being involved.

12.5 Marriage and childbearing

Choice of marriage partner is not of course a random matter. Numerous studies have demonstrated the phenomenon of assortative mating in which spouses tend to resemble one another more closely than members of the population at large in characteristics such as educational attainment, IQ, social class and height (see, for example, Harrison *et al.*, 1976). The norm in England is to seek a partner of a similar background, while, in general, relatives are avoided as partners, and of course marriage to close relatives is not legally sanctioned or permitted by the Church. Despite this, cousin marriage occurs and may be relatively frequent among wealthy families, where it can serve to preserve or enhance family fortunes. But empirical studies of consanguinity in English marriage are lacking (however, see Galton, 1869). It is of course of significance from a genetic point of view when a recessive disorder is carried by a kinship because it makes it much more likely that two carriers will produce a child together. In many societies – estimates run as high as 20% of them – marriage to a cousin may be the preferred option (see Tillion, 1983). This is the case among a number of ethnic minority communities in Britain, and concern about this has been expressed by geneticists (see Darr and Modell, 1988; Modell, 1991; Modell and Kulieu, 1992). Sometimes statements and actions by geneticists have been extremely ethnocentric, showing little concern, understanding or respect for the marriage patterns of other communities (see Chapter 14). Increasingly, however, the social advantages of cousin marriage are becoming better understood and, rather than simply suggesting to families that carry recessive disorders that they should marry out, other options of using carrier detection or prenatal diagnosis are being discussed.

Little is known of the extent to which genetic disorders affect the choice of marriage partners or decisions to marry or to have children (but see below). With some disorders such as neurofibromatosis I, where there may be disfigurement and intellectual impairment and self-image may be poor, marriage may be delayed or avoided altogether (Huson *et al.*, 1989). Interestingly, the relative fertility of affected males seems to be lower than for females (Crowe *et al.*, 1956); however, the extent of such effects is largely unknown. Others who know they carry a disorder may marry but avoid having children.

Where a person knows that they may be carrying a genetic disorder, such as Huntington's disease, the issue arises of telling a potential partner. This is illustrated in one of the personal accounts in Chapter 1 where, in the previous generation, a woman married someone who later proved to be carrying the Huntington's mutation without being told about the family history. This became a long-standing source of resentment. At least judging by a very small amount of anecdotal evidence, it seems relatively rare for a marriage not to proceed when a potential partner learns of the disorder carried by a family.

In some societies there have been attempts to intervene in marriage patterns to reduce births of children with serious recessively inherited disorders. In Cyprus the Greek Orthodox Church will not marry a couple unless their carrier status for beta-thalassaemia is known. It seems that marriages usually proceed even if both spouses are carriers, but the couple then use prenatal diagnosis and abortion to avoid births of affected children. Some migrants from Cyprus have carried this practice to other countries, including Britain. Schemes in Britain to promote carrier testing in non-pregnant populations for cystic fibrosis and Tay–Sachs disease have not attracted wide uptake (see Chapter 5). We have argued elsewhere that this may be because, unlike in Cyprus where thalassaemia is well known, there is little perception in Britain that either cystic fibrosis or Tay–Sachs disease pose a threat for most people (Richards and Green, 1993).

In some Jewish communities in the USA there has been acceptance of carrier detection schemes for Tay–Sachs disease that do set out to influence marriage patterns. In one Orthodox community where marriages are arranged by matchmakers, carrier status of young people is made known to the matchmakers but not to the young people themselves or their families. This is to avoid the creation of what has been called 'genetic wallflowers' or negative effects on self-image of carriers, which have been reported in some studies (e.g. Marteau *et al.*, 1992). Matchmakers then arrange marriages in such a way to avoid the marriage of two carriers (Merz, 1987). In another Jewish community where marriages are not usually arranged, young people undergo carrier detection but are not given the test results. Instead they receive a code number. If subsequently they are contemplating a serious relationship they can call a hot line with the two numbers and they will be informed if both are carriers.

Much genetic counselling has been directed at reducing the fertility of couples who are at risk of producing children with a genetic disorder, or at least suggesting this as a possible option. The evidence about the consequences of counselling is reviewed in Chapter 3. This suggests that it seldom results in more than small changes in fertility and may often serve simply to confirm pre-existing attitudes and intentions. In the case of Huntington's disease, where there seems to be a widespread professional belief that those at risk of carrying the mutation should not reproduce, there are, at most, indications of some modest reductions in fertility after counselling (Barette and Marsden, 1979; Tyler and Harper, 1983). Couples who want to have children and who know themselves to be at risk may avoid contact with genetic counsellors because they do not wish to hear the expected advice. But with Huntington's disease it has been suggested that the situation may be complicated by a lack of social inhibition, which may be an early symptom of the disease (Pridmore, 1993).

Although there has been a good deal of effort to try to gauge the 'effectiveness' of counselling in terms of subsequent births, there has been much less interest in trying to understand the ways in which those who carry genetic disorders reach decisions about having children. Some of the salient points are illustrated in the personal accounts in Chapter 1. Many factors may be involved. In families with Duchenne muscular dystrophy, for example, the direct experience of looking after a child at the stage when the disablement becomes serious seems to have a more powerful effect in reducing subsequent births than simply theoretical knowledge of the disorder (Green and Murton, 1993). With late onset conditions (e.g. Huntington's disease) or those where life expectancy is now extended to early adulthood (e.g. cystic fibrosis) the risk of having a child with the disorder is one that some couples are prepared to take. The decision to have a child in these situations may well be encouraged by statements from the research community about the possibility of future 'breakthroughs' in treatment or even cures. In the case of Huntington's disease there has been relatively little use made of prenatal diagnosis and abortion (see, for example, Adams *at al.*, 1993). Using this technique means either (1) the possibility of disclosing whether one or other parent is in fact a carrier or (2) the possible abortion of a healthy child, if exclusion testing was used. Understandably, many parents have rejected both of these options.

The psychological dynamics of these choices about children remain to be explored in detail. We do not know, for example, how far issues of genetic identity may be involved. How important is it to parents that in some of these situations they may have to consider aborting a child who carries the same mutation as themselves? We know, for example, that partners of those at risk for carrying the Huntington's mutation state that they are more interested in using prenatal diagnosis than is the at-risk partner. Does this issue help to explain why abortion for genetic reasons seems more traumatic than that for social reasons (see Richards, 1989)? There are several questions related to gender that await research. Are reproductive decisions different if it is the mother's family rather than the father's that carries the dominant disorder? In Chapter 3, Shiloh describes a study that indicated that husbands are more opposed to risking the birth of a child with an X-linked condition than their wives. Another observation that might suggest that gender differences may be very significant is that in counselling for hereditary breast and ovarian cancer it is very unusual for women who are at risk of carrying the gene mutation to seek advice about limiting their fertility or about the selection of the sex of their children. While these are, of course, late onset conditions, cancers do occur in some families in early adulthood and some women who come forward for counselling have had extensive experience of nursing family members, most often their mothers, through their terminal illnesses.

In some situations genetic disorders may influence families as a whole. Both the direct consequences of the disorder and the effects of providing care may affect life chances and so lead to downward social mobility in a condition like neurofibromatosis (see Spaepen *et al.*, 1992). There is historical evidence that families carrying Huntington's disease may have faced discrimination. In the seventeenth century many of those in New England accused of witchcraft came from families who carried Huntington's disease, and one might speculate that emigration to escape persecution was one of the reasons why the disease reached both South and North America so early in the process of European colonisation (see Vessie, 1932; Maltsberger, 1961; Klawans, 1988).

12.6 Family dynamics and emotional defences

Illness of any kind may cause changes in relationships within families, but in genetic conditions, because of their hereditary nature, these changes may have particular potency. There is the issue of who may be held to be 'responsible' for the condition. In recessive disorders, both parents of an affected child will be heterozygote carriers so the finger cannot be pointed at either side of the family. The same is not true of dominantly inherited conditions, but we have too little research on family processes to be able to cite more than anecdotes about this. There is the added complication of gender in sex-linked conditions. Duchenne muscular dystrophy affects only boys (except in rare cases), but where the gene mutation has been inherited it will have come from the mother. Certainly in some families these tangled issues of gender may play a part in conflicts between parents of affected boys (see Green and Murton, 1993). In couples who carry recessive disorders, women report more feelings of guilt than their partners. Of course, looking after a child with a serious and inevitably fatal condition causes great stress for all members of the family, as has been documented for cystic fibrosis by Burton (1975). It is hardly surprising that occasionally in such situations accusations of blighted inheritance will fly between family members.

For late onset conditions, elaborate psychological processes have been described and interpreted as ways that families may cope with the uncertainties of who may develop the condition. The best known of these has been termed preselection (Kessler, 1988; Kessler and Bloch, 1989) but there is little information about how commonly this occurs. This is a family process whereby a child or young person is singled out as the one who will eventually develop the condition. According to Kessler, this often happens in Huntington's disease at the stage where a parent first begins to show obvious symptoms, and it functions to reduce uncertainties inherent in the condition. He states that the process of preselection is unconscious and the individual selected may not be fully aware of what is being done. Family members may draw attention to behaviour which links the selected person to the affected parent and so the perception is reinforced: 'That's what mom used to do', 'Dad walked that way' (Kessler, 1988). Kessler states that the process serves to reduce anxiety, and the non-selected siblings may say that they 'know' they will not get the condition and for them it will help to resolve ambivalence toward the affected parent.

Not surprisingly, the process may have profound effects upon the selected individual who may speak of *when* they will develop the condition rather than *if* they will. In some cases these individuals may imitate the mannerisms and movements of an affected person. It has been suggested that sometimes the process of preselection may be the result of the recognition of very early signs of the disease, which those familiar with the disease may pick up long before a conventional clinical diagnosis is possible. But predictive testing shows this is not always the case: families may get it wrong. Sometimes the preselection begins very early, even at birth with a child being given the name of the affected parent. Of course, in due time, if a family member who has not been preselected develops the condition there may be a considerable upheaval in family relationships.

The phenomenon of preselection may be subsumed within a broader category of family scripts. This is a term that has been used by family therapists to denote a pre-existing system of beliefs about family relations into which a new member is integrated

(Byng-Hall, 1989). So, for example, a newly born child may be assigned some of the characteristics and potentially the role within family relationship that had been taken by a grandfather who had recently died. The importance of these ideas is that they draw attention to the fact that family relationships are not created *de novo* by each new family member but there are beliefs about roles, relationships and individual characteristics that precede them and may help to produce their identity and place within the family. Such beliefs may be particularly powerful when a family is burdened by a serious genetic condition.

Preselection is of obvious significance in the context of predictive testing. Other members of the sibship may put pressure on the selected person to get tested to confirm the situation or, conversely, the selected person may decide there is no point in being tested because they already 'know' what the result may be. If a preselected individual is tested and found not to carry the gene mutation there are likely to be significant family repercussions, assuming of course that the family members are told the result (see Bloch *et al.*, 1989; Tibbens *et al.*, 1990). Cases have been reported where the negative test result has resulted in the preselected individual being effectively ostracised from the family.

In any family, predictive testing is likely to have many implications for family members. As already mentioned, genetic investigation of a family may lead to contacts and discussions among family members. If linkage testing is involved, this may be intensified as DNA samples are collected from family members. And, of course, test results may carry implications for ancestors and descendants in the kinship. Obvious ethical dilemmas arise when, for instance, a young person seeks predictive testing for a late onset condition and the parent does not wish to know his or her own genetic situation. There is also the issue of informing other kinship members if a genetic condition is recognised in a family. Many families will take it upon themselves to inform other members of the kinship and to explain the implications to them. But what about branches of the kinship with whom they have lost contact? Should they try to track down those who, for example, may have emigrated to the Antipodes a generation earlier, or should clinicians undertake these tasks? Or should that branch of the kinship be left to find out for themselves they have inherited the condition? Certainly there are families who welcome help and support from professionals in informing kinship members. But what if they do not want to tell? In the UK the Nuffield Council on Bioethics has recently (1993) recommended that medical confidentiality might be broken and clinicians should intervene in situations where family members do not want to inform someone in the kinship for whom there may be implications of a genetic diagnosis (see Chapter 9). Elsewhere, we have argued against such clinical imperialism on the grounds that families may have good reasons for not passing on information to others and that to intervene might have disturbing consequences for family relationships. Additionally, to do so erodes the increasingly fragile concept of clinical confidentiality (Richards and Green, 1995).

Family members may adopt many other strategies to cope with genetic threat (see Wexler, 1979, 1984). Denial is perhaps the most straightforward – to try to ignore the issue, refuse to discuss it and to keep well away from geneticists. There have been sug-gestions, for example, that some women who are at risk of hereditary breast cancer avoid breast self-examination and breast screening clinics because these are a reminder of the potential threat to themselves. Others take exactly the opposite course, learning as much as possible about the disorder, seeking whatever clinical advice may be avail-able and becoming active in self-help and patient groups. Members of the same sibship may adopt differing strategies, which can complicate genetic counselling if they attend a joint session.

12.7 Knowledge about inheritance

As I have mentioned earlier, knowledge about the inheritance of family characteristics is part of our family cultures. Such knowledge covers physical features, characteristics of behaviour and personality, mannerisms and personal habits, as well as health and proneness to illness (see, for example, Davison, Frankel and Smith, 1989; see also Chapter 15). This lay knowledge holds that these characteristics are determined by what we inherit, with varying contributions from the environment. Such lay knowledge may be seen as part of the wider body of knowledge that is part of the notion of kinship (Strathern, 1992). So, for instance, as our kinship system is bilateral, it is usually assumed that, potentially at least, characteristics may be inherited from either side of the family. However, this bilaterality seems to be frequently modified by an assumption related to gender – that daughters will usually be seen to inherit particular female characteristics from their mother, and sons from their fathers. There is another association with kinship: the terminology we use for the lay discussions of family resemblances is the same as that we use when talking of how goods and wealth may be passed on at death. We may inherit our good looks, as well as the family fortune, from our forebears. A common language might suggest common processes, so that those who leave you things in their will may be the same kin who pass on inherited personal characteristics. Our own research has shown that lay discussions about genes may refer to them as objects somehow separate from the rest of the person, that may or may not be passed to children, just as our personal possessions might be.

When family members become aware that they may carry a genetic disorder, general discussions of characteristics that may be passed to children or resemblances between relatives suddenly take on a new salience and meaning – or so I want to suggest. Family members try to make sense of the pattern of occurrence of the disorder they can observe in their family in terms of their previously held knowledge about inheritance. Not all members of a blood line will develop a dominantly inherited condition, so further ideas are introduced to account for the intermittent appearance of the disorder: it skips a generation (another idea that can be traced back to classical Greek science), it appears only in first-born children, it is associated with particular physical or personal-ity types in the family, and so on. These additional assumptions may play a part in psy-chological defence systems as well as being an intellectual part of an attempt to understand what is happening. Both these functions may be important because they can provide strong reasons for holding to such lay knowledge even when contradictory scientific accounts are offered. This, I would suggest, makes it important to try to understand lay knowledge systems and to take account of them in genetic counselling (see Richards, 1993). When a counsellor provides a Mendelian account of a genetic disorder, the issues may not be new to the family member. Rather the counsellor may be challenging long-held knowledge that has served as an important understanding and psychological defence system, at least for some people. It follows from this that we might expect the reception of the counsellor's account to be rather different in situations where the counsellee has been aware that the disorder runs in the family, and so has developed specific knowledge, than others where family members had not previously realised that the condition was inherited. This latter situation will usually be true for recessive inherited disorders, whereas patterns of dominantly inherited conditions will have often been obvious to family members and knowledge and discussion about them may have become part of family talk.

Difficulties may arise in genetic counselling where there are conflicts between lay knowledge and the Mendelian account provided by the counsellor (see Richards, 1995). These situations may parallel others that have been described elsewhere in medicine, where a patient's belief or illness representation may determine the response to medical advice or information (see, for example, Janz and Becker, 1984; Leventhal, 1989). I would suggest that there are three aspects of the Mendelian explanation that seem particularly likely to conflict with lay knowledge and so, in that sense, are hard to understand for some people. Or, to put it in other terms, there seem aspects of the Mendelian account that do not appear to be part of lay knowledge systems. These involve notions that seem restricted to people who have received scientific genetic education.

1.   Lay knowledge does not often seem to include a notion of something that is inherited, aside from the condition itself. It is the condition that is inherited, or a proneness to it, rather than a gene mutation that may be expressed as the condition. This can lead to a number of difficulties in making sense of Mendelian explanations. If you do not have the condition, how can you pass it to your child? This makes recessive inheritance a puzzle. Perhaps that is one reason why carrier screening seems of so little interest to many people (see Richards and Green, 1993), or why some people who have received genetic counselling may be able to quote correctly the chances of producing an affected child without understanding the chances of producing a carrier child or one without the gene mutation (Snowdon and Green, 1994). If you, or members of your family, do not have the condition, how can it be that you might be able to produce a child with it? It is possible too that this is part of the explanation for the spoilt identity some carriers feel. Do they have the condition? If not, what is it they are carrying? If they are carrying the gene mutation, perhaps they do have at least 'a touch' of the disease, which could be why they may view their future health less positively (Marteau *et al.*, 1992). The same question may arise with dominantly inherited conditions that are only expressed in one sex such as the *BRCA1* gene mutations and breast and ovarian cancer. It seems to be the daughters or granddaughters of women who have the disease who see themselves prone to these cancers. Those who might have inherited the gene via a father seem to

find it more difficult to understand their situation. Penetrance[3] and gene expression may also seem rather obscure concepts. If the penetrance of a dominant gene is less than 100%, again there is a puzzle. What sense can be made of someone who is said to have a gene mutation which causes the cancer, but does not develop a cancer? These matters become clearer with a scientific model of a gene coding for amino acid sequences in a protein and a mutation disrupting the process so that the protein molecule cannot perform its usual function, but these, and notions of gene expression, are not the constructs of lay knowledge.

2. Mendelian ratios are easy to understand if you have a concept of chromosomes as paired structures, with gametes being produced by a reduction division. It then becomes clear why two carriers of a recessive disorder have a one in four chance of producing an affected child, or why a parent with a dominantly inherited disorder has a 50% chance of passing on the disorder to that child. Although genetic counselling seems relatively successful in enabling people to quote the correct chance, further probing shows that most people are unable to understand these ratios fully (Snowdon and Green, 1994). I would suggest that in addition to the difficulty noted above, this is because the ratios are not always anchored in a model of chromosomes as paired structures. Such models and notions of reduction division do not seem to be part of lay knowledge. Here the situation may be paralleled by the historical development of the scientific understanding of genetics. Mendel published the results of his breeding experiments in 1865, but the scientific world ignored his work. It was only at the turn of the century when the role of the chromosomes, and of mitosis and meiosis, became clearer, that hypotheses were developed that pointed to a link between the behaviour of chromosomes and data on inheritance. Then Mendel's work could be 'rediscovered', as it was independently by at least three biologists (Moore, 1993).

3. Geneticists talk in terms of risks and probability. As much research has demonstrated, some of their clients have a good deal of difficulty in making human sense of what they are talking about (see Chapter 3). It is not entirely clear why these concepts should present such problems. Notions of chance, in themselves, are not difficult to handle, as is indicated by the culture of the betting shop. But why, then, is it so easy to find examples such as these? A family is carrying a dominant condition. The two siblings are both able to say that they have a 50% risk of inheriting the condition. One sibling goes for predictive testing. She does not carry the gene mutation. She cannot tell her brother about the test because she says she feels it implies he has the mutation. She knows intellectually that this is not true and that his 50% risk figure is completely independent of her situation, but her actions are guided by her feeling, not her intellectual understanding. Or, with a recessive condition, where a couple have had one child who is affected, they believe that next time they will have a normal child. As Nancy Wexler (1979) has put it, we find it hard to believe that chance has no memory. But perhaps that is exactly where the betting analogy breaks down. You back a horse that you think has a good chance of winning. It does win. Obviously you back the same horse again: it is a good horse likely to win again. Chances in one race are related to chances in another. Or do we like to hold on to notions of winning or losing streaks? Or, to return to the genetic

---

[3] Penetrance is the extent to which a gene is expressed. If all those who carry a gene mutation develop the condition the penetrance is 100%. However, if some carriers of mutations do not develop the condition that penetrance will be less than 100%. For example, with *BRCA1*, the most common dominantly inherited gene associated with breast and ovarian cancer, about 85% of those who carry the gene are thought to develop the cancer: the penetrance is 85%.

example, that after the birth of a child with a recessive disorder, future children will be normal as lightning does not strike more than once in the same place? However, the inheritance of a genetic condition depends on a random event: which of each pair of chromosomes from each parent is passed via an egg or sperm to the embryo. But conception is not usually seen in these terms; rather as a process of blending of characteristics from each parent. Notions of kinship do not include the random element inherent in biological explanations. And even where we have an intellectual understanding of the situation, we may have emotional reasons for clinging to another view.

In summary, I am suggesting that we hold lay knowledge about inheritance that relates to our ideas of kinship and is not consistent with the scientific Mendelian account of the inheritance of single gene disorders. Despite the fact that the Mendelian explanation has been part of scientific dogma for almost a century, it has not replaced much older lay knowledge. Mendelian genetics is taught in schools and is now part of the National Curriculum. However, it is very rare among the man or woman in the street, including the younger generation who have experienced the National Curriculum, to find anyone who is able to provide a Mendelian account of inheritance (Lee and Ponder, 1995). I have suggested that those from families that are known to their members to carry an inherited condition may have cultural and psychological reasons for holding to their lay account of inheritance. More widely, though the vocabulary of scientific genetics (gene, DNA) is widely used and has become part of popular culture, there is little indication that the scientific meaning of these terms is generally understood (see Chapter 11, and Richards (1995)). A further reason why accounts of inheritance and genetic risk presented in genetic counselling may not be widely accepted is that we do not, it seems, find it easy to apply notions of chance and probability to our own reproduction. Again, there may be both cultural and psychological reasons for this resistance. The systematic exploration of lay knowledge about inheritance and reproduction needs to be extended beyond English families to the ethnic minorities in the UK and more widely to other cultures. To meet the needs of families who carry genetic disorders, genetic counselling must take more account of lay knowledge of inheritance and the psychological and family processes that govern our family relationships and reproduction.

12.8 Acknowledgements

I am grateful to Josephine Green, Nina Hallowell, Ginny Morrow, Frances Murton, Frances Price, Helen Statham and other members of the Centre for Family Research for their helpful comments on earlier drafts and other assistance with the preparation of this chapter. Marilyn Strathern has been a stimulating discussant on a number of occasions, as has Sarah Smalley. Jill Brown and Sally Roberts provided excellent technical support. Ray Price, Floyd Tillman and Charlie Walker gave inspiration when the writing flagged.

Our research on families and the new genetics is supported by grants from the Medical Research Council and the Health Promotion Research Trust.

12.9 References

Adams, S., Wiggins, S., Whyte, P. Bloch, M., Shokeir, M. H. K., Soltan, H. *et al.* (1993) Five year study of prenatal testing for Huntington's disease: demand, attitudes, and psychological assessment. *Journal of Medical Genetics*, **30**, 549–56.

Barette, J. and Marsden, C. D. (1979) Attitudes of families to some aspects of Huntington's chorea. *Psychological Medicine*, **9**, 327–36.

Bloch, M., Fahy, M., Fox, S. and Heyden, M. R. (1989) Predictive testing for Huntington disease II. Demographic characteristics, life-style patterns, attitudes and psychosocial assessments of the first fifty-one test candidates. *American Journal of Medical Genetics*, **32**, 217–24.

Burton, L. (1975). *The Family Life of Sick Children*. London: Routledge and Kegan Paul.

Byng-Hall, J. (1989) The family script: a useful bridge between theory and practice. *Journal of Family Therapy*, **7**, 301–5.

Crowe, F. W., Schull, W. J. and Neel, J. V. (1956) *A Clinical, Pathological and Genetic Study of Multiple Neurofibromatosis*. Springfield, IL: C. C. Charles.

Darr, A. and Modell, B. (1988) The frequency of consanguineous marriage among British Pakistanis. *Journal of Medical Genetics*, **25**, 186–90.

Davison, C., Frankel, S. and Smith, G. D. (1989) Inheriting heart trouble: the relevance of common-sense ideas to preventative measures. *Health Education Research*, **4**, 329–40.

de Regt, A. (1994) Inheritance and the relationship between family members. Paper presented at a seminar 'L'Europe des Familles Parenté et Perpétuation Familiale', Poitiers, France.

Evers-Kieboom, G., Swerts, A. and Van den Berghe, H. (1991) Partners of Huntington patients: implications of the disease and opinions about predictive testing and prenatal diagnosis. *Genetic Counselling*, **39**, 151–9.

Finch, J. (1989) *Family Obligations and Social Change*. Cambridge: Polity Press.

Finch, J. (1994) Kith and kin in the 1990s. In D. Clark (ed.), *The Jackie Burgoyne Memorial Lectures*, vol. 1 (1989–1993). Sheffield: PAVIC Publications.

Finch, J. and Hayes, L. (1994) Inheritance, death and the concept of the home. *Sociology*, **28**, 417–33.

Finch, J. and Mason, J. (1992) *Negotiating Family Responsibilities*. London: Routledge.

Firth, R., Hubert, J. and Forge, A. (1970) *Families and their Relatives*. London: Routledge and Kegan Paul.

Funder, K. (1991) *Images of Australian Families*. Melbourne: Longman, Cheshire.

Galton, F. (1869) *Hereditary Genius: an Inquiry into its Laws and Consequences*. London: Macmillan.

Gilby, R. and Pederson, D. (1982) The development of the child's concept of the family. *Canadian Journal of Behavioural Sciences*, **14**, 110–21.

Green, J. M. and Murton, F. E. (1993) *Duchenne Muscular Dystrophy: the Experiences of 158 Families*. Cambridge: Centre of Family Research, University of Cambridge.

Harris, C. C. (1990) *Kinship*. Milton Keynes: Open University Press.

Harrison, G. A., Gibson, J. B. and Hiorns, R. W. (1976) Assortative marriage for psychometric, personality and anthropometric variation in a group of Oxfordshire villages. *Journal of Biosocial Science*, **8**, 145–53.

Holloway, S., Mennie, M., Crosbie, A., Smith, B., Raeburn, S., Dinwoodie, D. *et al.* (1994) Predictive testing for Huntington's disease: social characteristics and knowledge of applicants, attitudes to the test procedure and decisions made after testing. *Clinical Genetics*, **46**, 175–80.

Huson, S. M., Compstan, D. A. S., Clark, P. and Harper, P. S. (1989) A genetic study of von Recklinghausen neurofibromatosis in South East Wales. *Journal of Medical Genetics*, **26** 704–11.

Janz, N. K. and Becker, M. H. (1984) The health belief model, a decade later. *Health Education Quarterly*, **11**, 1–47.

Kessler, S. (1988) Invited essay on the psychological aspects of genetic counselling. V. Preselection: a family coping strategy in Huntington's disease. *American Journal of Medical Genetics*, **31**, 617–21.

Kessler, S. and Bloch, M. (1989) Social systems responses to Huntington disease. *Family Process*, **28**, 59–68.

Klawans, H. L. (1988) *Toscanini's Fumble and other Tales of Clinical Neurology*. London: Bodley Head.

Lee, J. and Ponder, M. (1995) Knowledge about heredity and susceptibility to health risks among young people and their parents. (In preparation.)

Leventhal, H. (1989) Emotional and behaviour processes. In M. Johnston and L. Wallace (eds.), *Stress and Medical Procedures*. Oxford: Oxford Science and Medical Publications.

Macintyre, S. and Sooman, A. (1991) Non-paternity and prenatal genetic screening. *Lancet*, **338**, 869–71.

Maltsberger, J. T. (1961) Even unto the twelfth generation – Huntington's chorea. *Journal of the History of Medicine and Allied Sciences*, **1**, 1–17.

Marteau, T. M., van Duijn, M. and Ellis, I. (1992) Effects of genetic screening on the perception of health: a pilot study. *Journal of Medical Genetics*, **29**, 24–6.

Merz, J. (1987) Matchmaking scheme solves Tay Sachs problems. *Journal of the American Medical Association*, **258**, 2636–7.

Modell, B. (1991) Social, and genetic implications of customary consanguineous marriage among British Pakistanis. *Journal of Medical Genetics*, **28**, 720–3.

Modell, B. and Kulieu, A. M. (1992) *Social and genetic implications of customary consanguineous marriage among British Pakistanis. (Galton Institute Occasional Paper*, Second series, no. 4.)

Moore, J. A. (1993) *Science as a Way of Knowing. The Foundation of Modern Biology*. Cambridge, MA: Harvard University Press.

Nuffield Council on Bioethics (1993) *Genetic Screening: Ethical Issues*. London: Nuffield Foundation.

Pridmore, S. (1993) Reproduction and the Huntington's disease process. *Psychiatric Genetics*, **3**, 215–21.

Richards, M. P. M. (1989) Social and ethical problems of fetal diagnosis and screening. *Journal of Reproductive and Infant Psychology*, **7**, 171–85.

Richards, M. P. M. (1993) The new genetics: some issues for social scientists. *Sociology of Health and Illness*, **15**, 567–87.

Richards, M. P. M. (1995) Lay knowledge of inheritance and genetics. (In preparation.)

Richards, M. P. M. and Green, J. M. (1993) Attitudes towards prenatal screening for fetal abnormality and detection of carriers of genetic disease: a discussion paper. *Journal of Reproductive and Infant Psychology*, **11**, 49–59.

Richards, M. P. M. and Green, J. M. (1995) Genetic screening, counselling, consent and family life. In the *Proceedings of a Symposium on Genetic Screening*. London: Centre for Medical Law and Ethics and the Nuffield Foundation.

Schneider, D. (1968) *American Kinship: A Cultural Account*. Englewood Cliffs, NJ: Prentice Hall.

Shakespeare, J. (1992) Communication in Huntington's disease families. Paper presented at the Third European meeting on Psychosocial Aspects of Genetics. University of Nottingham, September.

Snowdon, C. and Green, J. M. (1994) *New reproductive technologies: attitudes and experience of carriers of recessive disorders*. Centre of Family Research, University of Cambridge.

Spaepen, A., Borghgraef, M. and Fryns, J. P. (1992) Von Recklinghausen–neurofibromatosis: a study of the psychological profile. In G. Evers-Kiebooms, J. P. Fryns, J. J. Cassiman and H. Van den Berghe (eds.), *Psychological Aspects of Genetic Counseling*, pp. 85–91. New York: Wiley–Liss, Inc., for the National Foundation – March of Dimes. (*Birth Defects: Original Article Series*, vol. 28.)

Strathern, M. (1992) *After Nature: English Kinship in the Late Twentieth Century*. Cambridge: Cambridge University Press.

Tibbens, A., Vegter, V. D., Vlis, M., Niermeijer, M. F., Kamp, J. J. P. V. D., Roos, R. A. C. *et al.* (1990) Testing for Huntington's disease with support for all parties. *Lancet*, **335**, 553.

Tillion, G. (1983) *The Republic of Cousins*. London: Al Saqi Books. [Originally published by Editions du Seuil (1966) as *Le Harem et Les Cousins*.]

Tyler, A. and Harper, P. S. (1983) Attitudes of subjects at risk and their relatives toward genetic counselling in Huntington's chorea. *Journal of Medical Genetics*, **20**, 179–88.

Vessie, P. R. (1932) On the transmission of Huntington's chorea for 300 years: the Bures family group. *Journal of Nervous and Mental Disease*, **76**, 533–70.

Wexler, N. S. (1979) Genetic 'Russian roulette': the experience of being 'at risk' for Huntington's disease. In S. Kessler (ed.), *Genetic Counselling: Psychological Dimensions*. New York: Academic Press.

Wexler, N. (1984). Huntington's disease and other late onset genetic disorders. In A. E. H. Emery and I. M. Pullen (eds.), *Psychological Aspects of Genetic Counselling*. London: Academic Press.

Wilson, P. and Pahl, R. (1988) The changing sociological construct of the family. *Sociological Review*, **36**, 233–66.

13

Ethics of human genome analysis: some virtues and vices

JANICE WOOD-HARPER and JOHN HARRIS

13.1 Introduction

The applications of human genome analysis (HGA) are numerous and diverse and are being implemented with increasing rapidity as progress in genetic technology and knowledge escalates. A predominant question, frequently being expressed in both Europe and the USA, is one that addresses the problem of whether advances stemming from HGA will be used with due regard to upholding established moral principles and human rights or whether, with insufficient awareness of the ethical issues involved and inadequate regulation, there is a real danger that these might be violated. There are undoubtedly important potential benefits to be gained by both individuals and society but, at the same time, there is some anxiety that serious, adverse social and psychological consequences might arise unless the new knowledge is used ethically.

This chapter will identify a wide range of issues that have already arisen or might feasibly be encountered in the future as the result of HGA. These will be discussed from the standpoint of ethical principles with the intention of illustrating that, overall, society has much to gain and that many of the possible hazards may be overcome provided that professional and public awareness of ethical implications is promoted and safeguards are instigated to ensure the amelioration of harm.

13.2 What is human genome analysis?

Human genome analysis is the resolution of genetic information that is encoded within the entire complement of human hereditary material. This is being achieved through the Human Genome Project (HGP), which is an international collaborative effort. Its main aim, in the short term, is to produce a detailed genetic and physical map of the human genome contained within the deoxyribonucleic acid (DNA) of the 23 pairs of chromosomes of a human cell. Ultimately it is expected that the entire genome, comprising an estimated 100 000 genes, which contain approximately three billion nucleotide base pairs, will be sequenced.

Progress in human genome analysis is being achieved by the identification and characterisation of an ever increasing number of genes, the determinants of inheritable traits. Variations in such traits arise as a consequence of genes existing in a variety of forms, or mutations, owing to alterations in base pair sequence. Mutations usually have deleterious effects, producing hereditary disorders that are often manifested as symptoms of disease. Genetic diseases can be defined as those 'in which the composition of the genome is of decisive importance in the development of the disease' (Danish Council of Ethics, 1993).

Human genome research is currently providing much information about the genetic basis of diseases, especially those caused by single genes. Over 4000 such diseases are now known and most of the more serious ones have already been mapped with an accuracy sufficient to permit their prediction within families (Harper, 1992*a*). Progress towards the identification of the exact locations of individual genes in the genome, together with their sequencing, will enable new understanding of the precise way in which each functions and how defects result in harmful physiological effects. A greatly enhanced understanding of the aetiology of diseases such as cystic fibrosis and Duchenne muscular dystrophy has already been made possible by HGA.

Such diseases, however, represent only a fraction of the total. The majority of inherited diseases, including coronary heart disease and various forms of cancer, are multifactorial, that is, they are caused by the interaction of several genes whose expression is influenced by environmental factors. As these diseases have a more complex pathology, it will be some time yet before HGA can provide an explanation of the genetic basis of their inheritance. Nevertheless, it is envisaged that eventually a good understanding will be gained of all genetically influenced disorders.

13.3 What is HGA likely to make possible?

As more genes are isolated through HGA, there will be an escalation in the development of diagnostic tests for an increasing range of genetic disorders.

Genetic tests are already available for many single-gene disorders and, with the increasing commercial interest of biotechnology companies in the mass manufacture of tests, these are likely to become cheaper and therefore more accessible. A more favourable cost–benefit analysis would also influence the accuracy of testing for specific conditions by providing justification for the implementation of tests to identify rarer mutations.

Genetic diagnostic testing can be distinguished from more conventional methods used in medicine for the detection of disorders in that it is dependent on neither a particular tissue type nor its clinical state. Almost all cells contain DNA and only a small sample is required. Usually blood is used but cheek cells in saliva, obtained by mouthwash which is less invasive, can also be suitable. Additionally, DNA is immutable and tests can therefore be conducted at any stage of life from conception (Harper, 1992*b*; Nuffield Council on Bioethics, 1993).

One of the most significant differences is that genetic tests permit the diagnosis of a disorder in the absence of disease symptoms and can therefore be predictive, that is, they can yield information about statistical expectations for the future. They enable the identification of individuals who are either presymptomatic of late onset disease or those with increased genetic susceptibility conferring predisposition to future disease symptoms. As tests become more reliable it is expected that their predictive value will increase in some cases so that the likely time of onset of a particular disease and the severity of symptoms can be ascertained with more certainty. Genetic tests can also identify carriers of recessive genetic diseases, who will not manifest any disease symptoms themselves but can transmit the genetic disorder to their offspring, who may be affected.

Expansion of genetic testing is prompting much debate, both in the USA and in Europe, about the feasibility, potential benefits and ethical implications of implementing population screening programmes for genetic disorders (Beaudet, 1990; Gilbert, 1990; Netherlands Scientific Council for Government Policy, 1990; Natowicz and Alper, 1991; Rosner, 1991; Fost, 1992; Holtzman, 1992; Nuffield Council on Bioethics, 1993). Population screening has particular relevance to preventive health policies, especially for those defects causing diseases that are serious but treatable and or for those whose onsets and degree of severity can be influenced by environmental factors. In both cases harms may be ameliorated through early detection and intervention.

The ultimate aim of HGA will be achieved only in the longer term. It is acknowledged that, in the immediate future, a serious 'gap' will remain between diagnosis of diseases and the ability to treat them (Friedmann, 1990; British Medical Association, 1992). However, with further elucidation of gene function and the biochemical and molecular pathology of disease, it is envisaged that more effective forms of traditional or conceptually new therapies, such as gene therapy, will eventually be developed (Friedmann, 1989). Gene therapy uses genetic engineering techniques to manipulate the human genome in order to rectify the adverse effects of defective genes. Although this form of treatment is currently only in its infancy and therefore its practice is restricted to a very small number of genetic conditions, it is expected to herald a new era in the treatment of genetic disease. Furthermore, gene therapy might not only provide benefits for those treated but it may make possible the elimination of genetic diseases from future generations.

In view of the fact that acquisition of knowledge from HGA is fast outpacing the development of therapeutic remedies, it becomes urgent, in advance of such therapies, to decide the ethics of using such tests. In particular, the information revealed by genetic testing and screening is of an intensely personal nature and can also be unique in its ability to be predictive of an individual's future health status. Consequently, its value to others, and therefore the potential for its misuse, is immense, and controls are essential to ensure that it is used ethically. The overall significance of HGA in this respect, however, is viewed not so much as its having raised a whole new set of ethical issues but rather as its having precipitated the need for them to be confronted (Beckwith, 1991; Collins, 1991; Fletcher and Wertz, 1991).

The implementation of any new procedure needs to be regulated to ensure that the most desirable ends are brought about in an ethical way for both individuals and society as a whole. To achieve such ends, not only should there be debate among geneticists, politicians, lawyers and ethicists but also, as clinical applications of genetic

knowledge become more widespread to potentially touch the lives of an increasing proportion of the population, the importance of an informed public must be emphasised.
Public awareness needs to be stimulated and its opinions on the social and ethical
acceptability of a particular new technology should be both sought and seen to be
taken into account before final decisions are made. The importance of this aspect is
becoming increasingly recognised, as was recently demonstrated by public debate and
media coverage regarding possible new developments in reproductive technology
(Laurance, 1994).

13.4 Ethical issues

Human genome analysis is a technique, a tool, and like all tools it may be used for good
or ill. There is a famous line in a classic Western film *Shane* in which the sentiment is
expressed that the world would be better if all guns were abolished and the particular
peaceful valley in which they lived was rid of all guns including the gun of the hero,
Shane himself. Shane's response is to the effect that a gun is just a tool, no better nor no
worse than the person using it. It is often thought that the ethical problems concerning
new technologies can be removed if the technologies are banned or restricted. This is
clearly naive. In this chapter it is our intention to show precisely this Janus-faced nature
of technology, that it is capable of being used for both good and ill and that we can
decide on the legitimacy of its use only by reference to an analysis both of the likely
ways in which it will be used, and of the consequences of denying the human community such benefits as it may make possible.

Human genome analysis is a tool, primarily for providing information, but of course
that information makes possible the development of further techniques for making
physical changes to human beings. Some of these physical changes would be therapeutic, others fall into the class of actions that are sometimes called 'eugenic'. Let us deal
briefly with the ethics of information management first and then the ethics of perfect-
ing physical changes in human beings.

Information management

Human genome analysis will yield unprecedented quantities of information about the
human individual. Much of this would be available as early as the identification of the
gametes that will form the resulting individual, but certainly most of it will be available
from the zygote stage onwards. Much of this information will be morally inert, giving
us details simply of the genetic constitution of the individual, but some of it will reveal
predictive data on a whole range of features concerning this individual, his or her life
expectancy; susceptibility to diseases, environmental features, pollutants, etc. Some of
the predictions based on these data will be highly accurate and determinate, but most
will be relatively indeterminate and statistical in nature. The ethical principles that
immediately arise are of course those that have to do with **autonomy** – the individual's
capacity for self determination – and **privacy** – the individual's rights to maintain

control over personal information. Equally, this information will affect the *rights* of others, from those who form personal relationships with the individual in question to employers, national agencies including government, and so on. On the one hand, the individual will require access to the maximum information about herself in order to make accurate predictions and informed decisions about her life. At the same time, and in so far as others also have access to this information, it is potentially a great inhibitor of her autonomy. Privacy will require that she is able to maintain control of others' access to this information or, failing that, is able to control their use of it. Technological advance is likely to make it increasingly difficult for individuals to limit the availability of information about themselves derived from human genome analysis. For this reason, if privacy is to be maintained, it is more likely to occur within a framework of legal prohibitions on the use of such information rather than in vain attempts to limit access to it. The balance that has to be struck is that between the individual's autonomy and privacy, the interests of third parties, and the public interest. We should bear in mind, however, that in a society in which individual rights and freedoms are regarded as important, the interests of third parties on the one hand and the public interest on the other should, except in the most exceptional of circumstances, be subordinate to the autonomy and privacy of the individual.

Physical changes to the human individual

The information revealed by human genome analysis will enable both therapeutic and other material changes to be made to human individuals. Where these changes are genetic they may be made either to the germ-line, in which case they will be passed on indefinitely to successive generations, or to somatic cells, in which case they will take effect as one-off changes to the genetic structure of that particular individual. Therapeutic changes, if they are understood as those required to restore some conception of normal functioning, are relatively unproblematic from a moral point of view, although many are still fearful of the implementation of such changes via modifications to the germ-line. However, non-therapeutic or marginally therapeutic changes are seen as more problematic. It is not necessary to consider such changes as might reduce or impair human functioning, but where they would enhance human functioning or indeed confer other sorts of advantages (such as protection from environmental pollutants or immunities to disease) they clearly would be regarded by many as highly desirable. Here, the relevant principles seem to be, on the one hand, *protection of individual autonomy* and, on the other, *public interest*.

A guiding principle is that we should not deny people the objects of their choice whether those objects are seen in terms of removing existing disease states or in terms of enhancing normal functioning or where they are morally neutral, unless it can be decisively shown that such changes are not self-regarding, that is, they are harmful to others or more generally harmful to the public interest. We should, however, be very wary of defining the public interests simply in terms of what a given generation of people happen to want or to feel comfortable with. Therefore an overriding principle should be that unless palpable harms can be demonstrated as the likely results of the use of technology or unless a clear and unequivocal public interest can be identified as being thereby damaged, *the presumption should be in favour of permitting the physical changes or modifications made possible by human genome analysis.*

With these very general principles before us we can now move on to examine in greater detail the particular applications of the new genetic knowledge.

The following discussion will identify these applications and, by reference to ethical principles, will emphasise the main concerns that are raised. It will concentrate predominantly on what is already or will imminently be possible, and consider briefly what is likely to be feasible in the more distant future. It will consider the probable desirable and undesirable outcomes of such applications; and, in the case of potential harms, forward suggestions as to how these could be minimised by policies or controls to ensure that genetic knowledge is used ethically for the overall benefit of both individuals and society.

13.5 The shorter-term applications of the new genetic knowledge: genetic testing and screening

In the shorter term, emphasis will be on the use of genetic knowledge to provide means of avoidance or prevention of genetic disease rather than on producing new therapies or cures. It is envisaged that there will be a massive increase in provision of facilities for genetic testing and screening.

Genetic testing, as distinct from screening, provides diagnostic information about an at-risk individual for whom there is prior indication of a genetic condition or defect being present (Nuffield Council on Bioethics, 1993). Testing is therefore usually performed at the request of an individual where there is a family with a particular genetic disease, or a couple who already have an affected child.

In contrast, a screening programme constitutes a search, in the absence of such indications, within a population where all members are screened. Its purpose is to identify individuals either affected by or susceptible to a serious genetic disease or those who are carriers, and results often require confirmatory diagnosis. Another distinction is that screening gives rise to particular ethical problems concerning informed consent and management of unanticipated information.

Expansion of the range of genetic diseases for which screening is appropriate is likely to be slower than increase in availability of genetic testing. The introduction of such programmes for specific diseases is not necessarily dependent upon their incidence in the population but rather upon factors such as their severity and the extent to which they are treatable. This is demonstrated by the existing programme for the detection of phenylketonuria (PKU), a condition that affects considerably fewer newborns than, for example, cystic fibrosis, for which testing is presently largely restricted to those at known risk. Factors such as efficiency of testing, that is the detection rate, and cost effectiveness will also be influential.

Genetic tests and screening programmes are at present usually performed at the prenatal or neonatal (newborn) stages, in later childhood or in adulthood (Nuffield Council on Bioethics, 1993). Pre-implantation testing of embryos, however, is in its initial stages and there is even the possibility of diagnosing female gametes before fertilisation (Collins, 1991).

Both testing and screening will generate a copious amount of genetic information and it is the consequences of the potential uses of such information that the following discussion, based upon the stages of life when testing is performed, will address.

Pre-implantation and prenatal

Pre-implantation and prenatal testing and screening each raise unique issues, which will initially be addressed separately. Thereafter the two procedures will be considered together as there is a commonality with regard to many of the social and psychological consequences of their practice.

Pre-implantation testing enables the embryos that carry abnormal genes to be identified using *in vitro* fertilisation (IVF) techniques together with pre-embryo biopsy. Embryos found to be defective are subsequently not transferred. This technique is appropriate for carrier couples or for older women. It is not currently widely available, however, and is restricted to a very few conditions, such as cystic fibrosis. One of its main advantages is that it may constitute the only acceptable means by which a couple that, for example, had only become aware of its at-risk genetic status as the result of already having had an affected child and felt unfit to cope physically, mentally or emotionally with a second who was similarly affected, could be ensured a child free of that particular risk and of others for which tests are available.

An ongoing factor that is likely to severely limit the availability of pre-implantation screening is its reliance on IVF techniques, which have low priority in terms of government funding in the UK. Facilities remain predominantly in the private health sector and therefore access is largely restricted to those who can afford to pay. Consequently, equity of treatment is not possible, a situation that is inherently unjust in that it does not permit all at-risk couples to benefit. However, this does not by itself necessarily constitute sufficient reason to prohibit its clinical use, because there are many other accepted forms of medical treatment that also are not allocated on an equitable basis.

Another reason that may preclude pre-implantation screening on a wide scale is that IVF itself currently has a relatively low success rate (20–30%). With continuing research, however, higher rates may be achieved, and expansion of the practice of pre-implantation screening, for a wider range of genetic diseases, would ensure that the techniques for identifying defective genes in the early embryo became more accurate. Not only would this increase the number of potential couples who could benefit from the technique, but together with knowledge from HGA and genetic engineering technology, it would facilitate the possibility of manipulating the genome of an embryo *ex utero*, thereby correcting its defect, avoiding its destruction and allaying moral objections to disposal of embryos.

Until this is feasible, however, destruction of a pre-embryo or non-implantation could be viewed as being preferable, by avoidance of agonising decisions relating to termination, and ethically more acceptable to both the couple and society as a whole, at a time when abortion is generally tolerated but not universally endorsed. Indeed, it may also be deemed as more tolerable to anti-abortion groups, in so far as the embryo had never been implanted, and to those who believe that the embryo does not achieve a moral status consistent with the conferral of full rights until that stage.

The advantage of prenatal screening and testing is that, to some extent, it overcomes the problems associated with equity of access in that the embryo has been conceived by natural means. The more recent technique of prenatal diagnosis by chorionic villus sampling (CVS) at 8–12 weeks' gestation, rather than by amniocentesis, enables termination at earlier stage of pregnancy, which can be a less traumatic experience for the woman in terms of both her physical and mental well-being. At present, it is used to detect genetic disorders of the fetus in cases where there are prior indications and is particularly desirable for those conditions that are serious, fatal or not amenable to treatment and are liable to cause extreme distress in families. Prenatal screening programmes are similarly restricted to high-risk groups living in particular locations, but a poignant example of the benefits to be gained for both individuals and society is illustrated by the contribution of the prenatal screening programme for thalassaemia in the London Cypriot community, which has resulted in a substantial reduction in the birth rate of affected children (Modell *et al.*, 1993). Screening programmes for thalassaemia have been even more successful in Cyprus and other Mediterranean countries, where they have resulted in the virtual elimination of such births. The success of any diagnostic testing or screening programme, however, is ultimately dependent upon the agreement of individuals to be tested. In the case of thalassaemia, there is a high demand for testing owing to the serious nature of the disease and the short life expectancy. For some other genetic diseases, however, which are either late onset, such as Huntington's disease, or for which improved methods of treatment are increasing life expectancy, such as cystic fibrosis, such a reduction in the number of affected births is less likely, owing to a higher proportion of high-risk couples who reject prenatal diagnosis.

The predominant aim of genetic testing, both at pre-implantation and prenatal stages, is to reduce the incidence of genetic disease in the next generation. This can be accomplished either by aborting affected fetuses or by identifying carriers, who can subsequently be offered appropriate counselling. Genetic testing can thereby enable the avoidance of damaging psychological effects to the parents of having a seriously disabled child, which can be manifested as feelings of guilt and misplaced responsibility, as well as the considerable burdens associated with caring for such a child, especially when life expectancy is very limited and of negligible quality. The savings to society can be immense in terms of financial costs and depletion of scarce resources related to the education, health care and community support of the disabled.

A considerable problem, however, is in selecting those genetic conditions, identified by testing, that would justify destruction of embryos and fetuses. Such selection necessarily requires that some forms of disability are considered to be more compatible with a worthwhile life than others. Disorders might be categorised on the basis of features such as whether the disorder is likely to cause fatality at birth; whether treatment or cure is possible; the predicted degree of disability and dependence; the time of onset of symptoms or, alternatively, the period of time during which the individual could expect to be symptom-free; and the extent to which quality of life would be impaired. Other factors, such as the cost to the healthcare system, might also be influential, but should not, by themselves, be decisive. Decisions would necessarily involve some sort of line to be drawn and particular difficulties would arise in the case of diseases, for example, of intermediate severity and for which effective therapy, although currently not available, might become feasible during the individual's lifetime. The demarcation is likely to become less arbitrary, however, as genetic tests become more comprehensive and reliable and if predictability becomes even more accurate than it is at present.

Another problem is presented by the question of who will make such decisions. It is important that geneticists' determination of what is feasible is moderated through debate between ethicists, policy makers and the public as to acceptability of possible outcomes. It is possible that, as prenatal diagnosis in particular becomes more commonplace, the protection of defective fetuses will be safeguarded by guidelines. In any event, the significance of these decisions should be placed within the context of ethical arguments and legislation that already permits both the disposal of stored healthy embryos created during IVF procedures after a statutory period and the abortion of healthy fetuses to preserve the woman's physical and/or mental welfare.

Despite controls, however, it remains probable that fears will persist concerning possible misuse of genetic testing for the prevention of 'undesirable' characteristics. Lippman (1991) viewed prenatal diagnosis as 'approach[ing] children as consumer objects subject to quality control . . . [and] presuppos[ing] that certain fetal conditions are intrinsically not bearable' so that, as proliferation of such conditions occurs with improving diagnostic capabilities, concepts of 'normality' will be restricted. She envisages a rise in eugenics emerging from what she terms 'geneticization', which refers to a reductionism of individual differences according to their DNA codes and the employment of genetic technologies as interventions in the management of health problems.

Such fears of eugenics becoming a social norm are probably unfounded. In a society where discrimination exists, certain attributes are, without doubt, considered to be highly desirable and an individual lacking such attributes can expect to be disadvantaged, to a greater or lesser degree. Thus, it can be argued that not all forms of eugenics are necessarily undesirable (Dawson and Singer, 1990; Harris, 1993). Selection of embyros or fetuses that possess certain characters might be ethically justified on the grounds of desire for the future welfare of the child. There is, for example, clear justification for selection based upon gender when genetic test results indicate the presence of sex-linked disease; skin colour in a society where discrimination adversely affects equality of opportunity; or intelligence, which influences choice of occupation and earning potential (Harris, 1992).

Some people have objected that to choose the skin colour, racial features or gender of children is an illicit form of parental preference. The phrase 'designer children' is often used pejoratively to describe the children of parents who are more concerned with fashion and pleasing themselves than with valuing children for the children's own sake. However, we should remember that choosing a same race or same race-mix child is also *designing* the child that you will have. This is no less an exercise of parental preference than is choosing a different race or race-mix, or, for that matter, colour. It is a truism that bears repeating, that once you have the capacity to choose and the awareness of that capacity, then choice is inevitable. It is no less an exercise of choice because the choice is exercised in a traditional way, nor because the choice may involve doing nothing at all. The best way to avoid totalitarianism and to escape the possibility of racial prejudice, either individual or social, dictating what sort of children people have, is to allow and support free parental choice in these matters, and to do so whether that choice is exercised by choice of procreational partner or by choice of egg or embryo. For such choice is likely to be as diverse as are the people making it (J. Harris and H. Wallace, unpublished results).

Whereas it will eventually become feasible to test for and selectively abort fetuses lacking desirable physical attributes, such as blue eyes, blond hair and white skin, or which are free of genetic disorders that would predispose them to genetic disease, in practice selecting for most characteristics that may better fit individuals for society, such as academic potential, abillity in sport or musical talent, would be much more problematic. The expression of such multigenic traits is influenced by complex interaction of the genes with environmental factors such as education, social class and upbringing. In any case, such traits are already, to some extent, effectively selected for by socially acceptable norms, such as choice of mating partner. The potential for advancing eugenics, such as through compulsory sterilisation of the mentally ill, already exists, yet its realisation is controlled by social and political policies, which will be equally determinant in preventing the misuse of genetic knowledge and technology (Dawson and Singer, 1990; Harris, 1992).

Consideration of the child's welfare also raises problems with consent and conflict of interests. The question of whether the potential parents have the right to consent on behalf of the embryo to its being genetically tested is principally dependent upon individual views on the moral status and therefore associated rights of the embryo, and in itself raises no new ethical issues different from similar consent to other interventions. Prenatal diagnosis does, however, present difficulties concerned with conflicts of interests. Conflicting interests of the potential parents, the child-to-be, geneticists, the healthcare system and society need to be prioritised so that the technology is used ethically.

Testing or screening at both pre-implantation and prenatal stages increases reproductive options. In order for couples to make an autonomous decision, it is essential that they are adequately informed and that the final choice, including that of refusal of testing and/or termination, is freely made, that is, not under coercive influences or to comply with societal pressures or expectations. The role of pre-test and post-test counselling is to fulfil the former requirement, and thus the importance of provision of facilities in this respect as testing becomes more widespread cannot be over-emphasised. There is, however, a danger that an increase in both the number of tests and disorders that can be diagnosed will augment women's perceptions of personal responsibility for the health of a child (Lippman, 1991). Thus women's freedom of choice would be restricted by their being made to feel irresponsible if they refuse available measures to avoid the birth of a defective child. This emphasises the need for safeguards against social control of pregnancy and abortion, through both effective counselling and public education, so that decisions are made voluntarily and without societal pressures.

Confidentiality and privacy of genetic information resultant from prenatal diagnosis should also be ensured, to prevent potential for discrimination (Billings *et al.*, 1992), which may influence decision-making. Discrimination can also be expected to arise as a consequence of abortion being justified on the basis of geneticists' decisions as to which particular disorders are to be tested for or included in screening programmes (Lippman, 1991). This could mean that, initially at least, a fetus with a common, untreatable disease determined by a single gene is more likely to be diagnosed pre-natally and selectively aborted than a fetus with a relatively rare, polygenic defect that may produce equally severe symptoms but for which screening is not feasible or cost-beneficial. In the latter case the fetus would be compelled to be born to a life of pain and suffering. This reinforces the need, previously addressed, to prevent a monopoly on decision-making, so that benefits, to both individuals and society, can be max-imised.

A widely perceived advantage of gaining prenatal diagnostic information is that it can provide reassurance during pregnancy, by allaying fears that the fetus may carry a genetic disorder. Although Lippman (1991) views this as being 'of a particular and limited kind . . . [because] most disabilities only manifest themselves after birth', support for this argument will recede as improvements in the reliability and accuracy of genetic diagnostic tests, performed at an ever earlier stage of life and for the detection of more disorders, will provide increasingly accurate predictions about onset and sever-ity of disability. She also suggests that the use of alternative 'low technology' approaches, provided by nutritional, social and other supportive services, can fulfil the need for reassurance. This is again debatable because, with an expected rise in public awareness and education as technology advances and test availability increases, it is doubtful whether these alternative approaches, alone, could always provide the same degree of alleviation of anxieties associated with the risk of carrying a defective fetus as could negative test results. Additionally, although confirmation of a positive result for a serious genetic disorder is, at the time, often devastating, it allows the couple to decide, at the earliest opportunity and with the help of sensitive, professional coun-selling, the best option for the child and for themselves. In the event of non-termina-tion, it can also provide a valuable period of time for psychological adjustment and making preparations for future lives. Consequently the 'wish to know' can be expected to be a strong, motivating influence on uptake of testing, as supported by evidence already of increasing public demand for prenatal testing services (Nelkin and Tancredi, 1989).

Another more indirect consequence of prenatal diagnosis is that it could further widen existent social divisions. First, screening programmes will, for some time, be available for the detection of only a few genetic conditions. Diagnostic tests, on the other hand, will initially be expensive and restricted in availability so that their use may be limited to those who are aware of their own risk of producing affected children, educated about testing facilities and can afford to pay. This inequity of access by which only a sharply defined sector of the public will be permitted to benefit from the advantages of prenatal diagnosis, apart from being inherently unjust, could lead to many genetic diseases becoming increasingly restricted to the poorer and less educated social classes. Secondly, it is these groups that are often already less advantaged, in terms of housing, financial pressures and leisure time, when faced with caring for a child with serious disability and these limitations might, of necessity, take priority over emotional or moral convictions about abortion or the desire for a child. A consideration of such factors would support the development of government policy to increase the provision of free prenatal testing and screening programmes, through the expansion of clinical genetic services, to all those at risk of serious genetic disease.

Neonatally or in later childhood

Screening for genetic disorders in the newborn has been practised for many years and is currently the only systematic and universal form of genetic screening. Until recently, detection of a genetic disorder relied upon indirect biochemical methods, involving the assay of either the abnormal enzyme produced by the defective gene or the products of the metabolic pathway under its control. Neonatal screening is primarily indicated for diseases associated with severe symptoms that are preventable if treatment is initiated immediately after diagnosis of the condition early in life. Established screening programmes for phenylketonuria and congenital hypothyroidism have been particularly successful in this respect.

HGA is resulting in a rapid increase in the range of genetic disorders that can be diagnosed by more precise methods that detect the gene directly. Also, it is expected that new therapies will gradually follow for diseases that are at present untreatable. It may then become feasible and beneficial to both individuals and society to screen all newborns routinely for a bank of genetic diseases that are preventable, or at least where symptoms can be alleviated, by early medical intervention. Looking further into the future, once the genetic basis of multigenic diseases has been identified, then early diagnosis of predisposition to disorders, such as heart disease and cancers, which have a high incidence in the population and are especially burdensome on society in terms of depletion of health resources, loss of working days and cost of sick benefits, will become increasingly important in an effort to increase the effectiveness of preventive health policies.

Although early diagnosis and therapeutic treatment of affected newborns does not raise any major ethical issues, information resulting from the use of new direct tests, which identify those with carrier status and therefore have implications for future reproductive decisions, would raise ethical dilemmas of a different order (Nuffield Council on Bioethics, 1993). There is a view, however, that carrier screening, because it can be of no obvious benefit at this stage, should be delayed until informed consent can be given (British Medical Association, 1992).

Testing of minors, who are unable to give informed consent, presents problems, similar to prenatal testing, regarding parental consent on their behalf. Such testing can enable parents to make future plans for the welfare of both the child and the rest of the family. It can also allay their own fears and uncertainties. Whereas it may be argued to be ethically acceptable for parents to request testing when it is clear that the child will derive some therapeutic benefit or other demonstrable advantage from early diagnosis of a genetic condition, the position is very different when there can no such benefit and/or when there is a likelihood of harmful consequences. For instance, there is general opposition to the testing of minors for Huntington's disease, on the grounds of not only the harm that may result from denial of the child's right to autonomy later in life (most adults reject testing for this condition) but also the possibility of psychological trauma during development, negative conceptions of self-worth, and social stigmatisation (Bloch and Hayden, 1990).

Genetic information revealed by testing of minors can thus have important implications for the child's future life. This requires that decisions both about whether or not the child should be tested and about when any information is best imparted to the child be made with care as to the child's best interests. For instance, the timing of disclosure might be dependent on the particular type of genetic disorder that has been detected. Information about carrier status needs to be communicated before there is a possibility of parenthood, whereas other conditions might require that the information is given before initiation of therapy or special education.

Adulthood

At present, genetic testing and screening in adulthood, as in other age groups, is restricted mainly to those in high risk categories, that is, those who have a family history of genetic disease and therefore an above average probability of either being affected or being a carrier.

Adult screening for carrier status in selected populations has, by increasing reproductive options, been effective in reducing the birth rate of affected children, and has prompted discussion on the most efficient ways of offering such programmes. Decisions will be dependent partly upon the extent to which the targeted population is informed and partly on its receptiveness. The importance of public interest can be illustrated by reference to screening for thalassaemia in Cyprus and Sardinia. Programmes, which were successfully initiated at the prenatal stage, have now progressed to the testing of young adults for carrier status, with the result that there is virtually universal uptake of prenatal diagnosis by couples where both members are diagnosed as carriers, and are moving towards screening of school children (Nuffield Council on Bioethics, 1993). Currently, however, most screening programmes, such as for cystic fibrosis, are still only in their pilot stages, and the degree of receptiveness of a wider population remains to be established before decisions are made regarding the advantages of more general implementation. Modell (1990) stresses the need to plan for future expansion of screening, by formulating national policy and improving the delivery of genetic services by the provision of an efficient infrastructure for screening, education for the public and health professionals, and a system to monitor services.

These developments, in conjunction with further knowledge acquired through HGA about the function of individual genes and their interaction both with other genes and with environmental factors, are likely to facilitate an increase in predictive testing for predisposition or increased susceptibility to disease and to accelerate considerations as to the feasibility of extending existing adult screening programmes to the more general population. This might be indicated most strongly for multifactorial diseases like cardiac heart disease, cancers and employment-related respiratory diseases. Screening is, in effect, already in existence for some of these. Knowledge of family history and linkage analysis are used to indicate predisposition or susceptibility to such conditions as breast cancer, bowel cancer and early onset Alzheimer's disease. The accuracy of these screening programmes is likely to improve with further identification of genes responsible for these diseases and the subsequent development of direct probe testing, both of which are likely to be made possible through HGA. The potential value of population screening for such multifactorial diseases lies in the following factors: many such diseases are widespread in society and responsible for the majority of untimely deaths; preventive measures and/or therapy can, at least to some extent, be effective in reducing their incidence or severity; and they are especially costly in terms of resources of the health system and loss of work hours.

Early identification of susceptibility to such diseases would permit, through regular medical check-ups, detection of symptoms and close monitoring of the progress of the disease at an earlier stage. Optimal, continuing health care would be expected to benefit the individual not only by increasing the effectiveness of therapeutic intervention but also by providing reassurance and emotional support. Substantial benefits might also accrue to the health system. The identification of individuals, previously unaware of their high susceptibility status, would present opportunities for preventive measures, related to diet, lifestyle, employment or geographical location. These could be effective, to varying degrees depending on the particular disease, in delaying the time of onset and/or lessening the severity of disease symptoms, both of which could result in savings of scarce health resources. It would also place greater responsibility on the individual for her own health status. This situation might be taken as strengthening the argument for health care's being rationed according to patients' readiness to accept such responsibility and act accordingly. Safeguards would therefore be required to prevent violation of equal rights to health care.

Another application of HGA is genetic monitoring. This involves the periodic examination of individuals 'to evaluate modifications of their genetic material . . . the premise [being] that such changes could indicate risk of future illness' (Office of Technology Assessment, 1990). Genetic monitoring could be most appropriately, though not exclusively, conducted in the workplace, where exposure to certain hazardous substances through the course of employment may result in chromosomal damage or the induction of gene mutations. Whereas such monitoring could benefit employees who are at high risk from occupational diseases, brought about by the work environment, and possibly also their future children, its practice could also be argued to be paternalistic and discriminatory or mainly in the interests of employers who wish to reduce their own costs. Controls therefore need to be implemented, particularly in relation to respect for autonomy and protection of privacy, if monitoring is to become more commonplace in the future (Harris, 1992).

Problems associated with violation of autonomy and confidentiality are also raised as a consequence of the genotypes of an individual's close relatives being implicated by test results. There is a danger that relatives, including the children, of the subject might be harmed by acquisition of genetic knowledge about themselves that they did not wish to know. It is especially important, in view of the predictive nature of some genetic information, that the 'right not to know' be protected. For this reason, there is general agreement that genetic testing and screening should be voluntary and require informed consent and also that the confidentiality of the information that is revealed is safe-guarded. Difficulties arise, however, where such information may be of value to rela-tives in preventing either the birth of a child with a serious genetic disease or an unnecessary termination of pregnancy (Nuffield Council on Bioethics, 1993). It must then be decided whether disclosure of personal genetic information is ethically justified to avoid a greater harm.

Psychological harms can also arise, within a family, as a result of genetic knowledge being acquired about individual members. It can create tension and be a source of resentment, hostility and recriminations. For the individual, knowledge of his genetic disorder can promote feelings of isolation and self-stigmatism. The severity of such harms is dependent not only on the nature of the particular disease that is diagnosed and society's attitudes towards it, but also upon the resilience of the individual. Diagnosis of diseases that are incurable, late in onset, or that affect mental health, such as schizophrenia, Alzheimer's disease and Huntington's disease, are likely to cause the most severe forms of distress. Testing for Huntington's disease presents particularly difficult ethical problems. Whereas a positive predictive test will arrest uncertainty and permit an individual to make practical and psychological preparations for future dis-ability and premature death, there is a strong likelihood that the knowledge can also be emotionally burdensome, risking serious psychological harm, manifested as symptoms of depression and suicidal tendency. Studies, which have demonstrated that such harms can be reduced by pre-test counselling and 'exclusion of subjects on the basis of psychological frailty', can be argued to provide some ethical justification for the use of minimal paternalism in decisions to be tested (DeGrazia, 1991). Any such violation of respect for autonomy must be carefully considered on a case-by-case basis. For some individuals, the psychological and emotional trauma might be preferable to living with uncertainty and continual anticipation of the onset of symptoms. Autonomous deci-sions to refuse testing should also be considered as paramount, with the possible excep-tion of when children are implicated.

To exercise the right to autonomy in decisions whether or not to be tested, an individual must be adequately informed. The dissemination of complex information related to genetic conditions and their inheritance can be particularly difficult, and the importance of effective counselling, including psychosocial support, must be emphasised. Because genetic information, compared with other forms of personal medical information, can be uniquely valuable, it is particularly important that individuals are made aware of all possible consequences of their decisions about testing. Such consequences do not relate exclusively to health. Information about an individual's genetic status can have much wider adverse implications for her financial security, employment, lifestyle, welfare of her family, and psychological well-being.

The potential for harm, through the misuse of genetic information, emphasises: the need for consent to testing and screening; that such consent should be voluntary and an informed decision to remain in ignorance respected; and that results should be confidential. The ease of obtaining samples of saliva for genetic diagnosis, however, enables testing to be carried out in the absence of consent. Additionally, individuals could be coerced into being tested or their genetic data could be accessed, without their consent, by interested third parties. These practices could result in an individual's being psychologically harmed by information about himself that he did not wish to know and being subjected to wrongful discriminatory practices. Incidences of genetic discrimination by social institutions, mainly insurers, resulting in stigmatisation and denial of services or entitlements to the 'asymptomatic ill' or those who will never suffer any significant impairment, have already been documented in the USA (Billings *et al.*, 1992). The case studies demonstrated the potential for discrimination resultant from refusal to undergo testing or on the basis of access to diagnostic data that indicated predisposition to genetic disease or carrier status. Genetic discrimination could emanate from the practice of pre-employment testing and the use of selection criteria by adoption agencies and assisted conception clinics. It could also reinforce already existent social discrimination in cases where a particular genetic disease is almost exclusively confined to a certain ethnic group, as is Tay–Sachs disease to Ashkenazi Jews.

Such discrimination on the basis of immutable genotypes is clearly wrong and potentially damaging to the substantial benefits to society that are possible through HGA. It would deter individuals from consenting to genetic testing or screening out of fear of serious, damaging repercussions, a decision that could result in harm to their own health or that of their families.

In the early 1970s, the poorly conceived mandatory screening programme for sickle-cell trait, which targeted the black community in the USA, had disastrous consequences. It resulted in discrimination in employment and insurance and the imposition of severe restrictions on reproductive choices, and as a result was widely perceived as being racially motivated (Wilfond and Fost, 1990; Collins, 1991; see also Chapter 14, by Bradby). Valuable lessons can be learnt from this programme, which failed to recognise the importance of public education, counselling provision, prenatal diagnosis facilities, and measures to prevent wrongful discrimination, in order to ensure that similar adverse consequences do not result from future screening programmes.

The development of policies is of major importance to minimise potential risks.

Among such policies that are necessary is the development of educational programmes for both professionals and the public so that they become more informed about genetics, its new applications and the opportunities that they offer. By this means, problems that arise, for example, out of the misinterpretation of genetic information, which can result in discrimination, ostracism and stigmatisation, could be avoided. More stringent controls could be implemented to reinforce privacy and so prohibit disclosure or access to genetic data, particularly if it is likely to cause harm, together with possible reform of anti-discrimination legislation.

However, despite such policies it could be argued that, with the predicted expansion in application of genetic technology, the imperative to explain all human 'health and disease, normality and abnormality' in terms of genetics (Lippman, 1991) will become stronger. There is also the possibility that it will reinforce adverse social attitudes and prejudices towards the disabled. The former need not necessarily produce a harmful consequence and may, in fact, reduce the likelihood of the latter.

Genetic research will, in all likelihood, discover that many disabilities have some genetic basis and, as genetic testing and screening proliferate, more people will learn of their genetic risks and become better informed in genetics. Gradually society could come to realise better that the 'luck of the genetic draw' falls indiscriminately, and that each and every person runs a risk of inheriting a genetic disorder, irrespective of race, colour or social status.

We all share, as human beings, in the order of 100 000 different genes. Ultimately, as society gains more understanding of some of the underlying causes of differences between individuals, manifested as disabilities that can sometimes be explained in terms of molecular differences in a single gene, then the distinction between 'normality' and 'abnormality' might be placed into a more realistic perspective. Societal perceptions of disability are often fuelled by fear stemming from ignorance. With genetic knowledge, these perceptions may change so that society becomes more tolerant and sympathetic towards those who are disabled merely through the misfortune of their genetic inheritance. Furthermore, the knowledge might satisfy a frequently experienced psychological need, not only of the disabled themselves but also of parents of disabled children, to be provided with some explanation of the cause of their disabilities.

In summary, provided that safeguards and policies are formulated to ensure that ethical principles are not violated, an expansion of genetic testing and screening, made possible through HGA, can in the shorter term be expected to facilitate significant benefits for society. These will be achieved through more accurate diagnosis of genetic disorders, earlier therapeutic intervention for the alleviation of symptoms and an increase in reproductive choices to avoid the birth of affected children. The ultimate advantages for society, however, are likely to be achieved further into the future. These will be mediated through new forms of therapy, which may raise particularly difficult ethical issues previously not encountered.

13.6 The longer-term applications of the new genetic knowledge

HGA, in facilitating new understanding of genetic diseases, may eventually permit the development of therapies for many diseases that are currently untreatable or incurable. These therapies may be of a traditional type or of a revolutionary kind made possible by advances in genetic technology. It is the latter that, by having the potential to effect changes in the genotypes of individuals as yet not conceived, presents the more difficult ethical problems.

Genetic engineering and gene therapy

It is now becoming feasible, through techniques of genetic engineering, to manipulate the human genome so that the adverse effects of defective genes are ameliorated. This is achieved by the insertion of 'healthy' genes into cells. By restoring 'normal' function, such techniques provide opportunities for effecting permanent cures for genetic diseases, rather than merely palliative measures, thereby considerably reducing the burdens of disease on society.

There are two types of gene therapy: somatic cell gene therapy, whereby the genetic content of cells of the individual being treated is engineered to allow the expression of the normal gene; and germline gene therapy, whereby the gametes or totipotent cells of pre-embryos are altered and the genetic modifications are inherited by future generations.

Somatic cell gene therapy is already feasible, and the first trials, for diseases like adenosine deaminase deficiency and cystic fibrosis, are currently in progress. The procedure, rather than raising any new ethical problems that are not already inherent in established medical practice, can be viewed as giving merely greater prominence to issues of, for example, safety, consent, the use of children in trials for the new therapy, and confidentiality (*Report of the Committee on the Ethics of Gene Therapy*, 1992). It is envisaged that, with further research, techniques will become more efficient and will be applied to many more genetic diseases. The benefits to be gained from this form of therapy could be immense in relieving both patients and their families of the suffering and distress frequently associated with a serious and incurable disease. The possibility of treating affected children would also mean that the often traumatic decisions involved in aborting defective fetuses could be avoided. Initially, because such therapy would be expensive, these benefits would be largely confined to the affluent rather than apportioned equally to all those who had need. However, with further refinement of the techniques and the probable large-scale commercial interest of biotechnology companies, this type of therapy could become much more common in the future.

Germline gene therapy, although not yet technically feasible, is likely to become so in the future and therefore requires that debate should already be taking place as to whether or not it should be permitted. Its distinction, from all other therapies, in changing the genetic constitution of future generations, raises complex ethical questions concerning its admissibility. These centre largely upon the difficulties, due to insufficient knowledge, in evaluating the magnitude of risks involved and the possibility of facilitating a slippery slope into enhancement of human traits and positive eugenics. Consequently, there is a consensus in many countries that its practice should not be contemplated at present (European Medical Research Councils, 1988; Committee of the Health Council of The Netherlands, 1989; Cook-Deegan, 1990; *Report of the Committee on the Ethics of Gene Therapy*, 1992; De Wachter, 1993).

One of the ethical issues is the problem of consent, on behalf of the yet unborn, to a procedure that might, in the long term, have harmful, unforeseen and irreversible consequences. It is important, however, to weigh such risks against the potential benefits. Germline gene therapy might constitute the only means, because of difficulties with delivery of genes to target cells, of curing diseases that affect many organs of the body. Its other advantage over somatic cell gene therapy is that it would be a one-time treatment. As the genetic correction would be inherited by the offspring of the patient, not only would repeated therapy for the disease in each generation be unnecessary, resulting in saving of health resources, but also some of the adverse social and psychological consequences of genetic testing and screening programmes could be avoided as some would no longer be required. The ultimate benefit for society, however, might be the realisation of the potential of germline gene therapy for eradicating genetic diseases from future generations (Harris, 1994; Wood-Harper, 1994). This is, however, looking far into the future, and before it could ever be achieved, there would need to be considerably more research and debate as to whether the use of such genetic technology could ever be ethically endorsed by society.

13.7 Final comment

There is wide recognition of both the potential benefits and dangers, to individuals and society, associated with the application of the new genetic knowledge. The concerns can be largely derived from the likelihood of new procedures and therapies infringing upon ethical principles and human rights but are, in the main, no different from issues already raised by established, and generally accepted, medical practices. Rapid progress in both genetic knowledge and biotechnology, however, makes it more imperative that the ethical issues are confronted and resolved by means of safeguards and strict regulation of practices.

To be dismissive of the new genetic technology on the grounds that it might result in unacceptable consequences would be to deprive society of the substantial gains that it makes possible. It is, however, necessary to be aware of the worst scenario and then to work towards a framework of regulation to ensure that it is avoided and that practices which are implemented function within defined limits of what is considered to be ethically acceptable.

An undoubted problem lies in the fact that these limits are frequently not clearly defined, causing difficulty in establishing what is, or is not, to be permitted on ethical grounds; nor are they static over time. As genetic knowledge increases with the likelihood of as yet unfeasible or unforeseen possibilities and as genetic techniques become more commonplace, then it is probable that the basis of what is considered to be permissible may shift. For instance, although germline gene therapy is currently viewed to be unethical, as more knowledge and further refinement of genetic engineering techniques are achieved, with the consequent reduction of risk, there may evolve a stronger case for arguing that the responsibility of the present generation, to use its knowledge and technical skills to secure the welfare of future generations by ensuring that they do not have to be burdened with genetic disease, outweighs any possible infringement of their rights.

Rigorous enforcement of controls formulated in the light of continual reassessment of the ethical acceptability of new genetic practices, in advance of their implementation, will be significant in ensuring that the 'vices' of human genome analysis can be avoided and the 'virtues' realised, to both the short-term and the longer-term benefit of society.

13.8 References

Beaudet, A. L. (1990) Invited editorial: carrier screening for cystic fibrosis. *American Journal of Human Genetics*, **47**, 603–5.

Beckwith, J. (1991) Foreword: the human genome initiative: genetics' lightning rod. *American Journal of Law and Medicine*, **17**, 1–13.

Billings, P. R., Kohn, M. A., and DeCuevas, M. (1992) Discrimination as a consequence of genetic testing. *American Journal of Human Genetics*, **50**, 476–82.

Bloch, M. and Hayden, M. R. (1990) Opinion: predictive testing for Huntington disease in childhood: challenges and implications. *American Journal of Human Genetics*, **46**, 1–4.

British Medical Association (1992) *Our Genetic Future*. Oxford: Oxford University Press.

Collins, F. S. (1991) Medical and ethical consequences of the human genome project. *Journal of Clinical Ethics*, **2**, 260–7.

Committee of the Health Council of the Netherlands (1989) *Heredity: Science and Society – On the Possibilities and Limits of Genetic Testing and Gene Therapy*. The Hague: Health Council of the Netherlands.

Cook-Deegan, R. M. (1990) Human gene therapy and congress. *Human Gene Therapy*, **1**, 163–70.

Danish Council of Ethics (1993) *Ethics and Mapping of the Human Genome*. Copenhagen: The Danish Council of Ethics.

Dawson, K. and Singer, P. (1990) The human genome project: for better or worse? *Medical Journal of Australia*, **152**, 484–5.

DeGrazia, D. (1991) The ethical justification for minimal paternalism in the use of the predictive test for Huntington's disease. *Journal of Clinical Ethics*, **2**, 219–27.

De Wachter, M. A. M. (1993) Ethical aspects of human germ-line gene therapy. *Bioethics*, **7**, 166–77.

European Medical Research Councils (1988) Gene therapy in man. *Lancet*, **i**, 1271–2.

Fletcher, J. C. and Wertz, D. C. (1991) An international code of ethics in medical genetics before the human genome is mapped. In Z. Bankowski and A. M. Capron (eds), *Genetics, Ethics and Human Values – Human Genome Mapping, Genetic Screening and Gene Therapy* (*Proceedings of the XXIVth CIOMS Conference*, Tokyo and Inuyama City, Japan, 22–27 July 1990), pp. 97–116. Geneva: Council for International Organisations of Medical Sciences.

Fletcher, J. C. and Wertz, D. C. (1992) Ethics and prenatal diagnosis: problems, positions, and proposed guidelines. In A. Milunsky (ed.) *Genetic Disorders and the Fetus – Diagnosis, Prevention, and Treatment* (3rd edn). Baltimore and London: Johns Hopkins University Press.

Fost, N. (1992) Ethical implications of screening asymptomatic individuals. *FASEB Journal*, **6**, 2813–17.

Friedmann, T. (1989) Progress toward human gene therapy. *Science*, **244**, 1275–81.

Friedmann, T. (1990) Opinion: the human genome project – some implications of extensive 'reverse genetic' medicine. *American Journal of Human Genetics*, **46**, 407–14.

Gilbert, F. (1990) Is population screening for cystic fibrosis appropriate now? *American Journal of Human Genetics*, **46**, 394–5.

Harper, P. S. (1992*a*) Editorial: the human genome project and medical genetics. *Journal of Medical Genetics*, **29**, 1–2.

Harper, P. S. (1992*b*) Genetic testing and insurance. *Journal of the Royal College of Physicians*, **26**, 184–7.

Harris, J. (1992) *Wonderwoman and Superman – The Ethics of Human Biotechnology*. Oxford: Oxford University Press.

Harris, J. (1993) Is gene therapy a form of eugenics? *Bioethics*, **7**, 178–88.

Harris, J. (1994) Biotechnology, friend or foe? Ethics and controls. In A. Dyson and J. Harris (eds), *Ethics and Biotechnology*, pp. 216–29. London: Routledge.

Holtzman, N. A. (1992) The diffusion of new genetic tests for predicting disease. *FASEB Journal*, **6**, 2806–12.

Laurance, J. (1994) Experts seek the public's view on 'womb-robbing'. *The Times*, 8 January, p. 4.

Lippman, A. (1991) Prenatal genetic testing and screening: constructing needs and reinforcing inequities: *American Journal of Law and Medicine*, **17**, 15–50.

Modell, B. (1990) Cystic fibrosis screening and community genetics. *Journal of Medical Genetics*, **27**, 475–9.

Modell, B. (1993) EC Concerted Action on developing patient registers as a tool for improving service delivery for haemoglobin disorders. In G. N. Fracchia and M. Theophilatou (eds.), *Health Services Research*. Amsterdam: IOS Press.

Natowicz, M. R. and Alper, J. S. (1991) Genetic screening: triumphs, problems, and controversies. *Journal of Public Health Policy*, **12**, 475–91.

Nelkin, D. and Tancredi, L. (1989) *Dangerous Diagnostics – The Social Power of Biological Information*, p. 64. New York: Basic Books.

Netherlands Scientific Council for Government Policy (1990) *The Social Consequences of Genetic Testing* (*Proceedings of a Conference on 16–17 June 1988*). The Hague: Netherlands Scientific Council for Government Policy.

Nuffield Council on Bioethics (1993) *Genetic Screening – Ethical Issues*. London: Nuffield Council on Bioethics.

Office of Technology Assessment, U.S. Congress (1990) *Genetic Monitoring and Screening in the Workplace*. Doc. No. OTA-BA-455. Washington, DC: Government Printing Office.

Report of the Committee on the Ethics of Gene Therapy (1992) Cm. 1788. London: HMSO.

Rosner, F. (1991) Screening for Tay–Sachs disease: a note of caution. *Journal of Clinical Ethics*, **2**, 251–2.

Wilford, B. S. and Fost, N. (1990) The cystic fibrosis gene: medical and social implications for heterozygote detection. *Journal of the American Medical Association*, **20**, 2777–83.

Wood-Harper, J. (1994) Manipulation of the germline – towards elimination of major infectious diseases? In A. Dyson and J. Harris (eds.), *Ethics and Biotechnology*, pp. 1121–43. London: Routledge.

14

Genetics and racism

HANNAH BRADBY

14.1 Introduction

Advances in scientific or technical knowledge are often accompanied by a concern about how the new knowledge will disrupt the existing moral order. The moral value of scientific knowledge depends on the use to which it is put, and the historical precedence of the abuse of theories of inheritance against various groups of people, including those distinguished by their ethnicity, has brought an urgency to concerns as to how the new genetics might be used against people. The continuing intolerance towards people of different ethnicities and religions, illustrated by the recent conflicts in Rwanda and the former Yugoslavia, indicates that the end of the Nazi regime in Germany did not mark the end of inter-ethnic genocide. The study of inheritance and human populations, since the Nazis were defeated in World War II, is often portrayed as neutrally progressing towards a 'truth', and not as an activity that has political implications. Decisions made about the type of research work to fund, what becomes defined as a clinical problem and the ends to which results are put, are guided by forces that are political in nature. An outcome of the supposedly neutral scientific research process, on which this chapter will focus, is the castigation of particular sections of society, because their habits, such as marriage patterns or diet, do not conform with what is considered to be genetically sound.

The ways in which genetics may be used against ethnic minority groups in the industrialised nations is not an 'ethnic minority problem', but a problem of the dominant ethnic group. This chapter is not about ethnic minorities and genetics, as this would imply that any dangers are somehow the responsibility of the minority group. Rather, it focuses on the possibility of racism being fuelled by procedures and research findings of the new genetics. To do this it is necessary to define the relationship between ethnicity, race and racism. What people call themselves and what they are called by others is of considerable importance and nowhere more so than when discussing genetics, where there is potential for confusion in the use of terms that have both a 'common sense' and a specialist meaning. Therefore, this chapter will start by discussing the use of language and the implications to be drawn from this. Three interrelated aspects of the practice of the new genetics that might affect ethnic or racialised minorities will be examined: screening, patient–professional relations and research.

14.2 Race

Until World War II, race was usually considered to be a scientific concept, a biological category used to 'measure' geographical, religious or skin-colour-based groupings, primarily associated with physical anthropology. Subsequently, scientific opinion has rejected race as a useful classificatory tool (Barkan, 1992). In terms of genetics there is no evidence for the existence of discrete, self-reproducing groups of people that could be called races. Lewontin succinctly sums up the evidence by stating that 85% of all identified human genetic variation is accounted for by differences between individuals, whereas only 7% is due to differences between what used to be called races (Lewontin, 1993). In some cases the term 'race' might be useful as an analytical category for the study of racism (Miles, 1993), but its use might confer analytic status upon something that is no more than an ideological construct (Phizacklea, 1984). In a society where racism is the official policy of a small minority of elected representatives (Pilkington, 1993; *Guardian*, 1993), the danger of giving race scientific credence by using it as a term in research is to be guarded against. The challenge is then to analyse the causes of racism while avoiding the implication that race exists (Miles, 1993).

14.3 Racism and racialisation

To define racism we need to acknowledge that, although expression of a belief in the existence of biologically distinct races is now rare, there continues to be a discourse of the signification of some real or alleged biological characteristic as a criterion of other group membership that also attributes to that group further, negatively evaluated characteristics (Miles, 1993). The process of racialisation, whereby meaning is attributed to particular biological features of human beings, as a result of which individuals may be assigned to a general category of persons that reproduces itself biologically (Miles, 1989), is important for an analysis of racism.

14.4 Institutional racism

It is relatively easy to identify racism where individuals or policies use a notion of humanity divided into a hierarchy of biologically distinct races. But what of instances where the expressed intention of the policy is not racist but the effect is to disadvantage one or more racialised groups? Institutional racism is defined as circumstances where racism is embodied in exclusionary practices or in discourses that are formally non-racialised but where it can be demonstrated that racism has or has had a determinant influence (Miles, 1989). In the current practice of genetics, the well documented evidence of racism in science, including genetics and anthropology (Gould, 1980; Rose *et al.*, 1984; Kevles, 1985; Jones, 1994) constitutes a racist discourse that fulfils the criteria of a determinate influence in claims about institutional racism. That is to say, where genetics draws attention to differences between ethnic groups it is difficult *not* to interpret such data in a framework that sorts people into a hierarchy with a moral dimension.

14.5 Ethnicity and racialisation

The groups that tend to be subject to racialisation in Britain (and elsewhere) are ethnic minorities, where ethnicity is defined as the real, or probable, or in some cases mythical, common origins of a people with visions of a shared destiny (Brah, 1993), which are manifested in terms of the ideal or actual language, religion, work, diet, or family patterns of that people. From this definition it can be shown that an ethnic group need not necessarily be racialised and that it is not always ethnic majorities who subject minorities to racialisation and discrimination, as illustrated by the powerful ethnic minorities in South Africa under apartheid and in the ex-colonies of the British Empire, who have racialised and discriminated against the ethnic majority. Racialisation can take place where the two ethnic groups have different skin colours, but this is not a necessary condition (Miles, 1989) as in the examples of anti-Jewish and anti-Irish racism (Cooper, 1984; Rossiter, 1992).

Since its earliest use in the nineteenth century, race is a classification that has been 'objectively' allocated to groups of people by outsiders. Ethnicity has been used as a complementary term that encompasses socially mediated differences between groups, such as language, food, dress, marriage patterns and people's self-identifications (van den Berghe, 1978). Here it is argued that the term race should no longer be used because of its historical associations, and that ethnicity should be defined to include the social and religious differences between groups and such physical and genetic differences that are associated with particular descent groups.

14.6 Ethnocentrism

In this chapter ethnocentrism will be taken to mean a belief in the superiority of one's own ethnic group and the corresponding dislike of other such groups. Although the belief does not rely on the existence of race, where ethnocentrism is discernible it is likely that racism might be in danger of developing (British Sociological Association, 1990).

14.7 Migration

Another term often associated with ethnic groups is 'migrants'. Once an ethnic minority has been in a country for more than one generation, it is erroneously referred to as an immigrant group, particularly as it implies that British-born minority ethnic people have incomplete rights as citizens (Donovan, 1984). Migration may have played an important part in both the history and identity of a group, as well as the evolution of its particular set of frequencies of various genes. Although populations cannot be divided up into distinct groups on the basis of their genetics, there may be differences in the frequencies of particular genes, which can be of clinical significance. Gene frequencies may be subject to environmental and evolutionary influences (e.g. sickle cell genes), and can be affected by social factors such as migration and marriage patterns.

14.8 Slippage between terms

The concept of ethnicity is difficult to grasp because it varies with political and historical context and with the passing of time. If the relationship between illness and ethnicity is to be studied, it must be accepted that it is likely to be complicated, interactive and time dependent (Cooper, 1993). It is important to be able to conceptualise ways in which peoples might differ from one another without slipping back into a 'common sense' idea of race as encapsulating some inherent disjuncture in humanity. It is also important to remember that ethnicity is something that everyone has, not just the minority ethnic groups, although there has been a dearth of research exploring what constitutes the majority ethnicity in Britain (see Frankenberg, 1993). The complexity of summing up the meaning of ethnicity in a single term, without falling back into essentialism, has led some to comment that the term 'ethnic' is woolly (Jones, 1994) and non-committal (Huxley and Haddon, quoted in Kevles, 1985). I would argue that the woolliness and lack of commitment is in the usage of the term, rather than in the term itself.

14.9 Different types of term

For evidence of the considerable confusion between race and ethnicity one need only look at the terminology used in articles about genetics, which not only use terms referring to ethnic divisions and so-called racial divisions as though they were synonymous, but also political and national terms. For instance, an academic article uses 'African-American', 'Caucasian', 'Northern Irish' and 'Israeli' as four comparison groups for a study of the cystic fibrosis gene in different populations. Each of these terms refers to a different type of division in humanity: African-American is a political/ethnic term, Caucasian derives from discredited race theory, Northern Ireland is a region (albeit disputed) of a nation and Israel is a nation (Cutting *et al.*, 1992). One source uses Caucasian and white as interchangeable terms (Lebel, 1983), another uses Caucasian and black as contrasting terms (Donegan, 1993), and a third describes the possibility of genetic differentiation between 'Caucasian' and 'Afro-Caribbean' people while noting the difficulty of distinguishing between 'subcategories of Caucasians such as Europeans and people of Indian and Pakistani origin' (Bown, 1993*b*). As an illustration of the reigning confusion, consider those Britons of Pakistani and Indian origin, who may identify themselves, and be identified by others, with the political term 'black' who, in this last article, are subsumed under 'Caucasian', which in other contexts is synonymous with 'white'.

14.10 Caucasian

The use of the term 'Caucasian' is another instance of the confusion and overlap between ideas of race, racialisation and ethnicity. It is used in both academic papers (e.g. Lebel, 1983; Williamson, 1993) and newspaper articles (e.g. Donegan, 1993), usually to refer to people of the majority ethnicity of the United States of America or the states of northern Europe. The term comes from the era when measuring skulls was thought to be a means of identifying races, and skulls from the Caucasus mountains were thought to represent the purest form of the white-skinned race (described as 'the most beautiful race of men' by Blumenbach (quoted in Jones, 1994), who came from this area). Given its association with race theory, the use of the term Caucasian as a scientific classification (Cutting *et al.*, 1992; Williamson, 1993; Jones, 1994) seems ill-judged, especially when it is being contrasted with other categories that are nationalities or political groupings.

Until the term 'ethnicity' is clearly dissociated from race it will be difficult to speak of ethnic differences in genetic health without racist overtones. The slippage between the two terms means that ethnic differences in genetic diseases may be interpreted as fixed, immutable and indicative of other differences. The fact that certain genetic diseases are associated with particular ethnicities can be, and is, interpreted within a racist frame work which in an extreme manifestation questions the legitimacy of residency of minority ethnic groups in Britain (see, for example, Anionwu, 1993). This has been attributed to 'the ancient impulses setting group against group' (Kevles, 1985), but whether or not this explanation is accepted, it should be a priority to minimise the possibility of its happening. This task is made more urgent by the presence of an audience that is receptive to genetic evidence of the existence of what are seen as inherent or fundamental differences between ethnic groups. Such evidence can be used to give racist beliefs and practices a gloss of scientific 'truth'.

14.11 Racism and eugenics

Whenever racism and the science of heredity have been linked in the past it has been called eugenics. The term was coined to indicate the science of improving the human race, before genetics existed, and has rightly earned a bad name, given the horrific means that have been used to this end (Kevles, 1985; Wilfond and Fost, 1990; Cooper, 1993). Eugenics, as applied to different ethnic groups, rests on a notion of humanity as divided into a series of biologically distinct lineages that differ in quality, with pale-skinned, narrow-nosed, straight-haired, northern Europeans seen as the pinnacle of development (Goodey, 1991; Jones 1994), that evolved in parallel with theories about human heredity and anthropological explorations. This racist notion was the justification for the annulling of all 'mixed' marriages in German South West Africa (Jones, 1994). In the USA, as well as Nazi Germany, eugenic doctrine was virulently deployed against minority groups, some of which were defined as minorities because of their alleged moral or intellectual disabilities (Gould, 1980; Rose *et al.*, 1984; Kevles, 1992), and many of whom were minority ethnic groups. The close historical relationship between eugenics and racism must be a good reason for guarding against any hint of eugenics in the practice, let alone the theory of new genetics (see Greer, 1994; Malik, 1994).

There is a suggestion that a form of eugenics exists that has no relationship with racism. It has been termed negative eugenics (Wertz, 1992) or, less favourably, genetic paternalism (Birke *et al.*, 1990). In contrast with the positive eugenics of the Nazi regime, negative eugenics refers to the prevention, or reduction, of births with genetic disease and possibly a wider promotion of well-being in individuals, but without reference to the welfare of the population (Congress of United States Office of Technology Assessment, 1988; Wertz, 1992). The link with old-style positive eugenics is refuted on the grounds that the well-being of the individual rather than that of the population has become the responsibility of geneticists (Kevles, 1985; Miringoff, 1991; Jones, 1994). Perhaps this individualistic approach is workable in a society that is free of individual and institutional racism, but we can only imagine such a society. Kevles's dictum that attention should be paid to the dissenters from the eugenic revival 'however shrill they may sometimes be' (Kevles, 1985) is worth repeating here, given that the exercise of looking for parallels between the historical precedent of the Nazi use of eugenics and current genetic practices has been dubbed sensationalism by others (e.g. Weatherall, 1985). Knowledge of the historical abuse of people in the name of eugenics lends an urgency to contemporary discussion of discrimination and genetics (Kevles and Hood, 1992; Wertz, 1992). The fact that the term eugenics, with all its historical connotations, is being reclaimed in the name of genetic science and of the promotion of the welfare of the individual (see Congress of United States Office of Technology Assessment, 1988) should make us feel uneasy. Ethnic origin has been the basis of discrimination in the past (Wilkie, 1993), and the desire to improve humankind has repeatedly formed the basis of social movements (Miringoff, 1991), which makes reclamation seem a risky ploy.

14.12 The new genetics

The issues discussed in this chapter, such as abortion and social stigma, date back to long before the coining of the term 'the new genetics', and the concern is not so much with the new ethical dilemmas that might be produced by the technologies (see Harris, 1992) as with the new opportunities for the mishandling of old ones (Wertz, 1992).

14.13 Specific conditions

There are certain single-gene disorders with relatively high penetrance that are associated with particular ethnicities. Sickle cell disease is an umbrella term that describes a group of inherited disorders of the haemoglobin in the red blood cells (Anionwu, 1993) in which the alterations to the haemoglobin molecule cause the red blood cells to change to a sickle shape under certain conditions, blocking blood vessels, causing pain and damage to the organs where the blockage has occurred (Prashar *et al.*, 1985) (see Chapter 1 for personal accounts of these conditions). It is mainly associated with African–Caribbean ethnicity.

Thalassaemia covers a heterogeneous group of disorders caused by the partial or non-production of one of the chains of the haemoglobin molecule, the consequences of which can vary from a dependence on blood transfusions and high risk of childhood death to a mild form of anaemia (Weatherall and Letsky, 1984). It is most clearly associated with people who can trace their origins to Mediterranean, middle Eastern and south Asian countries, although those of Chinese, Caribbean and African background also show a higher carrier frequency than the majority British population (Anionwu, 1993). Sickle cell and thalassaemia disorders are collectively referred to as haemoglobinopathies.

Cystic fibrosis describes the inability to make a protein that facilitates the movement of chloride ions through the cells lining the lung, resulting in dry airways and less protection from infection, among other disabilities (Geddes, 1994), and it is associated with northern European ethnicity. Tay–Sachs refers to the deficient activity of a lipid-degrading enzyme, which leads to the accumulation and deposition of the lipid in various tissues (Hecht and Cadien, 1984); it is associated with Ashkenazi Jewish ethnicity.

The reason for the strong association between ethnicity and a disease-causing allele is relatively clear with haemoglobinopathies: the carrier status offers protection against malaria and it is in areas where this disease is or was endemic that these disorders are found in the highest frequency. The movement of peoples has meant that the disease is also found in non-malarial areas. The existence of different polymorphisms indicates that the mutations happened more than once. It is more difficult to account for the association between Tay–Sachs and cystic fibrosis and ethnicity because there is no obvious selective advantage offered by carrier status that could account for the spread of the gene. It has been suggested, without firm evidence, that protection from tuberculosis in the past could account for the distributions of both conditions (Hecht and Cadien, 1984; Cutting *et al.*, 1992). Where a mutation has occurred in a relatively isolated population, a founder effect and random genetic drift can account for the establishment of an allele, despite carrying some selective penalties.

Table 5.1 (p. 124) gives estimated frequencies of affected births for the conditions most likely to affect different ethnic groups in Britain.

Although associated with particular ethnicities, these conditions also occur in other ethnic groups; for instance, Tay–Sachs occasionally occurs in non-Jewish populations (Hecht and Cadien, 1984), sickle cell disease is thought of as a 'black disease' (Kirk, 1987) but occurs also in those of northern European origin (*American Journal of Nursing*, 1993) and cystic fibrosis, which is thought of as a 'white disease', can occur in those of African (Owen *et al.*, 1993) and of Pakistani (Anionwu, 1993) descent. Within the broad ethnic group of 'African' there are important differences in prevalence of sickle cell disease according to local ethnic variations (Aluoch and Aluoch, 1993). Thus, the association between a particular ethnicity and a single-gene condition is not such that the two are coexistent: the relationship is statistical, not absolute. This is due both to spontaneous mutations and to movement between ethnic groups; where migration has brought two ethnic groups into proximity it is social mores that promote or discourage the mixing of the two gene pools. Where there is strong discouragement, taboos are always broken by some people, for instance the incidence of sickle cell is higher among Americans of northern European descent from the south of the USA, compared with those of the same ethnicity elsewhere. This points to mixing with the gene pool of African Americans and northern European Americans, which blood group studies confirm. In the future, associations between ethnic groups and diseases will probably become 'looser' owing to increasing numbers of children with parents of different ethnicities (Horn *et al.*, 1986).

The conditions shown in Table 5.1 are the diseases with a genetic component most strongly associated with particular ethnic groups, but there is a wide range of other conditions whose distributions vary with ethnicity, for instance peptic ulcers, diabetes mellitus (Emery and Rimoin, 1990*b*), albinism and multiple sclerosis (Emery and Rimoin, 1990*a*). The variation of prevalence of disease with ethnicity exists for different conditions, in complex association with other variables. The relationship between ethnicity and genetic disease is further complicated by the variable expression of genetic material owing to interaction with the enviornment (Davison *et al.*, 1994). For instance, the course of sickle cell disease in an individual is influenced by climatic and socio-economic factors (Weatherall and Letsky, 1984) and can be exacerbated by physical strain (Lewontin, 1993) and deprivation (Horn *et al.*, 1986). In Britain, people with cystic fibrosis can now expect to live to about 25 years if in receipt of appropriate treatment (Geddes, 1994), whereas until recently few reached adulthood. Therefore the progress of a disease is dependent on many more factors than ethnicity alone. Many diseases do not show an association with ethnicity, and although this chapter concentrates upon disorders that disproportionately affect racialised minorities, conditions exist that affect racialised minorities less than majorities, including certain dominant genetic disorders, such as Huntington's disease.

Upon the diagnosis of one of the autosomal recessive disorders discussed above, genetics can offer the individual screening and counselling followed, in some cases, by remedial therapy, but in most cases by screening future pregnancies and the possibility of aborting fetuses with the condition. In the long term, research is needed to find better preventative and remedial therapies.

14.14 Screening

A useful starting point for discussing screening is its definition as the identification, among apparently healthy individuals, of those who are sufficiently at risk of a specific disorder to justify a subsequent diagnostic test or procedure (Wald, 1984) (see Chapters 3 and 6). One example would be ascertaining the age of all first-time pregnant women in order to perform amniocentesis or chorionic villus sampling on older women, for the detection of Down syndrome in the fetus. Universal screening is where every individual is tested for a condition, such as cystic fibrosis. One of the simplest screening methods is to ascertain the ethnicity of the individual, to judge whether they have a higher risk of carrying any particular recessive disorders. There are important questions about the level of costs and benefits to the individual and to society that justify the introduction of screening programmes and about the optimal time in a person's life cycle at which to offer them information on their genotype (Hecht and Cadien, 1984; Serjeant, 1985; Horn *et al.*, 1986; Kirk, 1987; Dodge, 1988; Williamson, 1993; Green and France-Dawson, 1994). While these questions are still being debated in the literature, there is a preceding question about the use that is made of the information obtained by screening. There is disagreement as to whether the prime purpose of screening is the reduction of the incidence of affected births (the criterion by which the success or failure of screening programmes is often judged), or helping individuals to take decisions about their own procreation (Wilfond and Fost, 1990), an aspect that is emphasised by those who feel that non-directive counselling is an important part of the screening process. Opinions vary, and some writers put no emphasis on the individual's interests whatsoever, for instance: 'The medical objective of genetic screening is to predict and prevent disease of heritable nature' (Zeesman *et al.*, 1984).

One aspect of this question is the uses to which individuals put information about their own genotype, and although it is not a well-explored area (Weatherall, 1985) there is some research into the psychological impact of screening during pregnancy on women (see Green, 1992; Green and France-Dawson, 1994; see also Chapter 6). Another issue is the use to which other individuals and institutions put the information revealed through screening. Where early knowledge of a homozygous state offers the possibility of better prophylactic care, as with phenylketonuria and sickle cell, screening is of obvious benefit to the individual in terms of their health and physical well-being (Horn *et al.*, 1986). Screening can identify individuals who are heterozygotic as well as homozygotic for recessive conditions. Individuals may want this information to plan their reproductive careers, but the possession of information carries hazards as well as benefits. There is evidence to show that discrimination on grounds of heterozygous and homozygous status for recessive conditions can occur (Duster, 1990). Given that ethnic minorities are often racialised and discriminated against, they might be particularly vulnerable to the risk of further discrimination.

It has been suggested that with some diagnostic tests, the costs of stigma outweigh any possible benefits (Wertz, 1992). The questions of what information people want about their recessive genotypes, how they can use it, how others can abuse it and whether any of this can be affected by the type of genetic counselling that is offered, are under-researched. The lack of research does not prevent the expression of opinions about the programmes that have been run. With respect to ethnic minorities and screening, the US sickle cell programme is held as an example of how not to screen, and some Tay–Sachs and thalassaemia programmes are seen as examples of good practice.

14.15 Sickle cell

The National Sickle Cell Anaemia control act, enacted by the US Congress on 16 May 1972, called for mandatory screening for the sickle cell trait in all people who were 'not of the Caucasian, Indian or Oriental race' (Kevles, 1985, p. 278) and marks the start of a screening programme that is held up as the benchmark for failure. The act had a serious effect on the lives of US citizens of African descent; in some states marriage licences were not issued to those of African descent unless the sickle cell test was taken (Kevles, 1985), in others a test was a condition of entry to school (Duster, 1990). The presence of the sickle cell trait could lead to loss of employment, increased insurance premiums, inappropriate medical therapy, delay in adoption of children and sometimes problems in the revelation of 'non-paternity' (Serjeant, 1985). It was as recently as 1981 that the Air Force Academy finally lifted its ban that prevented Americans of African descent, with the sickle cell trait (but not disease), from flying. In the fervour to stop the trait from being passed on to subsequent generations, one scientist suggested that 'there should be tattooed on the forehead of every young person, a symbol showing possession of the sickle-cell gene' to prevent them from 'falling in love with one another' (quoted in Miringoff, 1991, p. 50).

Although initially welcomed as an initiative to meet the long-neglected health needs of African Americans, and as a means by which they could take control of their own lives (Duster, 1990), the US sickle cell disease programme was subsequently described as racist, as a form of anti-black eugenics and even a step towards genocide (Kevles, 1985). There is widespread agreement that the programme was highly stigmatising (Kevles, 1985; Serjeant, 1985; Wilfond and Fost, 1990; Miringoff, 1991), and that the treatment of carriers of the sickle trait, both homozygous and heterozygous, was punitive (Duster, 1990). Attempts to explain the failure of the programme include the suggestions that the 'objectives of the programmes were not clearly stated or appreciated, there was confusion between the significance of the sickle cell trait and of sickle cell disease, and educational programmes were not always available to interpret the results of screening' (Serjeant, 1985, p. 375), and that it was due to inadequate provision of genetic counselling and the absence of a decision as to what to do with the information once the population had been screened (Weatherall, 1985). Only one author (Jones, 1994) raises the issue of the programme being directed at a racialised minority already subject to the stigma of racism, such that implications of inferiority chimed in with a long tradition of viewing those of African descent as less fit and less intelligent than other groups (Rose *et al.*, 1984).

In place of research to corroborate the suggestions for the reason for failure, reference is made to other genetic screening programmes that are judged to have worked, in as much as the frequency of affected births has fallen, and 'stigmatisation' does not appear to accompany knowledge of carrier status. In some cases these suggestions are obviously wrong; one author attributes the success of thalassaemia and Tay–Sachs programmes to intensive education campaigns before screening and to the geographical or ethnic criteria that linked and defined the populations concerned (Dodge, 1988). However, there are instances of unsuccessful programmes where one or both of these criteria were also fulfilled (Kevles, 1985; McKie, 1988).

The difficulty of assessing or even summarising the impact of the screening programmes on ethnic minorities is compounded by the assumptions and contradictions running through the writing about them. One is the issue of the purpose of screening: although non-directional counselling (Birke *et al.*, 1990; J. A. Raeburn, 1994) is seen as part of the screening process, the success of a programme is judged by the reduction in the number of affected births (Horn *et al.*, 1986; Modell and Petrou, 1989). Therefore, in the measurement of success there is no room for the idea that someone might be screening and counselled and then make use of this information to decide to proceed with an affected birth. The success of other programmes is judged by the number of people who take up screening, disregarding all those who make an informed decision to refuse to be screened. It is implied that there is a straightforward equation between the supply of information to individuals and the likelihood of their choosing to avoid producing offspring with genetic disease (see, for example, Haan, 1994). A programme is felt to have failed where stigmatisation of those who have been screened positive for heterozygote or homozygote status occurs. Unfortunately, what is meant by stigma or stigmatisation is not made clear, and the following examples will show that it is difficult to deduce what is meant from its usage. One writer feels that sickle cell is stigmatising 'as a consequence of its racial link' (Kirk, 1987), implying that the disease is only liable to attract stigma because it mostly affects black people. If this is the case, then the stigma arising from screening for the diseases that disproportionately affect racialised minorities is simply another aspect of racism. This does not seem to be the case because not all screening among racialised groups has given rise to stigma. The use of the term stigmatise is of note in the following passage:

> It is essential that society does not stigmatise or penalise individuals who are shown to be at high risk. After all they are normal at the time of testing and may not develop symptoms for several decades. *(Haan, 1990, p. 177)*

The statement that it is unfair for 'normal' people without symptoms of the disease to be stigmatised carries an implicit message that stigmatisation is only to be expected for those with symptoms. This statement could be taken as meaning that people who are 'abnormal' through being of a minority ethnicity could also expect to be stigmatised. The following statement by a clinical scientist makes a distinction between two types of stigma:

> Stigmatisation of heterozygotes, whether real or perceived, would disappear with universal understanding that we are all heterozygotes for a number of serious conditions. *(Dodge, 1988, p. 673)*

It is not clear how the author defines real as opposed to perceived stigma, but it may approximate to the distinction developed by Scambler (1989) drawing on Goffman's work (1990), between felt and enacted stigma. In this case 'felt stigma' refers to a spoiled or contaminated identity that can be felt by an individual, but hidden from others. Stigma that is enacted comes close to an idea of discrimination that may or may not be legitimate (Scambler, 1989). Deciding whether or not to run a screening programme, or how to maximise the acceptability of a programme to individuals, depends upon minimising the enacted stigma, or at least minimising people's perception that enacted stigma will occur. The perceived and the enacted stigma may differ considerably. Some research suggests that concern about stigma from public knowledge of carrier status dissuades people from being screened (Hecht and Cadien, 1984). The concept of stigma as used in the study of other medical conditions, such as cancer and epilepsy, needs developing in the context of genetic screening, as at present it is not clear exactly what people are keen to minimise during the process of genetic screening.

A second assumption betrayed by the quotation from Dodge (1988) is that stigma is likely to disappear once lay people are given enough information about the 'real' genetic situation. This is similar to the assertion that the crucial element for the success of a screening programme, in terms of minimising stigma and the reduction of affected births, is intensive community education (Hecht and Cadien, Zeesman *et al.*, 1984; Weatherall, 1985; Horn *et al.*, 1986; Dodge, 1988). Again, it does not allow that 'the community' or the individual might take the education programme that delivers information about the 'real genetic situation' very seriously, and then decide that screening is not something of which they wish to make use. There is evidence that education associated with screening programmes can increase knowledge about the disease in question, can bring about attitudinal change, but still be followed by a very low uptake of screening (Hecht and Cadien, 1984).

The fact that education programmes do not always lead to the outcome desired by the educators is illustrated by work on racism in the health services. For instance, increased knowledge about the stigmatised, racialised group among racists can be used to render the racism more informed, accurate and therefore cutting, rather than having a dissipatory effect (Currer, 1984). The fact that education does not necessarily prevent stigmatisation is illustrated by the case of a Greek village, where a vigorous community education programme was pursued before screening the whole village in order to reduce the number of births with sickle cell disease. Although the disease was not distributed along pre-existing lines of social cleavage, such as membership of a racialised group, those who were found to be carriers of the sickle cell gene were nevertheless socially rejected by those who were not. Carriers had no choice but to marry one another, and they continued to have children so the number of affected births did not fall (McKie, 1988).

Some of the US Tay–Sachs programmes are felt to have succeeded because of the system of arranged marriage that allowed only the marriage brokers access to the information about carrier status. The brokers would not proceed with a match between two carriers, but no one else ever needed to know genotypic information (Wilkie, 1993). However, the Greek village described by McKie also had an arranged marriage system, so this in itself did not guard against stigmatisation. The example of sickle cell shows that more research, both historical and prospective, is needed into the reasons why screening works in some cases but not in others (Weatherall, 1985). Lay interpretations of the connections between kinship, marriage patterns and inheritance are not well understood for minority ethnic groups, and research in this area would have implications for the majority groups as well (see Chapter 12).

14.16 Thalassaemia

A British screening programme for thalassaemia in London is judged a success for Cypriots because an 80% reduction in affected births resulted, whereas the reduction among Indians and Pakistanis was less (Petrou *et al.*, 1990). The success of this programme with Cypriots has been attributed to community education (Weatherall, 1985), and research that addresses the apparent failure of the programme either to reach or to persuade Britons of South Asian origin suggests that religion, previous bad experiences and lack of awareness of the availability of prenatal diagnosis were all part of people's reluctance to use prenatal screening (Anionwu, 1993). The thalassaemia screening programme in Cyprus is cited as a success (Dodge, 1988; Modell and Petrou, 1989), but unlike sickle cell disease in the US, thalassaemia in Cyprus is not confined to an ethnic or a racialised minority of the national population, which might suggest that the risks of stigma and discrimination would differ from the experience of US citizens of African descent or Britons of Cypriot descent. Widespread experience of caring for a sufferer of thalassaemia with the accompanying financial and emotional burdens are factors that have been associated with the success of thalassaemia screening programmes (Richards and Green, 1993).

14.17 Tay–Sachs

Although the Tay–Sachs programmes in US cities are often held up as successful in comparison with the US sickle cell programme, one commentary differs. It states that in contradiction to the claims of a *Lancet* editorial (*Lancet*, 1980), 'screening less than 15 per cent of the at-risk persons per community is not a measure of success', and that 'the United States has a long way to go' (Hecht and Cadien, 1984). If the US has a long way to go, then the rest of the world has even further, with Canada, Israel and South Africa screening fewer of their Ashkenazi Jewish population than the US, and the UK has furthest to go, being pronounced a 'dramatic failure' having screened only 0.1% of British Ashkenazi Jews (Hecht and Cadien, 1984). This worldwide failure is attributed to Tay–Sachs being a rare condition, to a lack of education about medical genetics and particularly Tay–Sachs, and to the greater interest in treatment for genetic disease than prevention (Hecht and Cadien, 1984). Another factor may be that the relationships between identifying oneself as Jewish, being thought of as Jewish by others, and racism vary between different groups of Jews. The success of Tay–Sachs screening in certain US cities may be due to the strong identification with orthodox Judaism and a lack of felt discrimination in these areas, which may affect the uptake of screening, but again this is a matter that could be illuminated with further research. There is a suggestion (though not backed by research evidence) that the relative success of the voluntary screening for Tay–Sachs in US cities, compared with the mandatory screening for sickle cell, was at least in part due to the structural privilege of American Jews compared with African Americans, in terms of education, employment and lack of racism (Duster, 1990).

14.18 Cystic fibrosis

Currently there is debate about whether or not universal screening for cystic fibrosis should be introduced. Lessons from screening for other diseases are being drawn, including worries about how the information of positive screening will be used (Williamson, 1993). Given the relative rarity of the condition, the high estimated cost per case of cystic fibrosis avoided (Wilfond and Fost, 1990) is also a consideration. The British and American ethnic minorities who suffer disproportionately from recessive autosomal conditions tend to be concentrated in urban areas (Brozovic *et al.*, 1989; Petrou *et al.*, 1990; Ballard and Kalra, 1994), so that service provision should, in theory, be easier and cheaper, compared with a proposal to screen universally for cystic fibrosis.

We know very little about the acceptability of screening programmes to minority ethnic groups (Weatherall, 1985). Such research as has been done has been rendered less useful by the small numbers involved and the even smaller numbers within the sample who had made decisions such as whom to marry and whether to procreate, when they might have made use of information obtained through screening (Zeesman *et al.*, 1984).

14.19 Counselling

If there is little research on screening, there is even less on counselling as part of the screening process. It is claimed that people are capable of assimilating data and making choices after non-directive counselling with relative ease (Williamson, 1993). However, this is more complicated with cystic fibrosis than, for example, sickle cell, owing to the high incidence of mutations that are felt to be too expensive to detect in the context of a screening programme (Wilfond and Fost, 1990; Cutting *et al.*, 1992; Owen *et al.*, 1993). This means that probabilistic information must be communicated during counselling. It is claimed that this is likely to be even more difficult with ethnic minorities (Wilfond and Fost, 1990). This may be true when the counsellor and the patient do not have a common language, but not for any other reason.

14.20 Resources

It is the absence of appropriate screening as well as the presence of inappropriate screening that alarms members of ethnic minorities in Britain and the USA (McKie, 1988). Although the US government was willing to fund a genetic screening programme, this has to be seen in combination with the lack of resources devoted to other aspects of black health care, and sickle cell disease care in particular (Miringoff, 1991). The prevalence of the sickle cell trait in people of African Caribbean background is twice that of cystic fibrosis in those of northern European descent (Horn *et al.*, 1986; Dodge, 1988; Franklin, 1988), yet there are fewer resources devoted to its prevention and treatment (Prashar *et al.*, 1985; Black and Laws, 1986; Birke *et al.*, 1990). Among the problems that have been highlighted are: despite accounting for up to 40% of all haematological hospital admissions, there is no good epidemiological study of the natural history of sickle cell disease, to establish prognosis, patterns of morbidity or life expectancy (Brozovic *et al.*, 1989; Birke *et al.*, 1990); the lack of a comprehensive plan of care for sufferers (Horn *et al.*, 1986); and, more generally, the reluctance of the British National Health Service to respond to the needs of sufferers of sickle cell disorders (McKie, 1988; Anionwu, 1993). It has been suggested that the lack of interest and resources is due to the fact that white people rarely suffer from sickle cell (Anionwu, 1993). Evidence from different sources certainly points to racism being at work (Prashar *et al.*, 1985; Black and Laws, 1986; Franklin, 1988; Anionwu, 1993). Despite the genetic basis of sickle cell having been established for 25 years, no therapy or cure has yet been developed (Wexler, 1992). It has been suggested that this is due to racism in the allocation of research funds preferentially towards those conditions that effect the ethnic majority (Black and Laws, 1986).

14.21 Patient–professional relations

If those ethnic minorities at risk of particular genetic diseases are screened, it is likely to lead to clinical decisions. A concern that is not specific to ethnic minorities, nor to genetic conditions, is whether the differing backgrounds of patient and clinician jeopardise the possibility of the patient's best interests being understood and promoted through ethnocentricity. This is part of the wider problem in which the professional or medical interests and the interests of the patient do not coincide. Consider news coverage from early in 1994 of two genetic issues.

The first news item centred on a private clinic in London that claimed to be able to offer parents the ability to choose the sex of their child. There was alarm expressed by politicians and journalists that British Asians might use the clinic to exercise a preference for boy babies (Greer, 1994; Milhill, 1994). Since amniocentesis became a widespread procedure it has been possible to detect prenatally the sex of a fetus, and therefore to perform an abortion if the child is not of the required gender. The practice of selecting male fetuses and aborting female fetuses is relatively widespread among certain social strata in various countries (Davis, 1990). New technologies provide new opportunities for pre-implantation sex selection without abortion (Pergament, 1991), but do not help to answer any of the moral dilemmas about the practice. There is no consensus on what are acceptable grounds for abortion. A US survey found that 60% of medical geneticists surveyed would test a fetus simply for its sex, or would refer to someone who would (*Medical World News*, 1987), yet it is asserted by another American that whatever their views on abortion, no reputable doctor would perform an abortion knowing it was simply to get rid of a female foetus (Davis, 1990). Sex selection has been called a precedent for eugenics (*Medical World News*, 1987), whereas others, although rejecting this claim, have nevertheless condemned sex selection as morally wrong owing to its contribution to the oppression of women (Wertz and Fletcher, 1993). Others have outlined the possible benefits of sex selection including the birth of fewer babies unwanted because of their gender, and that the birth rate might fall if parents achieved the desired sex composition of their family more quickly (Birke *et al.*, 1990).

The criteria that the majority ethnicity consider to be adequate grounds for refusing to continue with a pregnancy, such as a prenatal diagnosis of Down syndrome, may be unacceptable to others. Down syndrome is not a condition like cystic fibrosis which necessarily involves physical suffering. Although there are some congenital conditions associated with Down syndrome, many people who exhibit 21 trisomy do not have the associated problems. It could be argued that having an XX constitution is a genetic disability (Stacey, 1994) compared with the XY constitution. So we could ask why it is acceptable to abort on grounds of Down syndrome, but not on grounds of gender. Just as the meaning of genetic disability varies enormously between societies (McKie, 1988), so too do the implications of gender. Although there is evidence from Britain that there is no strong preference for the sex of children, either in the ethnic majority or ethnic minorities (Statham *et al.*, 1993; Malik, 1994), the question of what constitutes legitimate grounds for abortion is an important one to be considered in the context of genetics, given that the only clinical intervention for many of the conditions for which screening is possible is a so-called 'therapeutic' or 'medical' abortion (Richards, 1989; Cowan, 1992).

The acceptability of abortion varies between and within religions, societies and nations (Weatherall, 1985; Modell and Petrou, 1989; Galjaard, 1990; Panter-Brick, 1991; Anionwu, 1993; Williamson, 1993). In racist societies ethnicity and abortion become linked as women from certain racialised groups, perceived to be 'over fertile' and to have too many children, are encouraged to abort more often than the non-racialised women (Phoenix, 1990). The new genetics does not shed any light on the old ethical dilemmas surrounding abortion, but it can create novel situations where they are brought to the fore (Davis, 1990).

The second news item concerned a woman described as 'black' who was undergoing *in vitro* fertilisation and who, owing to a shortage of black donors, was using a white donor's egg (Donegan, 1993; Timmins, 1993; Fletcher and Muir, 1994; MacKinnon, 1994; McGourty, 1994; *The Times*, 1994). There was considerable debate over whether it was acceptable for the woman to be implanted with a white donor's eggs, but much of this was not so much a discussion of the ethics, as of the alarm among medics at the 'unnaturalness' of a black woman giving birth to a white baby, some commentators taking this as evidence of racism (Malik, 1994; Wilkie, 1994). The clinician interviewed in connection with the decision was quoted as explaining a 'genetic rationale' for this case rather than a moral one: the partner of the woman was of so-called 'mixed race'; therefore any of his progeny would be a mixture, regardless of the colour of the donor. He rejected the notion of implantating a 'black' woman with a 'white' egg under circumstances that he considered frivolous. The language used in the news coverage of this episode revealed a notion of ethnicity as something that is immutable and biological (Cooper, 1993).

Another area where there is evidence of ethnocentricity in some research is that of the practice of cousin marriage (Benallegue and Kedji, 1984; Krishan, 1986; Dunlop and Winter, 1987; Reddy, 1987; Rajeswari *et al.*, 1992). There is an implication that people ought not to indulge in a practice that carries genetic risks, but the social advantages are less often explored. It is suggested that once those who practice cousin marriage understand the genetic risks involved, they will be bound to desist (Feinmesser *et al.*, 1989; Kingman, 1993). There is evidence of good practice (B. Modell, unpublished; Darr and Modell, 1988), as well as bad in the treatment of families who practice cousin marriage and who have genetic disease (Anionwu, 1993).

14.22 Conclusion

This chapter has pointed to areas where there is evidence of racism or ethnocentrism in the practice or theory of the new genetics. The thoughtless use of certain terms means that biological determinism clings to discussions of ethnicity and can cloud the issues under examination. This has implications, particularly for racialised minorities, in the ways in which research findings may be interpreted (Lewontin, 1993). For instance, genetic fingerprinting has been presented in the media and in courts of law as a means of unambiguously finding the 'race' of a criminal, whereas the results from genetic fingerprinting should only be interpreted probabilistically (Bown, 1993*a*). Such misunderstandings would be less likely to arise if contingent and dynamic ethnicities were the focus of discussion, rather than 'races', which are thought to have fixed boundaries and characteristics. Research into ethnicity, racialisation and the practice of the new and the old genetics could have implications for majorities and minorities alike, but this depends on viewing ethnicity as a characteristic that is associated with racialised and non-racialised groups alike.

There has been a lack of systematic analysis of the reasons behind the outcomes of screening programmes, many of which have been directed towards ethnic minorities in Britain and the US. The possibility of repeating the same mistakes in future programmes, owing to a lack of research, is a concern at the time of writing, as the question of how and when to screen for cystic fibrosis is under discussion (S. Raeburn, 1994). There is evidence of ethnocentricity (which is not necessarily the same as racism) in the reported views of doctors who offer counselling to patients coming to decisions about prenatal techniques and IVF. This is of note, given the importance attached to non-directional genetic counselling as a means of helping people to come to their own decisions (Birke *et al.*, 1990). It is doubtful that the purported ideal of non-directional counselling could be given in Britain on the matter, for instance, of prenatal selection for eye colour, whereas the possibility of prenatal sex selection, which is acceptable in some cultures, has already been ruled out in Britain. This example is not used to recommend prenatal sex selection, but to show how the beliefs of the majority or dominant ethnic group are often assumed to be 'normal' and therefore advice along these lines is taken as 'non-directional'. Such assumptions are brought to light by an encounter with a group that does not share them. Such encounters can be instructive as to the nature of the beliefs of the counsellors and professionals and should be treated as an opportunity for critical reappraisal of the beliefs of the majority as well as the minority. In the context of genetic counselling a thorough examination of lay and professional beliefs about inheritance, marriage patterns and the uses of predictive screening in majority and minority ethnic groups is timely.

14.23 References

Aluoch, J. R. and Aluoch, L. H. M. (1993) Survey of sickle disease in Kenya. *Tropical and Geographical Medicine*, **45**, 18–21.

American Journal of Nursing (1993) Screening urged for all newborns. *American Journal of Nursing*. August, 12–13.

Anionwu, E. N. (1993) Sickle cell and thalassaemia: community experiences and official responses. In W. I. U. Ahmad (ed.) *'Race' and Health in Contemporary Britain*. Buckingham: Open University Press.

Ballard, R. and Kalra, V. S. (1994) *The Ethnic Dimensions of the 1991 Census: A Preliminary Report*. Manchester: Census Microdata Unit, University of Manchester.

Barkan, E. (1992) *The Retreat of Scientific Racism: Changing Concepts of Race in Britain and the United States Between the World Wars*. Cambridge: Cambridge University Press.

Benallegue, A. and Kedji, F. (1984) Consanguinité et santé publique: étude algérienne. *Archives Françaises de Pédiatrie*, **41**, 435–40.

Birke, L., Himmelweit, S. and Vines, G. (1990) *Tomorrow's Child: Reproductive Technologies in the 90s*. London: Virago.

Black, J. and Laws, S. (1986) *Living With Sickle Cell Disease: An Enquiry into the Need for Health and Social Service Provision for Sickle Cell Sufferers in Newham*. London: East London Branch Sickle Cell Society.

Bown, W. (1993a) Doubts over DNA evidence 'exaggerated'. *New Scientist*, 23 January, p. 6.

Bown, W. (1993b) Race, crimes and genetic fingerprints. *New Scientist*, 23 January, p. 6.

Brah, A. (1993) Re-framing Europe: en-gendered racisms, ethnicities and nationalisms in contemporary Western Europe. *Feminist Review*, **45**, 9–28.

British Sociological Association (1990) Anti-racist language: guidance for good practice. Lexicon of terms and guidelines for their use. British Sociological Association.

Brozovic, M., Davies, S. C. and Henthron, J. (1989) Haematological and clinical aspects of sickle cell disease in Britain. In J. K. Cruikshank and D. G. Beevers (eds.), *Ethnic Factors in Health and Disease*. London: Wright.

Congress of United States Office of Technology Assessment (1988) *Mapping Our Genes: Genome Projects: How Big, How Fast?* Baltimore: Johns Hopkins University Press.

Cooper, R. (1984) A note on the biologic concept of race and its application in epidemiologic research. *American Heart Journal*, **108**, 715–23.

Cooper, R. S. (1993) Ethnicity and disease prevention. *American Journal of Human Biology*, **5**, 387–98.

Cowan, R. S. (1992) Genetic technology and reproductive choice: an ethics for autonomy. In D. J. Kevles and L. Hood (eds.), *The Code of Codes: Scientific and Social Issues in the Human Genome Project*. Cambridge, MA: Harvard University Press.

Currer, C. (1984) Pathan women in Bradford – factors affecting mental health with particular reference to the effects of racism. *International Journal of Social Psychiatry*, **30**, 72–6.

Cutting, G. R., Curristin, S. M., Nash, E., Rosenstein, B. J., Lerer, I., Abeliovich, D., Hill, A. and Graham, C. (1992) Analysis of four diverse population groups indicates that a subset of cystic fibrosis mutations occur in common among Caucasians. *American Journal of Human Genetics*, **50**, 1185–94.

Darr, A. and Modell, B. (1988) The frequency of consanguineous marriage among British Pakistanis. *Journal of Medical Genetics*, **25**, 186–90.

Davis, J. (1990) *Mapping the Code: The Human Genome Project and the Choices of Modern Science*. New York: John Wiley.

Davison, C., Macintyre, S. and Smith, G. D. (1994) The potential social impact of predictive genetic testing for susceptibility to common chronic diseases: a review and proposed research agenda. *Sociology of Health and Illness*, **16**, 340–71.

Dodge, J. A. (1988) Implications of the new genetics for screening for cystic fibrosis. *Lancet*, **ii**, 672–4.

Donegan, L. (1993) 'Designer' baby sparks race choice row. *Guardian*, 31 September.

Donovan, J. L. (1984) Ethnicity and health: a research review. *Social Science and Medicine*, **19**, 663–70.

Dunlop, L. S. and Winter, R. M. (1987) Analysis of perinatal mortality by ethnic group: does consanguineous marriage contribute to mortality due to congenital malformations? *Journal of Medical Genetics*, **24**, 241.

Duster, T. (1990) *Backdoor to Eugenics*. London: Routledge.

Emery, A. E. H. and Rimoin, D. L. (eds.) (1990*a*) *Principles and Practice of Medical Genetics*, 2nd edn, vol. 1. Edinburgh: Churchill Livingstone.

Emery, A. E. H. and Rimoin, D. L. (eds.) (1990*b*) *Principles and Practice of Medical Genetics*, 2nd edn, vol. 2. Edinburgh: Churchill Livingstone.

Feinmesser, M., Tell, L. and Levi, H. (1989) Consanguinity among parents of hearing-impaired children in relation to ethnic groups in the Jewish population of Jerusalem. *Audiology*, **28**, 268–71.

Fletcher, D. and Muir, H. (1994) Doctors defend mixed-race baby for black woman. *Daily Telegraph*, 1 January.

Frankenberg, R. (1993) Growing up white: feminism, racism and the social geography of childhood. *Feminist Review*, **45**, 51–84.

Franklin, I. M. (1988) Services for sickle cell disease: unified approach needed. *British Medical Journal*, **296**, 592.

Galjaard, J. (1990) Challenges for the future. In A. A. H. Emery and D. L. Rimoin (eds.), *Principles and Practice of Medical Genetics*, vol. 2. Edinburgh: Churchill Livingstone.

Geddes, D. (1994) Cystic fibrosis: gene therapy trials come to the UK. *MRC News*, Spring, 13–16.

Goffman, E. (1990) *Stigma: Notes On the Management of Spoiled Identity*. London: Penguin.

Goodey, C. F. (ed.) (1991) *Living in the Real World: Families Talk About Down's Syndrome*. London: Twenty-One Press.

Gould, S. J. (1980) *Ever Since Darwin; Reflections in Natural History*. Harmondsworth: Penguin.

Green, J. M. (1992) *Calming or Harming? A Critical Review of Psycholgical Effects of Fetal Diagnosis On Pregnant Women*. London: Galton Institute.

Green, J. M. and France-Dawson, M. (1994) Women's experiences of screening in pregnancy: Ethnic differences in the West Midlands. Paper prepared for meeting entitled 'Culture, Kinship and Genes', Abergavenny, 28–30 March.

Greer, G. (1994) Racism written into the gender agenda. *Guardian*, 21 March.

The Guardian (1993) Back to blackshirts and bullyboys. *Guardian*, 28 September.

Haan, E. A. (1990) Ethics and the new genetics. *Journal of Paediatric Child Health*, **26**, 177–9.

Haan, E. A. (1994) Screening for carriers of genetic disease: points to consider. *Medical Journal of Australia*, **158**, 419–21.

Harris, J. (1992) *Wonderwoman and Superman: The Ethics of Human Biotechnology*. Oxford: Oxford University Press.

Hecht, F. and Cadien, J. D. (1984) Tay–Sachs disease and other fatal metabolic disorders. In N. J. Wald (ed.), *Antenatal and Neonatal Screening*. Oxford: Oxford University Press.

Horn, M. E. C., Dick, M. C., Frost, B., Davis, L. R., Bellingham, A. J., Stroud, C. E. and Studd, J. W. (1986) Neonatal screening for sickle cell diseases in Camberwell: results and recommendations of a two year pilot study. *British Medical Journal*, **292**, 737–40.

Jones, S. (1994) *The Language of the Genes*. London: Flamingo.

Kevles, D. J. (1985) *In the Name of Eugenics: Genetics and the Uses of Human Heredity*. Harmondsworth: Penguin.

Kevles, D. J. (1992) Out of eugenics: the historical politics of the human genome. In D. J. Kevles and L. Hood (eds.), *The Code of Codes: Scientific and Social Issues in the Human Genome Project*. Cambridge, MA: Harvard University Press.

Kevles, D. J. and Hood, L. (eds.) (1992) *The Code of Codes: Scientific and Social Issues in the Human Genome Project*. Cambridge, MA: Harvard University Press.

Kingman, S. (1993) Why cousins can be just too close. *Independent*, 6 July.

Kirk, S. A. (1987) Sickle cell disease and health education. *Midwife, Health Visitor and Community Nurse*, **23**, 200–6.

Krishan, G. (1986) Effect of parental consanguinity on anthropometric measurements among the Sheikh Sunni Muslim boys of Delhi. *American Journal of Physical Anthropology*, **70**, 69–73.

Lancet (1980) Population screening for carriers of recessively inherited diseases. *Lancet*, **ii**, 679–80.

Lebel, R. R. (1983) Consanguinity studies in Wisconsin. I. Secular trends in consanguineous marriage, 1843–1981. *American Journal of Medical Genetics*, **15**, 543–60.

Lewontin, R. C. (1993) *The Doctrine of DNA*. London: Penguin.

MacKinnon, I. (1994) Tougher rules urged for 'designer baby' clinics. *Independent On Sunday*, 2 January.

Malik, K. (1994) Children of a confused society. *Independent*, 3 January.

McGourty, C. (1994) Dark genes dominant factor in skin colour. *Daily Telegraph*, 1 January.

McKie, R. (1988) *The Genetic Jigsaw: The Story of the New Genetics*. Oxford: Oxford University Press.

Medical World News (1987) Fetal sexing: is it ethical? When? *Medical World News*, 28 September, 18–19.

Miles, R. (1989) *Racism*. London: Routledge.

Miles, R. (1993) *Racism after Race Relations*. London: Routledge.

Milhill, C. (1994) MPs seek ban on clinics 'selling babies like hamburgers'. *Guardian*, 16 March.

Miringoff, L. (1991) *The Social Costs of Genetic Welfare*. New Brunswick, NJ: Rutgers University Press.

Modell, B. and Petrou, M. (1989) Thalassaemia screening: ethics and practice. In J. K. Cruikshank and D. G. Beevers (eds.), *Ethnic Factors in Health and Disease*. London: Wright.

Owen, P. P., Elias, S., Woods, D., Hanissian, A. S., Schoumacher, R. A. and Bishop, C. (1993) Cystic fibrosis mutations in white and black Americans: an approach to identification of unknown mutations with implications for cystic fibrosis screening. *American Journal of Obstetrics and Gynaecology*, **168**, 1076–82.

Panter-Brick, C. (1991) Parental responses to consanguinity and genetic disease in Saudi Arabia. *Social Science and Medicine*, **33**, 1295–302.

Pergament, E. (1991) Preimplantation diagnosis: a patient perspective. *Prenatal Diagnosis*, **11**, 493–500.

Petrou, M., Modell, B., Darr, A., Old, J., Kin, E. and Weatherall, D. (1990) Antenatal diagnosis: how to deliver a comprehensive service in the United Kingdom. *Annals of the New York Academy of Sciences*, **612**, 251–63.

Phizacklea, A. (1984) A sociology of race relations or 'Race Relations'? A view from Britain. *Current Sociology*, **32**, 199–218.

Phoenix, A. (1990) Black women and the maternity services. In J. Garcia, R. Kilpatrick and M. Richards (eds.), *The Politics of Maternity Care: Services for Childbearing Women in Twentieth Century Britain*. Oxford: Clarendon Press.

Pilkington, E. (1993) Bark of the dogs of war. *Guardian*, 16 September.

Prashar, U., Anionwu, E. and Brozovic, M. (1985) *Sickle Cell Anaemia – Who Cares? A Survey of Screening and Counselling in England*. London: The Runnymede Trust.

Raeburn, J. A. (1994) Community genetics across cultures. Paper prepared for meeting entitled 'Culture, Kinship and Genes', Abergavenny, 28–30 March.

Raeburn, S. (1994) Screening for carriers of cystic fibrosis. *British Medical Journal*, **308**, 1451–2.

Rajeswari, G. R., Busi, B. R., Murty, J. S., Rao, V. V. and Narahari, S. (1992) Selection intensities and consanguinity in the Yadava and Vadabalija of Visakhapatnam, Andhra Pradesh, India. *Social Biology*, **39**, 316–19.

Reddy, P. C. (1987) Consanguinity and inbreeding load in two Mala populations of Andhra Pradesh. *Man in India*, **67**, 357–63.

Richards, M. P. M. (1989) Social and ethical problems of fetal diagnosis and screening. *Journal of Reproductive and Infant Psychology*, **7**, 171–85.

Richards, M. P. M. and Green, J. M. (1993) Attitudes toward prenatal screening for fetal abnormality and detection of carriers of genetic disease: a discussion paper. *Journal of Reproductive and Infant Psychology*, **11**, 49–56.

Rose, S., Lewontin, R. C. and Kamin, L. J. (1984) *Not in Our Genes: Biology, Ideology and Human Nature*. Harmondsworth: Penguin.

Rossiter, A. (1992) 'Between the Devil and the deep blue sea': Irish women, Catholicism and colonialism'. In G. Sahgal and N. Yuval-Davis (eds.), *Refusing Holy Orders – Women and Fundamentalism in Britain*. London: Virago.

Scambler, G. (1989) *Epilepsy*. London: Tavistock/Routledge.

Serjeant, G. R. (1985) *Sickle Cell Disease*. Oxford: Oxford University Press.

Stacey, M. (1994) Political and social issues. Paper presented at meeting on 'Culture, Kinship and Genes', Abergavenny, 28–30th March.

Statham, H., Green, J., Snowdon, C. and France-Dawson, M. (1993) Choice of baby's sex. *Lancet*, **341**, 564–5.

The Times (1994) Unnatural childbirth. *The Times*, 1 January.

Timmins, N.(1993) Couples barred from choosing race of babies. *Guardian*, 31 December.

van den Berghe, P. (1978) *Race and Racism: A Comparative Perspective*, 2nd edn. New York: John Wiley.

Wald, N. J. (ed.) (1984) *Antenatal and Neonatal Screening*. Oxford: Oxford University Press.

Weatherall, D. J. (1985) *The New Genetics and Clinical Practice*, 2nd edn. Oxford: Oxford University Press.

Weatherall, D. J. and Letsky, E. A. (1984) Genetic haematological disorders. In N. J. Wald (ed.), *Antenatal and Neonatal Screening*. Oxford: Oxford University Press.

Wertz, D. C. (1992) Ethical and legal implications of the new genetics: issues for discussion. *Social Science and Medicine*, **35**, 495–505.

Wertz, D. C. and Fletcher, J. C. (1993) Feminist criticism of prenatal diagnosis: a response. *Clinical Obstetrics and Gynaecology*, **36**, 541–67.

Wexler, N. (1992) Clairvoyance and caution: repercussions from the human genome project. In D. J. Kevles and L. Hood (eds.), *The Code of Codes: Scientific and Social Issues in the Human Genome Project*. Cambridge, MA: Harvard University Press.

Wilfond, B. S. and Fost, N. (1990) The cystic fibrosis gene: medical and social implications for heterozygote detection. *Journal of the American Medical Association*, **263**, 2777–83.

Wilkie, T. (1993) *Perilous Knowledge*. London: Faber and Faber.

Wilkie, T. (1994) Colours of morality. *Independent On Sunday*, 2 January.

Williamson, R. (1993) Universal community carrier screening for cystic fibrosis? *Nature Genetics*, **3**, 195–201.

Zeesman, S., Clow, C. L., Cartier, L. and Scriver, C. R. (1984) A private view of heterozygosity: eight-year follow-up study on carriers of the Tay–Sachs gene detected by high school screening in Montreal. *American Journal of Medical Genetics*, **18**, 769–78.

15

Predictive genetics: the cultural implications of supplying probable futures

CHARLIE DAVISON

15.1 Introduction

The phrase 'predictive genetic testing' refers to the examination of a sample of genetic material with the aim of producing information about the health-related future of the person from whom it was taken. Such tests can be carried out on genetic material collected from individuals as fetuses, or as born people of any age. At present, predictive genetic testing has several routine applications in ante-natal care, such as the use of chorionic villus sampling in the early identification of fetuses with chromosomal abnormalities. The use of predictive genetic testing in people already born has, until recently, been restricted to a handful of relatively rare genetic disorders, such as Huntington's disease or muscular dystrophy. This chapter, however, is principally concerned with the extension of predictive testing into a wide range of very common illnesses and conditions, which is being facilitated by current rapid developments in human genetics (Wilkie, 1993).

In ante-natal testing, the process creates the opportunity for prospective parents to make decisions about the termination of an affected fetus, or to be forewarned of the birth of a child with special needs. In individuals already born, predictive genetic testing allows for the early start of therapeutic or prophylactic regimes (if any exist for a given disorder), and for life decisions to be informed in a way that would not otherwise be possible – the decision of someone who has inherited Huntington's disease not to reproduce, for example. These new directions in medicine have been and continue to be the focus of intense ethical and policy debates (see, for example, Dunstan, 1988; Murray, 1991; Takagi, 1991; Muller-Hill, 1993).

It is difficult to forecast future developments in applied medical genetics with great precision, but it is generally expected that it will soon be possible to test large numbers of individuals for genetic predisposition to a wide range of cancers, respiratory diseases and cardiovascular disorders, as well as other common physical and mental conditions such as diabetes, alcoholism and schizophrenia. Apart from this involvement in mainstream medical fields, the supposed genetic basis of what Murray (1991) calls 'a vast range of ethically significant behaviour', such as sexual orientation, are also under investigation (e.g. Baron, 1993; Hamer et al., 1993). In the view of one Nobel laureate in the field, the supply of personal genetic profiles by high-street slot machines could be commonplace within the next 30 years (W. Gilbert, quoted in Christie (1992)).

Predictive genetic testing for adult-onset chronic disorders is currently at an exciting stage in its development (Harper and Clarke, 1990; British Medical Association, 1992; Wilkie, 1993). The speed of developments in molecular genetics poses some pressing questions for policy-makers and regulatory bodies and also for society as a whole. In a rapidly changing scientific field, technical knowledge may oustrip the development of the ethical, cultural and political infrastructure required to control it. In modern genetics, clinical techniques may become available before society has had the opportunity to consider fully and decide upon the appropriate uses to which they may be put and any controls that might be deemed necessary. As Harper (1992) has observed:

> At this stage, careful discussion and planning can insure that the powerful new tools of genetics can be made useful servants for our benefit rather than become weapons of control and restriction. . . . For a short while only, the choice is still ours to determine.

15.2 The inherent uncertainty of assessing future events

A major branch of contemporary social theory (see, for example, Giddens, 1991; Beck, 1992; Bauman, 1993 (in which see especially Chapter 7)) has become concerned with the cultural infrastructure within which people make decisions about events and time in a 'post-industrial' or 'late-modern' era. Central to this concern are the processes by which events yet to happen become defined as legitimate elements of present thoughts and decisions. In Giddens's (1991) memorable phrases, we have embraked upon 'the colonisation of the future', a cultural development that was bound to spring from a 'system geared to the domination of nature and the reflexive making of history'.

Within this overall picture, the invention of scientific techniques to 'read' the human genome and furnish people with accurate assessments of the chances and conditions of the future onset of illnesses is plainly important. In the context of casting individual futures, predictive genetic testing can be seen as the applied cartography of the colonisation movement. Indeed the term 'mapping' is often applied to developments in the codification of human DNA. This apparently precise and quintessentially scientific field of endeavour carries with it a strong image of accuracy and certainty, largely bound up with the notion of genetic inheritance being a 'programme' or 'master design' that governs the subsequent development of an organism. Because of this particular attribute of the science of genetics, it is important to clarify the elements of predictive genes testing whose essential function is the production of uncertainty.

In spite of the identification of the post-modern condition, for most of us the future remains less concrete than the past or present and necessarily retains a sketchy or provisional element. Prognostications and forecasts must, by their very nature, contain more 'ifs' than 'whens', and assessments of things to come demand to be conceptualised in terms of possibility, probability and risk.

Almost entirely, genetic information concerning susceptibility or predisposition is concerned with possible rather than defined futures. A much used and telling analogy is that of playing cards. Each person is dealt a genetic hand, but only in very rare cases does this automatically define a player as a winner or loser. Rather, any outcome depends on how the cards are played, which in turn depends on how that play interacts with the playing of cards from other hands beyond the original player's control. As Bauman (1993) pointed out, risk as a cultural category

> belongs to the discourse of gambling, that is to a kind of discourse which does not sustain a clear-cut opposition between success and failure, safety and danger, one which recognises their co-presence in every situation . . . risk, therefore, is resonant with the post-modern view of the world as a game, and the being-in-the-world as play.

This theoretical view is supported by our own fieldwork data from South Wales (Davison *et al.*, 1989, 1991, 1992). Here, assessments of both the risk and the distribution of misfortune were consistently articulated in terms of an interplay between attempted control and a range of forces beyond that control, including the workings of randomness and chance. The cultural idiom most suited, therefore, was that of game playing, and informants regularly used images such as 'the luck of the draw', 'dicing with death' or 'a bit of a lottery' to explain the health or safety situation in which they and others found themselves.

In the particular field of predictive genetic testing for future illness, this game-like uncertainty means that, even when a genetic defect is found in an individual, the expected disorder may never develop, and that in most circumstances the time of symptom onset will be unknown. Such uncertainties flow directly from two fonts: variations in expression, and the intervention of other (i.e. non-genetic) misfortunes.

Variations in the expression of genetic material means that, in any given case, the inheritance of a particular sequence of DNA does not automatically imply that the individual concerned will go on to develop the attribute (or ailment) indicated by the sequence concerned. A major reason for differences in expression is that genes and mutations vary in terms of their 'penetrance'. Some material (such as the gene for Huntington's disease) exhibits very strong penetrance, with all individuals who inherit it developing the disorder. Even in such cases, the age a person reaches without suffering symptoms of illness can vary greatly. In other situations (the inheritance of *BRCA1* gene for breast/ovarian cancer, for example), women who inherit the genetic material in question not only show variation in the timing of illness onset, but also whether they become ill at all.

The examples of Huntington's disease and *BRCA1* both concern relatively simple single-gene scenarios. The genetic components of many of the most important chronic diseases in industrial societies, however, are probably due to the influence of several different genes. In such circumstances, the expression (timing, severity, causation of illness or not, etc.) of genetic vulnerability will depend on relatively complex combinations of the expression of all the different genes involved.

Although variations in penetrance and the expression of polygenic conditions are clearly linked to the production of uncertainty in predictive genetic testing, a yet more important process is the interaction of genetic endowment with environment. As Jones (1993) pointed out:

For many people, the role of inheritance compared to that of experience is an obsession. It is an obsession that goes back long before genetics. Even Shakespeare had a say: in *The Tempest*, Prospero describes Caliban as 'A devil, a born devil, on whose nature nurture can never stick'. There are still endless discussions about whether musicality, criminality or – most fashionable of all – intelligence is inherited or acquired. More seriously, there is debate about the role of genetics and environment in controlling illnesses such as cancer and heart disease.

In fact, most disorders with a genetic component rely on the interaction of gene(s) and environment to bring about a clinical manifestation of disease. It has been estimated that 4000 disorders are brought about by such interactions; indeed, it is difficult to think of many diseases for which this is not the case (Takagi, 1991). In those illnesses characterised by highly penetrant genes, environmental influences may be expected to be of little importance, as the pathogenic effects of genetic inheritance will be manifested whatever conditions are experienced. Even in these cases, however, the timing and severity of the onset of illness are thought to be wholly or partly determined by the interaction of individuals and their environments (Summers, 1993).

Even for those diseases with obvious and strong environmental causes, genetic make-up can be shown to play some role. As Rose (1985) pointed out, if everyone smoked 20 cigarettes a day then lung cancer would appear to be a genetic disease. Even in the situation where this does not pertain, the role of genetic make-up in the risk of lung cancer could be seen to work either through determination of the susceptibility of the lungs of the smoker to inhaled carcinogens, or through genetic influences on smoking behaviour itself (Carmelli *et al.*, 1992).

Conversely, the risk of diseases with a strong genetic component can be greatly influenced by environment. For example, non-insulin-dependent diabetes mellitus (NIDDM) shows 90% concordance in monozygotic twins in studies of homogeneous groups (Barnett *et al.*, 1981) and studies are under way to identify candidate genes (Elisin *et al.*, 1986). The prevalence of the disease within a community, however, can change dramatically when environmental developments take a hand. Rapid economic development in the South Pacific, for example, has caused some places that were virtually free of NIDDM to rank among the highest-prevalence countries in the world (Zimmet *et al.*, 1978). Such cases are clear demonstrations of the often relative nature of the designation of a disease as either environmental or genetic in origin (Orchard *et al.*, 1986; Coughlin, 1992; Phillips, 1993).

In the sense that predictive genetic testing is 'only' a process whereby people are furnished with probabilistic statements about the risk of suffering from future illness, the process is not new. Many types of screening, risk factor analysis and other medical testing can produce similar information (Davison *et al.*, 1994). The implementation and management of, for example, mass screening for hypertension (see Collins *et al.*, 1990; Macdonald *et al.*, 1984) or cholesterol (see Brett, 1984, 1991) are in some senses akin to predictive genetic testing. The similarities lie in the gathering of data derived directly from samples taken from an individual that are analysed in such a way that they indicate that person's potential for future ill health. Such techniques have a range of both personal and collective effects, including options for early treatment, the adoption of prophylactic drug or 'lifestyle' regimes and more effective calculations of future service needs. Like genetic screening, non-genetic types of medical screening have also been associated with the production of stigma, anxiety and a deterioration in the quality of life of those taking part (Holland and Stewart, 1990).

There are, however, particular attributes of predictive genetic testing that make its potential impact of special importance. In the first place, clinical genetics addresses a vast range of illnesses and conditions, representing such an increase in the relative importance of prediction in medicine that many observers speak of a 'revolution'. Furthermore, screening will produce relatively complicated information relating to the various vulnerabilities and risks pertaining to any one individual that might, for example, be interactive (or not) or environmentally sensitive (or not). Such an increase in the level of complexity of possible futures indicates that many small and large life choices will become subject to the calculus of what has been called the 'actuarial mindset' (Nelkin and Tancredi, 1989).

Predictive genetic testing is capable of supplying information at an extremely early stage of human development. Data with a strong image of accuracy about a person's susceptibilities and risks in terms of, for example, heart disease, will become available much earlier than current measurements of blood pressure, body mass, serum lipid levels, or smoking behaviour. The extreme earliness that can characterise DNA-based information is closely related to genetic screening's most distinctive feature, the fact that it largely concerns aspects of an individual that are transmissible through generations. It is from these twin sources, on the one hand the intensely personal nature of a DNA 'blueprint' and on the other the more diffuse group aspects of shared inheritance and generational transmission, that the culture impact of predictive genetic testing flows.

15.3 Risk, identity and the management of lifestyle

> It seems clear that virtually all diseases will have some component of genetic predisposition, so that some individuals in some families are more at risk than others . . . ultimately, great social and economic benefit will be gained through the development of risk modification strategies specific to each individual. *(Summers, 1993)*

The widespread introduction of predictive genetic testing will dramatically widen the range of information available to people who want to plan their 'lifestyle' specially to avoid illness. The logical progression of this type of development is a situation in which it would become common for people to know about their own personal genetic risk profile across a range of disorders, and for them to design an 'individually tailored' set of behaviours. Someone with an inherited susceptibility to coronary thrombosis and musculo-skeletal problems, for example, may decide never to eat high-fat foods nor play impact or contact sports. Another person with a quite different 'genetic read-out' may become particularly wary of entering smoky rooms, or being exposed to bright sunlight. As foreseen by the theorists of late and post modernity, the construction of a unique self through individual 'bricolage' may come to supersede membership of population aggregates (such as social class) as the source of defining behaviours and personal identification.

Just as individual identities are increasingly defined by consumption in general, the way we 'consume' risk information and our 'consumption' of prophylactic behaviour and products will become increasingly important in defining who we are and who we appear to be. This tendency is, of course, already discernible in the ways in which individuals and groups accomplish the cultural management of non-genetic risk and prophylaxis. Crawford (1984), for example, illustrated how attitudes to personal responsibility for the maintenance of health and safety are of supreme importance to the definition of self in a world where

> Americans have acquired a sense of somatic vulnerability . . . [and] have been exposed to a virtual media and professional blitz for a particular model of health promotion that emphasises lifestyle change and individual responsibility . . . the message is clear: living a long life is essentially a do-it-yourself proposition.

In our own research (Davison *et al.*, 1991; Backett and Davison, 1992), the definition of a person in relation to this movement was illustrated by informants' tendencies to typify self or others as, for example, 'the kind of guy who never wears a seat-belt'. This information, as part of a cultural set mutually recognised by interlocutors, communicates more about the person in question than their habits concerning a particular piece of safety equipment.

Those, like Summers, who foresee 'great social and economic benefit' in the production of individually tailored lifestyle regimes, assume, of course, that individual recipients of the results of predictive genetic testing will decide that health maintenance (or, more accurately, illness avoidance) is a high priority. They also assume a relatively unproblematic relationship between information given and action taken. As Otway and Wynne (1989) pointed out, like much contemporary risk communications, their visions 'rest on unexamined and unarticulated assumptions about who is communicating what, to whom and in what context'.

Within the essentially utilitarian ideology shared by clinical genetics and health promotion, it is axiomatic that a person receiving positive test results will decide 'I have inherited a particularly high risk of lung disease, so I must not smoke'. It should be born in mind, however, that popular cultures of risk and misfortune may also legitimate a quite different response: 'I am genetically programmed to get lung disease, so it doesn't much matter what I do' (Davison *et al.*, 1989, 1992).

This range of responses highlights the fact that risk assessments can only derive social and cultural meaning from the specific context within which they are set. This contextual relativity of risk has been illustrated by several qualitative and anthropological studies of health-related behaviour. For example, Graham (1987) found that working-class British mothers who smoked balanced the physical risks of tobacco smoking against the emotional and mental risks of stress and what they saw as bad child care (excessive anger, beating) and opted, in this context, to smoke. Connors, in her work with intravenous drug users in the USA, produced similar findings. Here, in the context of social events centred on needle sharing, it was preferable for informants to risk HIV infection from sharing a needle than to risk the stigma and suspicion attendant on washing the needle in bleach or using a new one (Connors, 1992).

The existence of similarly 'unorthodox' or 'anti-establishment' sub-cultures among wider social groups is illustrated by Balshem's account of how inhabitants of a poor inner-city neighbourhood in the USA developed a form of cultural resistance in reaction to a behaviour-based cancer prevention programme (Balshem, 1991). In this case, local cultural understanding of illness causation was such that health promotion advice was simply categorised as inappropriate and irrelevant.

The primacy of cultural context in the genetics field is underlined by Rapp's work in genetic counselling sessions. This research showed that decisions to seek prenatal genetic information were specific to the cultural context of the lives of the potential consumers and were

> based on all the assumptions, fears and norms concerning healthy and sickly children ... the meaning of illness in family history, the shame or pride attached to the bearing (or non-bearing) of children; beliefs about fertility, abortions, femininity and masculinity, and the social consequences and prejudices surrounding disability. *(Rapp, 1988)*

Risk information for preditive genetic testing, as we have seen, is characterised by its possible production at a very early stage in human development. This, in conjunction with the familial content of the information (risks are only present because they have been inherited), will have immense implications for future constructions of parenting, family responsibility and what may be considered adequate child care:

> One can envision law enforcement or child protection agencies looking for children with genetic conditions to make sure that their parents are providing them with proper medical care or prevention strategies. Although it seems far-fetched today, assuming that a gene that predisposes a person to skin cancer is discovered in the future and that such cancer is preventable if one stays out of the sun, it would be possible to search DNA databases to identify children at risk and require that their parents protect them from this genetic hazard by keeping them out of the sun and away from the beach. *(Annas, 1993)*

Such developments in clinical genetics have certain implications for the future of family life. Somewhat ironically, the very information that could facilitate a trend towards genetic individualism would concern a person's membership of a biological descent group (the vast majority of significant genetic material would be 'passed on in the blood'; see Section 15.5, and Chapter 12). If genetic tests relied on the use of 'markers', the provision of meaningful information would rely on the testing of several members of a biological descent group. Such a situation could conceivably provide extra reasons for people to maintain contact with an extended family, because accurate genetic information would require that a social group opt for testing, rather than any one individual. Genetic tests based on single genes, however, would put less of a premium on material from biologically related individuals. In these cases, the tracing of other people who have (and may transmit) unwanted genetic material would be a matter for the public health authorities rather than an individual citizen who knew little and cared less about a range of biologically related, but socially insignificant, others.

15.4 Predictive information and the potential for social exclusion

Predictive genetic testing, as we have seen, provides information, much of it probabilistic, concerning the health-related futures of specific individuals. Much of this information will have a direct input to estimations of a person's longevity, need for medical attention and potential for work and earning. Such information has the power to transform the personal aspects of employment, insurance and financial credit (Roscam Abbing, 1991; Billings *et al.*, 1992). As Jonsen *et al.*, (1991) pointed out:

> the constant stream of information that inundates citizens of modern industrial democracies profoundly affects their views of themselves, their opinions of society and their modes of behaviour. Those who lack information, or do not know how to obtain it or use it, are radically disabled.

In health and life insurance, for example, it can be argued that it is only fair for companies to make actuarial assessments about prospective clients before they set premiums. Otherwise, clients with little likelihood of making claims are placed in the position of subsidising the premiums (and pay-outs) of those who are much more likely to claim. From the point of view of the person with an undesirable genetic test result, risk-related premiums appear as grave injustice. From the position of the genetically average client, premium adjustment and even exclusion in cases of genetic risk will appear as sound commercial reasons for choosing one company over another. As the director of one large American health insurance company stated: 'We don't think it's fair for someone to get insurance at standard rates when they represent more than standard risk' (J. Payne, quoted in Cherskov, 1992).

Medical information and other data (place of residence, age, occupation, etc.) are currently used in this connection. The potential for individual premium setting and the fine-tuning of risk assessment that predictive genetic testing allows, however, can be seen as violating the principles of shared liability that underpin the very idea of insurance (Daniels, 1990; Light, 1992). One US lawyer, has even warned that the Human Genome Project was in danger of becoming 'a welfare program for insurance companies' (T. Morelli, quoted in Cherskov, 1992).

The field of employment displays similar tensions between individual rights and goals and the interests of companies, institutions and the wider society. Which interests will be best served by the introduction of predictive genetic testing depends largely on the balance that is struck between the unregulated use of genetic information in the organisation of work and its control by state or collective authorities (Council on Ethical and Judicial Affairs, 1991; Draper, 1991). For example, a predictive genetic test indicating vulnerability to smoke and dust, such as the currently available test for alpha1 anti-trypsin deficiency (Gustavson, 1989), could be used in two basic ways. On the one hand, it could be implemented with the aim of reducing company costs due to absence through illness and compensation claims from workers with lung damage. To achieve this, the test could be administered to all staff and those found to be genetically vulnerable to dust and smoke could be removed from work on the grounds of unsuitability. On the other hand, the test could be implemented with the aim of protecting or enhancing individual rights and freedoms in the labour market. In this case, workers found to exhibit potential vulnerability could be redeployed to non-dusty environments. Pre-employment testing, or the carrying of susceptibility information by all workers as a matter of routine, could allow for the running of a system of reserved employment for workers whose employment potential is limited by their genetic profile.

Such choices are, essentially, political rather than scientific or clinical. The example of alpha1 anti-trypsin deficiency illustrates how current debates centred on the exclusion of people with HIV infection (Daniels, 1990) and other medical risk factors (Light, 1992) will be intensified if predictive genetic testing becomes widespread.

15.5 Predictive genetics and the culture of family and inheritance

Because of the potential detail that predictive genetic testing gives assessments of the future, we have so far concentrated on the social and ethical impacts of this aspect of the new genetics. Genetic information concerns the biochemical functions that in a very basic sense determine the nature of each of us as organisms. Thus, it is highly likely that developments in the production and distribution of information will have a considerable impact on our relationship with the highly personal areas of shared culture that are our kinship systems. Because of the diversity of modern populations in terms of class and ethnic origin, it has been pointed out that it is impossible to talk of 'British kinship' as a unitary system (Strathern, 1992). Although accepting this point of view in terms of the subtlety and cultural context of kinship as a lived experience for any individual, we hope to show here that, at a certain level of abstraction, discussion of the potential impact of the new genetics on Western or 'Euro-American' kinship is possible. Our aim in this is to illustrate the kinds of cultural issues that the advent of widespread predictive genetic testing will raise.

In the vast majority of settings in the modern world, a normal fact of social existence is that most people have a notion of who they are in terms of where they came from (technically, that some aspects of their social identity are derived from their membership of a descent group). This rudimentary aspect of kinship systems does not suggest, of course, that socially defined descent is the same as biologically defined descent; indeed, the two often vary (Needham, 1960; Gellner, 1987). This variation exists not only on the theoretical level, but also on the practical, in that it is estimated that quite large numbers of British people (up to 10%) are not the biological offspring of the couple they have always thought of as their mother and father (Macintyre and Sooman, 1991).

One aspect of reckoning shared by most contemporary Euro-American systems of reckoning descent, however, is the idea that any individual belongs to a family with different 'sides'. This notion arises from the fact that, in common with many other European systems, British kin reckoning tends to be bilateral, that is to say that an individual belongs to both a mother's and father's family groups.

In terms of the charting of the inheritance of ailments, weaknesses, susceptibilities and the physical and personal attributes that may be associated with them, British 'sides' systems have much in common with the descent models of clinical genetics. There is a common expectation that children will resemble parents and grandparents ('take after') and that similarities will be discernible between related people of the same generation ('spit', 'image'). Because the 'sides' system implies a dendritic structure (my parents also each came from two sides, so did theirs, etc.) the charting of traits and attributes even within a relatively limited generational depth takes on an aspect of speculation and even lottery (Davison *et al.*, 1989). Further complication is added by the tradition mentioned above that 'inheritance' can either be essentially physical ('a chip off the old block' – nature) or social ('following in her father's footsteps' – nurture). British people speak equally readily of ingrowing toe nails and diamond rings as being 'inherited'.

Because of the superficial similarity between scientific genetics and bilateral kinship reckoning, it is tempting to believe that the findings of predictive genetic tests will be relatively easily assimilated into family life (see Richards's dicussion in Chapter 12). Such an expectation is illustrated by the fact that genetic counsellors tend to see the communication of accurate probability and odds calculations (rather than conceptual issues concerning inheritance) as their main professional role (Harper, 1988; Parsons and Atkinson, 1992). In terms of the interaction of predictive genetic testing with popular cultures, two issues are of particular importance: first, that British popular culture expects to find character traits such as nervousness or wit following the same sort of route as hair or eye colour; secondly, that popular systems are much more ready to link traits and attributes into significant groups, giving primacy to a relatively 'holistic' appreciation of self and inheritance (Davison *et al.*, 1989). The importance of this divergence in terms of the impact of predictive genetic testing is best illustrated by an example.

The daughter of a family, as she moves into her late thirties, is recognised by relations, friends and neighbours alike as developing a build (thick arms, strong shoulders) and a gait (slight rolling) that make her the 'image' of her grandmother at that age. She has also acquired the habit of flicking the hair from her eyes that makes her just like her mother. The fact that she has started to (absently mindedly) miss sections of conversations and then goes on to interject inappropriate opinions is so like her mother that it is the cause of much mirth amongst family and friends. Because of these inheritances, it is widely assumed that the woman in question will very probably suffer with arthritis in her late fifties and sixties, because so many of the women ('on both sides, actually') in the family have, and particularly because the mother and the grandmother (so strongly 'taken after' in other ways) both had a lot of trouble with their joints in later years.

Closer family members have quite serious concerns about the severe depression suffered by the woman's mother when she was in her mid-thirties and again in menopause.

In this example, a range of different physiological and personal traits are charted through three generations. Among the protagonists and observers they make a set and it 'stands to reason' that the inherited body shape, the absent-mindedness and the slightly humourous conversational quirk should go together. This easy relationship with linkage leads naturally to the conclusion that arthritis is a distinct possibility, and that depression may never be far away. An extreme form of such processes (termed 'preselection') has been charted in families with Huntington's disease (Kessler, 1988).

The example can be hypothetically extended into a time in which predictive genetic testing is available and the woman, as a youngster, is provided with a personal genetic profile. As this service is part of the health system, information given is concerned with vulnerability and susceptibility, and the emphasis is placed on the benefits of early prophylactic action. The principal finding, in this case, is that a marked susceptibility to arthritis is identified, and passed on with the recommendation that great care is taken with respect to choice of sports activities, heavy lifting, and repetitious arm and hand movement (in a job, for example).

When this information is placed, as an overlay, on to the popular culture of inheritance, the linkage process is played out in reverse. The intimation that arthritis is 'on the cards', is now a starting point, and may stand as official scientific evidence that the woman 'takes after' her mother and grandmother. Because family inheritance lore tends to link joint trouble with depression, a strong potential exists for the advanced 'labelling' of the subject as someone to be treated in a particularly kind (or harsh) manner. While predictive genetic testing in Huntington's disease had the potential to prevent damage to a person through inaccurate 'pre-selection' (Kessler, 1988), its application more generally may have the opposite effect. Because of the relative complexity and sophistication of the 'sides' system, medically produced knowledge (which will inevitably consist of a series of 'bits' of information) will be easily assimilated. But through the action of popular 'linkage' concepts, the results of predictive genetic testing will often mean more than those who produced and communicated the information intended. Where such information is involved with what may be some of an individual's most important life decisions, such as whether or not to have children, its potential impact on real lives is considerable.

15.6 Conclusion

As part of a much broader orientation towards managing probability and risk, predictive genetic testing arises in a pre-existing landscape of social and cultural trends. In many fields of life the general tendency towards basing action on future-oriented information has already, had an impact, bringing into question the 'newness' of some cultural aspects of the 'new genetics'.

The medicalisation of symptomless at-risk states, for example, has already occurred in several fields, most notably blood pressure. Many different calculations of the probability of illness and death have long been used in the insurance and loans business, the most recent disputes and dilemmas having been centred on HIV testing. Genetics will not be the first link between employment, fitness and future health. It simply joins an already turbulent area of institutional struggle and individual dispute. Lifestyle movements with both official and commercial elements have already attempted to incorporate knowledge of heart disease risk into everyday cultural life. Twenty-first century health promotion campaigns based on predictive genetic testing will, in all likelihood, be similar to 1980s encouragements for people to 'know their cholesterol number' and then lower their personal risk. A similar 'measure-and-manage' culture has long surrounded body weight and shape.

The advent of predictive genetic testing, though, brings these disparate strands together and thus has the potential to make a serious impact both on individual lives and on collective organisation. Although augmenting the ethical, political and social issues listed above, genetic prediction is likely to pose new ones by virtue of its ostensible accuracy and completeness, its concern with information about transmission across generations and its possible supply at a very early stage of human development. Unavoidable decisions, choices and dilemmas will arise in any or all of a wide range of personal, group and institutional activities. If genetic assessments of an individual's probability of developing an adult onset disease become widely available, current cultural struggles and transformations concerning the construction of moral values, the nature of normality and what constitutes acceptable behaviour can only intensify.

15.7 References

Annas, G. J. (1993) Privacy rules for DNA databanks – protecting coded 'future diaries'. *Journal of the American Medical Association*, **270**, 2346–50.

Backett, K. and Davison, C. (1992) Rational or reasonable? Perceptions of health at different stages of life. *Health Education Journal*, **51**, 55–9.

Balshem, M. (1991) Cancer, control and causality – talking about cancer in a working-class community. *American Ethnologist*, **18**, 152–72.

Barnett, A. H., Eff, C., Leslie, R. D. G. and Pyke, D. A. (1981) Diabetes in identical twins – a study of 200 pairs. *Diabetologia*, **20**, 87–93.

Baron, M. (1993) Genetic linkage and male homosexuality – reasons to be cautious. *British Medical Journal*, **307**, 337–8.

Bauman, Z. (1993) *Postmodern Ethics*. Oxford, and Cambridge, MA: Blackwell Publishers.

Beck, U. (1992) From industrial society to the risk society – questions of survival, social structure and ecological enlightenment. *Theory, Culture and Society*, **9**, 97–123.

Billings, P. R., Kohn, M. A., de Cuevas, M., Beckwith, J., Apler, J. S. and Natowicz, M. R. (1992) Discrimination as a consequence of genetic testing. *American Journal of Human Genetics*, **50**, 476–82.

Brett, A. S. (1984) Ethical issues in risk factor intervention. *American Journal of Medicine*, **76**, 557–61.

Brett, A. S. (1991) Psychologic effects of the diagnosis and treatment of hypercholesterolaemia – lessons from case studies. *American Journal of Medicine*, **91**, 642–47.

British Medical Association (1992) *Our Genetic Future*. London: British Medical Association.

Carmelli, D., Swan, G. E., Robinette, D. and Fabsitz, R. (1992) Genetic influence on smoking – a study of female twins. *New England Journal of Medicine*, **327**, 829–33.

Cherskov, M. (1992) Fighting genetic discrimination. *ABA Journal*, June, 38.

Christie, B. (1992) On the brink of a molecular miracle. *The Scotsman*, 20 October, p. 15.

Collins, R., Peto, R., MacMahon, S., Hebert, P., Fiebach, N. H., Eberlein, K. A., Godwin, J., Qizilbash, N., Taylor, J. O. and Hennekens, C. H. (1990) Blood pressure, stroke and coronary heart disease. Part 2. Short term reductions in blood pressure – overview of randomised drug trials in their epidemiological context. *Lancet*, **335**, 827–38.

Connors, M. M. (1992). Risk perception, risk taking and risk management among intravenous drug users – implications for AIDS prevention. *Social Science and Medicine*, **34**, 591–601.

Coughlin, S. S. (1992) Applications of the concept of attributable fraction in medical genetics. *American Journal of Medical Genetics*, **43**, 1049.

Council on Ethical and Judicial Affairs, American Medical Association (1991) Use of genetic testing by employers. *Journal of the American Medical Association*, **266**, 1827–30.

Crawford, R. (1984) A cultural account of health – control, release and the social body. In J. B. McKinlay (ed.), *Issues in the Political Economy of Health Care*. London: Tavistock.

Daniels, N. (1990) Insurability and the HIV epidemic – ethical issues in underwriting. *Millbank Quarterly*, **68**, 497–525.

Davison, C., Frankel, S. and Davey Smith, G. (1989) Inheriting heart trouble – the relevance of common-sense ideas to preventive measures. *Health Education Research – Theory and Practice*, **4**, 329–40.

Davison, C., Frankel, S. and Davey Smith, G. (1991) Lay epidemiology and the prevention paradox – the implications of coronary candidacy for health education. *Sociology of Health and Illness*, **13**, 1–19.

Davison, C., Frankel, S. and Davey Smith, G. (1992) The limits of lifestyle – re-assessing fatalism in the popular culture of illness prevention. *Social Science and Medicine*, **34**, 675–85.

Davison, C., Macintyre, S. and Davey Smith, G. (1994) The potential social impact of predictive genetic testing for susceptibility to common chronic diseases – a review and research agenda. *Sociology of Health and Illness*, **16**, 340–71.

Draper, E. (1991) *Risky Business – Genetic Testing and Exclusionary Practices in the Hazardous Workplace*. Cambridge: Cambridge University Press.

Dunstan, G. R. (1988) Screening for fetal and genetic abnormality – social and ethical issues. *Journal of Medical Genetics*, **25**, 290–93.

Elisin, S. C., Corsetti, L., Ullrich, A. and Permutt, M. A. (1986) Multiple restriction fragment length polymorphism at the insulin receptor locus – a highly informative marker for linkage analysis. *Proceedings of the National Academy of Science, USA*, **83**, 5223–7.

Gellner, E. (1987) *The Concept of Kinship*. Oxford: Basil Blackwell.

Giddens, A. (1991) *Modernity and Self-Identity – Self and Society in the Late Modern Age*. Cambridge: Polity Press.

Graham, H. (1987) Women's smoking and family health. *Social Science and Medicine*, **25**, 47–56.

Gustavson, K.-H. (1989) The prevention and management of autosomal recessive conditions. Main example – alpha1-antitrypsin deficiency. *Clinical Genetics*, **36**, 327–32.

Hamer, D. H., Hu, S., Magnuson, V. L., Hu, N. and Pattatucci, A. M. L. (1993) A linkage between DNA markers on the X chromosome and male sexual orientation. *Science*, **261**, 321–7.

Harper, P. S. (1988) *Practical Genetic Counselling*, 3rd edn. London: Wright.

Harper, P. S. (1992) Genetic testing and insurance. *Journal of the Royal College of Physicians of London*, **26**, 184–7.

Harper, P. S. and Clarke, A. (1990) Should we test children for adult genetic diseases? *Lancet*, **335**, 1205–6.

Holland, W. W. and Stewart, S. (1990) *Screening and Health Care – Benefit or Bane?* London: The Nuffield Provincial Hospitals Trust.

Jones, S. (1993) *The Language of Genes – Biology, History and the Evolutionary Future*. London: HarperCollins.

Kessler, S. (1988) Preselection – a family coping strategy in Huntington's disease. *American Journal of Medical Genetics*, **31**, 617–21.

Light, D. W. (1992) The practice and ethics of risk-related health insurance. *Journal of the American Medical Association*, **267**, 2503–8.

Macdonald, L., Sackett, D., Haynes, R. and Taylor, D. (1984) Labelling in hypertension – a review of the behavioural and psychological consequences. *Journal of Chronic Disease*, **37**, 933–42.

Macintyre, S. and Sooman, A. (1991) Non-paternity and prenatal genetic screening. *Lancet*, **338**, 869–71.

Muller-Hill, B. (1993) The shadow of genetic injustice. *Nature*, **362**, 491–2.

Murray, T. H. (1991) Ethical issues in human genome research. *FASEB Journal*, **5**, 55–60.

Needham, R. (1960) Descent systems and ideal language. *Philosophy of Science*, **27**, 96–101.

Nelkin, D. and Tancredi, L. (1989) *Dangerous Diagnostics – the Social Power of Biological Information*. New York: Basic Books.

Orchard, T. J., Dorman, J. S., La Porte, R. E., Ferrell, R. E. and Prash, A. L. (1986) Host and environmental interactions in diabetes mellitus. *Journal of Chronic Disease*, **39**, 979–99.

Otway, H. and Wynne, B. (1989) Risk communication – paradigm and paradox. *Risk Analysis*, **9**, 141–5.

Parsons, E. P. and Atkinson, P. (1992) Lay constructions of genetic risk. *Sociology of Health and Illness*, **14**, 437–55.

Phillips, D. I. W. (1993) Twin studies in medical research – can they tell us whether diseases are genetically determined? *Lancet*, **341**, 1008–9.

Rapp, R. (1988) Chromosomes and communication – the discourse of genetic counselling. *Medical Anthropology Quarterly*, **50**, 143–57.

Roscam Abbing, H. D. C. (1991) Genetic predictive testing and private insurances. *Health Policy*, **18**, 197–206.

Rose, G. (1985) Sick individuals and sick populations. *International Journal of Epidemiology*, **14**, 32–8.

Strathern, M. (1992) *After Nature – English Kinship in the Late Twentieth Century*. Cambridge: Cambridge University Press.

Summers, K. M. (1993) Genetic susceptibility to common diseases. *Medical Journal of Australia*, **158**, 783–6.

Takagi, K. (1991) Genetic screening – policy making aspects. *Genetics, Ethics and Human Values*. (*Proceedings of the XXIVth CIOMS Conference*, Tokyo, 1990), pp. 117–25. Geneva: CIOMS.

Wilkie, T. (1993) *Perilous Knowledge – The Human Genome Project and its Implications*. London: Faber & Faber.

Zimmet, P., Arblarster, M. and Thorma, K. (1978) The effect of westernisation on native populations. Studies on a Micronesian community with a high diabetes prevalence. *Australia and New Zealand Journal of Medicine*, **8**, 141–6.

16

The new genetics: a feminist view

MEG STACEY

> ... women ... form the central focus of the family-based problems that genetic
> diseases create. *Peter Harper, clinical geneticist, Chapter 1, this volume*
> Without an awareness of the oneness of things science can give us only nature-in-
> pieces; more often it gives us only pieces of nature. *Barbara McClintock,*
> *cytogeneticist and Nobel prizewinner (quoted in Keller, 1983, p. 205)*

16.1 Introduction

Over the past 25 years the new reproductive technologies (NRTs) such as *in vitro* fertilisation (IVF) and the new genetics together have aroused my concerns. This may seem
odd to molecular biologists, geneticists, embryologists and obstetricians, whose spe-
cialisms differ one from the other and who for the most part work independently of
each other; but a woman could not help noticing that the first applications of the new
genetics impinged on pregnancy and birth – in prenatal screening and the offered abortion for fetuses found to be impaired. This represents a great change from the way in
which pregnancy, birth and possible disability were previously understood.
Consequent psychological problems for women were likely and seemed obvious. Yet
what thought was given to the problems was commonsensical rather than systematic or
scientific – and derived from men's, not women's, common sense.

To a sociologist it is axiomatic that a birth is a social as well as a biological event; the
arrival of a new member of society changes the social arrangements around it. New
relationships (mother, father, son or daughter) are established, and these changes
spread to other kin members – mothers become grandmothers overnight, cousins are
created. Altering the way in which children are conceived inevitably affects these relationships; although relatively few parents are directly involved in using the new techniques, these have implications for all the established mores about parenthood. Other
relations are changed: what had been a private event – conception – involves third
parties. This alters power relations between professionals and their clients.

However, none of these arguments seemed to weigh much: applications of both the
new genetics and the NRTs went ahead, with few calls being made for social science
research in parallel with the genetic, embryological and obstetric research. In the case
of the NRTs, such as IVF, Marilyn Strathern, a social anthropologist, elucidated why
this might be. She demonstrated an asymmetry in the understanding of what birth is:

those who thought of birth as biological did not think of it as also social; those who did think of it as social also thought of it as biological (Strathern, 1992*a*). Perhaps for some people biology is a sole determinant of life events. The talk around the new genetics increasingly began to look like that. Biological sex was discussed, but gender was left out along with the social context.

Some recognition was given to social consequences: scientists and practitioners were aware that the new techniques might give rise to religious and ethical objections and legal difficulties. Consequently to a greater or lesser extent lawyers, theologians and ethicists were called in to give their opinions. There seemed little understanding that systematic study of social and psychological consequences might be possible. Nor did women's experiential understanding apparently carry weight.

Increasing recognition is now being given to the psychological implications of the new human genetics, as a number of chapters in this volume testify. For example, behavioural science research has been recognised by the Medical Research Council in the UK and the Human Genome Project Ethical Legal and Social Implications Program in the USA. Cultural implications, particularly for kinship, have not only been studied by social anthropologists (Strathern, 1992*b, c*; Edwards *et al.*, 1993), but have become the topic of joint work between geneticists and social scientists.

Scientists and clinicians appear to have found social structural aspects harder to recognise than behavioural or psychological ones, including responsibility for their own input into structural change. Thus, the Clothier Report (HMSO, 1992) failed to understand the new ethical implications of the new genetics, so ably displayed by Derek Morgan in Chapter 9. Social change is often accompanied by political action and social movements that can add embarrassment and even apparent obstruction to the clinician's work. This then raises doubts on the part of scientists and clinicians about the possibility of systematic and detached analysis.

In this chapter, after explaining the notion of social structure, how it arises, is maintained or changed by social action (developed from Stacey, 1994), I shall illustrate how, in the past, scientific and medical developments have been implicated in changing social structure, including the gender order, and changing culture, including ways of knowing. Drawing attention to major contemporary structured divisions in industrial societies, I shall focus particularly on gender, both in the science of the new genetics and in the ways the new technologies are applied to the human body. Failure to recognise or systematically address the inevitability of social consequences will itself be taken as part of the phenomenon requiring analysis, including the power the new technologies put in the hands of a few.

In conclusion I shall argue that pointing out possible social consequences does not necessarily imply opposition to or obstruction of medical developments, but rather is a plea for therapeutic activity to be understood in its social context more widely and systematically than has hitherto been the case. Underlying my argument is the belief in the undesirability, indeed immorality, of causing further unnecessary suffering in the attempt to remove or ameliorate that which already exists.

16.2 Social structure and social action

'Structure' implies all those social arrangements that appear to be outside us, that exist before we arrive on the scene, that make social life possible, providing 'the means, media, rules and resources for everything we do' (Bhaskar, 1989, pp. 3–4; quoted in Porter, 1993, p. 593). In the sense that social structures were there before those of us currently alive were born, we do not create society. However, we do, in our everyday activities, continually reproduce or transform the structures that compose it. No society can exist without people; that is without human agency. There is more to society than 'individuals and their families' – the only aspects of the social world that Margaret Thatcher admitted the existence of. Society is very much there as something, albeit abstract, that both enables and constrains human action.

Social structures and the associated ideas are continually reinforced or changed by human agency. The sciences, including medical science, and medical practice themselves constitute a part of those social structures. Through the agency of genetic screening and later on through genetic manipulation, the new genetics and the ideology it purveys are making changes in the social structure and in the meanings we give to our lives. What changes are wrought will depend on how the procedures are applied, by whom, to what and to whom. There will be unintended consequences, whether beneficent or malevolent, conservative or radical, affecting differently those in different parts of the society: rich and poor, women and men, elites and masses, majority and minority ethnic groups. Along lines such as these major structural divisions are found in our society.

Although a reality, 'society' is not a being with a mind that it can make up – although medical practitioners frequently cast it in such a role – 'society' is derived entirely from human actions. 'Society making up its mind' frequently means 'government should decide'. This its members do by paying attention to opinion leaders, the media, parliamentary constituents, civil servants and members of parliaments or senates. Many people, many human agents in many interactions, will lead to the government's decision. However, only limited sections of the total society will have been involved; views of the dominant members are likely to prevail, research scientists and medical practitioners among them. Power and influence, their ownership or not by individuals and groups, are crucial aspects of social structure and social change.

16.3 Overriding nature

Persons have always, so far as we know, attempted to ameliorate the unwanted aspects of natural processes, to cure illness and to relieve pain and suffering, through ways that are many and varied. From time immemorial women have sought to prevent or encourage pregnancy and to terminate unwanted pregnancies. Elders, religious leaders, church or state have in various ways tried to regulate mating, overseen birth and the entry of the emergent new social being: death and its implications have always attracted collective interest. Such activities are not new and have involved a range of different authorities, not necessarily the healers.

The major new feature is that medicine, following science, has sought not just to modify the consequences of a quixotic natural order, but to control the order itself (Crombie, 1993). The new medicine from its earliest beginnings challenged and changed existing belief systems. Touching the untouchable began when corpses were dissected (illegally, initially); questioning the unquestionable has continued ever since. No given of nature retains its former status. Among recent examples has been the over-riding of the hitherto unquestioned 'given' that menopause signals the end of a woman's fertile life; the new genetics overthrows the inevitable 'naturalness' of heredity (Strathern, 1992*b*).

16.4 The social context

These medical achievements did not arise in a social vacuum simply as a result of changed ways of thinking, of the freedom from the old 'superstitions' which the enlightenment made possible. The developments depended on the availability of par-ticular social and economic arrangements. In the case of medicine, economic changes associated with the breakdown of feudalism and the development of capitalism led to the extension of urban areas and dispossessed many of their livelihoods and abodes in the countryside. Populations became available for the hospitals whose character was also changed by the new medical 'gaze' (Foucault, 1973). Medicine based on observa-tion, and increasingly on experiment, was born.

16.5 Gendered medicine

During the industrial revolution, the division of tasks and responsibilities between women and men was redefined. A new gender order, male dominated, was established by the nineteenth century, derived from the long-standing patriarchal familial and kinship system, but different in a number of important respects. Work and responsibil-ities for public affairs outside the house extended and an enlarged male public world developed to which women had restricted access. To this, medicine contributed.

Before the nineteenth-century professionalisation of medicine, women, along with men, played a major part in health maintenance and restoration. The physicians, all men and few in number, treated only the upper class. With particular interests in repro-duction and relationships among family and kin, responsible for feeding and looking after their families, housewives were the first resort when health care was needed for the majority of the population. Some had particular healing knowledge and skill called on by others beyond their family members; some were skilled in midwifery.

Quite profound changes in the social order and beliefs went along with the develop-ment of the new medicine and its associated division of labour. Midwifery (as its name indicates) had been exclusively women's work, probably seen as polluting for men. Such inhibitions were overcome as male doctors began to supervise childbirth. Women wishing to be seen as very up-to-date, and hoping no doubt for better outcomes, accepted the idea of medical birth attendants (Stacey, 1988, p. 53). Reformed nursing served medical practitioners in old and newly established hospitals, also providing a livelihood for the many spinsters of the mid-nineteenth century in an 'intermediate zone' (Stacey and Davies, 1983) between public and private domains. The new medicine had succeeded in creating a thoroughly male-dominated profession and thereby played its part in the establishment of the new male-dominated gender order in the nineteenth century, one that is now, in the late twentieth century, being again rearranged.

Women did not entirely abandon their healing role to the health professionals. Women, and especially housewives, continued and continue to play the informal role of family health workers; much of this now involves orchestrating professional health care for their families and effecting the interface between health and welfare professionals on their behalf (Graham, 1979, 1984; Stacey, 1988, pp. 206–209). This in part accounts for the distinctive relationship that women have with the new genetics, to which I shall refer later.

16.6 Gendered science

The conventional wisdom is that science is neutral, assured, scientists believe, by the strict application of rationality and logic and the associated procedures whereby its empirically based findings are always subject to falsification. This ideology is challenged by claims that science is socially constructed (Wright and Treacher, 1982) (see the dispute at the 1994 conference of the British Association for the Advancement of Science (Anon., 1994)). The challenge has nowhere been stronger than to science's lack of gender neutrality. Hilary Rose, an early contributor (Rose and Hanmer, 1976; Rose, 1978, 1983), gives a useful account (Rose, 1994) of the feminist charges about the masculinist bias of science drawing on writers such as Birke *et al.* (1980), Keller (1985) and notably Harding (1986).

The masculinism emerges in two ways: first, in the gender of the scientists; secondly, in the ideology that supports science. As to the first, men established modern science and still dominate it. Women scientists are mostly allocated to unsung laboratory work; those whose work is of prime importance get no or tardy recognition for it, whether as Fellows of the Royal Society or Nobel prizewinners (Rose, 1994, chs. 5–7). The topics taken for study, the problems set for solution, the methodologies used, and the public presentations reveal the masculinist ideology of science. Nature is not only to yield its secrets but to be controlled: it is to be worked against, not with. Women have traditionally been seen as part of nature and also to be controlled – for which Rose (1994, pp. 231–232) blames the Judaeo-Christian inheritance. The Cartesian dualism of intellect and body and the exclusion of feeling and emotion from the scientific enterprise are key feminist targets. Abby Lippman's critique of the biased presentation of prenatal genetic testing (Lippman, 1992, pp. 143–8) is an example of the lack of neutrality in the public presentation of scientific data. As Harding (1986, p. 63) concluded, 'the cultural stereotype of science . . . is inextricably intertwined with issues of men's gender identities . . . "scientific" and "masculine" are mutually reinforcing cultural constructs.'

The biography of the American cytogeneticist Barbara McClintock illustrates aspects of both of these points (Keller, 1983; Rose, 1994). The geneticists whose names are familiar as Nobel prizewinners – such as Crick and Watson, Monod and Jacob – are all men. McClintock was not awarded a Nobel prize until 1983, when she was 81, although in 1959–1960 she had published papers showing close parallels between her findings and those of Monod and Jacob. For 30 years mainstream male scientists ignored, perhaps because they could not understand, her careful work on transposition in genetic regulation. In common with other leading women scientists (Rose, 1994) but in contrast to, for example, Longino and Doell (1983) Haraway (1985), Hubbard (1979), Lowe and Hubbard (1983), McClintock did not claim feminism.

Mainstream biologists have said much about sex differences (Harding, 1986, ch. 4), for example explaining the contemporary domination of men over women by the evolutionary development of 'man the hunter'. Their enquiries have concentrated on biological differences between the sexes rather than on the *similarities* between them. Detached (because neutrality seems socially impossible) enquiry could have been expected to do that (Bleier, 1984). I do not here propose to go into whether the male bias manifests simply in examples of bad science or whether it imbues science-as-usual (see Haraway, 1981, 1985; Harding, 1986, p. 102ff). However, Harding's comment (1986, p. 103) about 'the seamlessness of science's participation in projects supporting masculine domination' chimes with the evidence.

16.7 Social sciences – also gendered

Modern social sciences, also children of the post-enlightenment era and the upheavals around the rise of capitalism, were concerned to apply the methods of the emergent natural and physical sciences to social life, perhaps offering means to control it, even suggesting social engineering, now generally seen as a hazardous route. Sometimes associated with reform, less often with revolution, more often with helping to sustain domination of a social order, sociology was particularly concerned with the question of order, how it came about, was sustained or fragmented. But sociology focused on the *public* order; the social order of the domestic domain was a minor appendage (Stacey, 1981).

Over the years the inappropriateness of methods that objectify the human phenomena of their enquiry has increasingly been borne in upon social scientists and the use of narrowly positivistic methods questioned. Sociologists, like other scientists, have sometimes themselves accepted unquestioningly the dominant values of their own societies. The discipline was also male dominated.

Failure to recognise the crucial importance of gender to analyses of social structure and process was characteristic until 20 years ago. From 1974, when women sociologists joined together to challenge the gendered nature of the profession and of sociological knowledge, the policy of the British Sociological Association has been to encourage teaching and research on gender issues to redress hitherto unrecognised gender bias in sociological knowledge. Similar develoments took place across the Atlantic. Despite this and major works such as Dorothy Smith's *The Everyday World as Problematic* (1987), the full implications of the feminist challenge have not been accepted and serious theoretical papers may emerge that are gender blind (e.g., Maseide, 1991). Interdisciplinary challenge to the gendered nature of the new genetics and the new reproductive technologies came from feminists, including biologists, social anthropologists, psychologists and sociologists.

The sociological enterprise recognises and analyses in a sustained and rigorous fashion the many connections between different aspects of the social world, including those between 'apparently narrow aspects of applied reproductive technologies [including genetics] and broader questions about the nature of the social order' (Stacey, 1992, p. 5): interrelations between one set of social actions and another, one part of the social structure with another. In the case of science and gender, the task is to observe and identify the many threads that make up the apparently seamless connection between science and masculinism of which Harding (1986) complained.

16.8 Scientific tunnel vision?

Gender divisions represent one important way in which our society is structured. These seem to have been ignored in discussions about the application of the new genetics, as have other major aspects of the social structure: the increasing gaps between rich and poor, between those who have work and those who do not, those who have somewhere to live or do not. Such divisions tend to be reduced to individual attributes: an unemployed man, a lower-class woman, a homeless person. However, these attributes represent the impact upon individuals of a divided society, the result of the many actions and interactions of individuals and groups that sustain or change the differences.

16.9 The disappearance of women

As obstetrics developed and, in contrast to midwives' perception, the focus increasingly came to be upon the fetus *in utero* (Jordanova, 1980, 1985; Arney, 1982; Oakley, 1984); the goal now was a healthy baby and less focus was on the woman carrying the child except in so far as she was a vessel for the production of that baby. Jordanova (1985) showed, through a series of illustrations from contemporary medical books, how the focus narrowed in the eighteenth century from the whole birthing woman to a section of her body from abdomen to mid-thighs. Gradually the woman disappeared from view, except perhaps for her uterus. Oakley has taken this theme further, reproducing Gruenwald's illustration of 'the fetal cosmonaut' floating free outside the woman and her uterus altogether (Oakley, 1984, p. 175). Petchesky (1987) argues convincingly that how fetal images are presented is *de facto* a political act.

Since the late 1960s the way in which women have been overlooked in academic research has been extensively explored from Sheila Rowbottom's *Hidden from History*

(1973) to Carol Gilligan's *In a Different Voice* (1982). Sarah Franklin's (1993) analysis of the UK debates on the Human Fertilisation and Embryology Bill underlines the remarkable way in which the embryo became an isolated object. Mike Mulkay (1994*a*, *b*) demonstrates how in these debates women did not exist as people for the male majority in both Houses of Parliament. Women spoke as women and from experiential knowledge; men from specific occupations (cf. Gilligan, 1982). Most women were simply concerned about whether embryo research and technologies would benefit *women*; on the whole they thought it would. Women parliamentarians did not support their few male colleagues who had spoken about the negative side effects for women (Spallone, 1989). Women and men shared an advocacy of family life and assumed that 'the interests of embryos, children, mothers, and fathers, naturally converged' (Mulkay, 1994*a*, p. 21). Women parliamentarians did not concur with the views of feminists cited earlier, who argue that the new genetics do not necessarily work in women's interests.

16.10 Gender blindness

As well as ignoring social structural components, most writing about the new genetics tends to be gender blind, as were the Clothier report (1992) and the Nuffield report (Nuffield Council on Bioethics, 1993). The broad-ranging and valuable *Human Genome Project: Ethics*, published by the Spanish Foundation BBV following an international workshop that had as its main aim to stimulate international cooperation in the ethical aspects of the human genome project, fails to discuss gender despite 37 papers from many disciplines and nations (Fundacion BBV, 1992). The Canadian Royal Commission on the new reproductive technologies was remarkable in that the five commissioners were all women. In the chapter on prenatal diagnosis (PND) their report says, 'we strongly reject the non-medical use of PND or genetic technologies, or their use in discriminatory ways that devalue being female' (*Proceed with Care*, 1993). National cultures strongly influence how the NRTs and the new genetics are received.

Although the Human Fertilisation and Embryology Authority document on sex selection mentions the concept of gender throughout (HFEA, 1993, para. 36), there appears to be a lack of understanding of the fundamental distinction between (biological) sex and (social) gender (Oakley, 1972). This distinction is not a matter of opinion but of observation. The notion of the gender order includes more than mere biological differences or the simple differential allocation of tasks between the sexes: it implies authority relations and arrogance–deference relations in the gender roles that are the norm in any one society. The order is one of gender, not of sex as such; it is socially constructed and may be constructed in a variety of ways that may change over time. Most societies we know about have a male-dominated gender order, the men being accorded a superior position (Stacey, 1988, p. 7). Anthropologists are now discovering that this conceptual division may be specifically Western (Moore, 1994), but it is with such societies that I am dealing here.

Despite all the equal opportunities legislation and initiatives, the gender order of British society continues to be male dominated, although the characteristics of that domination vary from one part of the society to another, not only between cultural ethnic groups but also within the majority ethnicity, including geographical and age-related variations. For none of the ethnic groups in the UK, including the white indigenous, can the assumption be made (as the HFEA document (1993) presumes) that a decision to go for a child of a particular sex is necessarily made for love. Nor can this be assumed even if the couple say they are agreed on the sex they wish. It may be that one partner has been over-persuaded by the other. Women may voluntarily accept decisions that support the male-dominated gender order, thus making their personal life more comfortable but contributing to the continued oppression of women. More empirical data are needed about matters such as these.

16.11 Gendered views

It is interesting to note how the possibility of sex selection initially roused different fears in women and men. John Harris, for example, was led to imagine a world in which women had been reproduced in larger numbers than men because it is their task to bear children, just a few men being kept rather like stud bulls (Harris, 1985). The fears of some feminists went the other way, imagining that men's desire for sons to keep the patriarchy going would work to the disadvantage of women (see Arditti *et al.*, 1984, pp. 236–77; Corea *et al.*, 1985, pp. 201–6; Rothman, 1988, pp. 133–46; Spallone, 1992, pp. 223–8; Steinbacher and Holmes, 1985).

Hannah Bradby, in Chapter 14, has surveyed data about sex selection. In some parts of India, where infanticide of girl babies has long been practised, amniocentesis is now being used by the wealthy to the same ends (Spallone, 1992, pp. 225–7). In the West the fear was expressed as a desire to have first a boy and then a girl, which would lead not to a numerical sex imbalance, but to a subtle underscoring of the superiority of the male over the female (Corea *et al.*, 1985, pp. 204–205). Research suggests that in Britain the wish is for children of mixed sexes, although the research of Carr-Hill *et al.* (1982) cannot tell us whether boy babies are preferred as first-borns. Statham *et al.* (1993) report data from pregnant women that suggest they have no particular preference for one sex or the other, nor a wish to know the baby's sex before birth – perhaps particularly in the case of first-borns. This research tells us nothing about men's preferences or the agreement/disagreement between women and men in general or between partners.

Harris's view seems to demonstrate the problems that philosophers may run into when they do not take empirical data about the real world into account. Maybe in his mind women are strong – indeed we are – and could achieve this end of severely reducing the male population. However, that reckons without the male-dominated gender order in which men still occupy most of the strategic power positions. The feminist view, probable or not, at least recognises the existence of the present gender order.

In a more recent book, Harris (1992) seems to suggest that the best plan would be to permit parents free choice of the sort of baby they want, not only sex, but other characteristics as well. The trouble is that choices are *not* free, they are informed by the beliefs and ideas of the society, which influence both conformists and rebels who both suffer social and economic constraints (see Rose and Hanmer, 1976). An understanding about such matters can be achieved only by looking at the empirical realities of what people do and think.

16.12 Not gender alone

Empirically, as we have seen, there are varied views among women as among men. Anne Finger, disabled by polio, a voluntary worker in a US abortion clinic (constantly under violent attack from extreme anti-abortionists), found she was facing the possibility that her newborn and wanted son might have been seriously damaged by a birth trauma. Finger, an active defender of the rights of the disabled to a full life, recognises the difficulties that impairment – whether by illness, accident or genetic inheritance – entails and the further disablement imposed by an unsympathetic world. In her life experience while waiting to see how her son would develop, she found herself in a constant internal debate, praying that he would not be disabled, yet somehow feeling a traitor that she should pray thus. This testimony of her own experiences (Finger, 1990), in common with those published at the outset of this book, reveals vividly the contradictions latent in us all, although not many are tested as Finger or our contributors were.

16.13 Sex, gender and inheritance

In this volume, Shiloh (Chapter 3), Green and Statham (Chapter 6) and Richards (Chapter 12) have discussed aspects of the intersection of sex and gender, particularly in cases of sex-linked genetic disorders such as Duchenne muscular dystrophy, fragile-X and some cancers. Among the particularly difficult and poignant problems that arise, some occur only because of knowledge provided by the new genetics. Knowledge may be constraining as well as liberating – sometimes both at once.

In addition to being the childbearers and child rearers, women seem always to have been 'kin-keepers' (Young and Willmott, 1957; Stacey, 1960; Firth *et al.*, 1970) and crucial unpaid health workers for their family (Section 16.5 above); now they are 'genetic housekeepers' (Chapter 12). Yet on account of previous gender blindness there is still so much we do not know.

In genetic clinics offering testing for inherited breast and ovarian cancer (e.g. *BRCA1*), where most clients are women who have referred themselves (Richards *et al.*, 1995), women may

> come to the clinic on their own or with other family members. Sometimes a mother and daughter, sisters, or more than one couple from the same family will attend the clinic together. Some male partners remain outside during sessions. *(Richards* et al.*, 1995, p. 3)*

Is this because they are afraid of what may be learnt or because such things are 'women's business'? It would be interesting to know.

Much of the responsibility for attempting to understand the genetic information they are being offered also falls upon women who have to negotiate with clinician or counsellor; then, on the basis of what they have understood, to negotiate (or not as they may decide) with partner, family and kin. Upon the women consequently fall many of the communication and decision problems that arise from the gap between clients' and geneticists' ways of knowing. The world of the clients is a different world from that of the genticists; furthermore the genetic account is light on the information that clients urgently want. Women experiencing the procedures find this particularly difficult, as Parsons's and Atkinson's research on a sample of women, mothers and daughters, medically defined as having a specific 'risk' of carrying and transmitting a gene mutation for Duchenne muscular dystrophy (Parsons and Atkinson, 1992; see also Parsons, 1990) shows. Simplifying statistical probabilities when explaining their chances to women did not always help. One woman, for example, whose medically defined carrier risk was 33%, told the researchers:

> They gave me odds, I cant remember the actual odds, I was classed, I think, as high risk . . . he did tell me about the blood test and what they looked for, the abnormalities and then how they calculated my risk . . . they said I've got to have eight children to have one affected, I thought – 'Do I have to have eight children to prove them wrong?'
> *(Parsons and Atkinson, 1992, p. 442; omissions in original)*

The authors indicate that the women's problems in understanding what they were told could not be accounted for by innumeracy on the women's part. The problems arose from difficulties in making sense of the information in their own terms, a problem shared by women whose personal accounts open this volume. The women tended to translate the risk assessments they were given into 'descriptive statements that were recipes for genetic decision making' (Parsons and Atkinson, 1992, pp. 447 and 454). Problems of this kind have echoes throughout this volume. In its difference from the scientific discourse this evidence echoes, but in a very specific way, the distinctive discourse of parliamentary women compared with the men.

16.14 Power and the new genetics

Throughout this chapter, power as a factor in social relations has been just beneath the surface of the discussion. Power, along with most other resources, is unequally distributed. Medical power can be exercised individually in relation to particular patients or collectively through public health programmes.

Clients and geneticists both bring knowledge to the consultation, but knowledge of different kinds. The geneticist's scientific knowledge has been accorded higher status in the society at large than the experiential knowledge of the client/patient, although people in general will recognise the relevance and worth of those dimensions of experience, feeling, meaning that the client expresses. Furthermore, the medical geneticist has access to the resources that the patient hopes may resolve the problem she has presented with – or that the client has been informed she faces. But knowledge of cure does not necessarily follow scientific knowledge of cause.

How the powers that higher status accords are exercised by the doctor or counsellor in the consultation is a matter of her/his understanding, sensitivity, training and personality. Where the doctor is a man talking with a woman, research evidence suggests that he (whatever his intention) will be more directive than a woman doctor would be (West, 1993). This comes apparently 'naturally' – and also unconsciously – derived from having been brought up in a male-dominated society. Consciously or unconsciously, practitioners may influence patients towards the solution that they think is correct. Marteau (1989) has shown, for example, how the way in which information is framed may influence decisions taken by both doctors and patients. Tuckett *et al.* (1985) have spelt out how complex are factors affecting patients' understanding of what they are told in consultations.

Some forms of influence go further. Josephine Green has recently compared obstetricians' views on prenatal diagnosis and termination of pregnancy in 1993 with those reported to Wendy Farrant in 1980 (Farrant, 1985). Some of her findings have already been reported (Green, 1993; see also Chapter 6). With regard to access to prenatal diagnosis, obstetricians were asked, 'Do you generally require that a patient should agree to the termination of an affected pregnancy before proceeding with amniocentesis/chorionie villus sampling (CVS)?' In 1980 three quarters of the sample (323 consultants) had answered 'yes'. In 1993 a third (34%) of 375 consultants also said 'yes' (J. Green, personal communication, 1994; see also Chapter 6).

This exercise of power denies knowledge to those women who do not wish to choose abortion. In that case a woman might reasonably wish to know the status of her fetus. Knowledge would give her time to learn about the impairment and to prepare for what she has to face. Insisting on agreement to termination before allowing access to amniocentesis or CVS seems an inappropriate use of power: denying this chance to a woman *reduces her choices*.

Pressure to abort impaired fetuses also reduces the collective choice to accept children as they are (Rothman, 1988). Questions arise about social attitudes towards those who do carry an impaired fetus to term in a society where their abortion has become the norm. It is all very well to refer, as the HFEA document (1993, para. 31) does, to the respect due to the choice of someone who refuses available sex selection for medical reasons. Those women who carry to term a baby with a known or high risk of impairment, refusing abortion, may in practice suffer disrespect for their actions; further, others may resent helping with the cost of the child if it is born impaired. Knowledge of that kind was part of what was in Finger's distressed mind referred to earlier (Finger, 1990). The notion that such a choice is worthy of respect only has meaning if all possible support is provided for the person who made it. Failure to provide support implies disrespect for the choice and the offspring.

16.15 Exercise of collective power

The right of doctors to treat patients has been accorded to them through the state-recognised medical registration system. Certain other powers and responsibilities are also laid upon them by the state: certification of birth, death and sickness, for example. Rights to practise relate to the treatment of individual patients. Matters such as general screening of the population, or of specified sections of the population, school children or pregnant women, for example, raise wider issues.

Offering a test that an individual has a legal right to accept or reject is by no means unproblematic. Inequalities of power and knowledge add to the difficulties of knowing whether consent is really fully informed (Alderson, 1990, 1993). In general screening for any inherited disease, with the implied intention of reducing the incidence of the pathological condition, the social action moves into a different territory. Although the overt intention is to offer a choice that comes from greater knowledge, a normative statement is in practice being made about the undesirability of the existence of people with such a condition and of giving birth to and bringing up such people. This is also the implication of individual actions and decisions, as we have seen. The geneticists' promotion of screening programmes for inherited disorders cannot help but carry a moral message about the worth of the lives of those individuals who carry the disorders that are the targets of such programmes.

In their sex selection document, the HFEA say 'that all human beings have intrinsically equal value and on this basis deserve equal respect' (HFEA, 1993, para. 29). As a moral statement this is impeccable; the statement that follows, that the diseased or disabled can make an important contribution to society, is empirically correct. However, despite some improvements, the disabled in British society are not in practice accorded equal status or respect compared with the so-called 'normal' members of the population.

A case can be made for saying that scarce resources would be better used by providing adequate services for the impaired than by trying to eliminate a relatively small number of sex-linked diseases. Whatever is done genetically, there will continue to be impaired persons – any one of us may become impaired at any time by reason of illness or accident. Each of us would then hope for care and continued support.

16.16 A potential distortion of the medical mission?

The killing mission of the Nazi doctors, an exercise of collective power, developed from public health programmes of the Weimar Republic (Lifton, 1987; Weindling, 1993; see also Chapter 10). Are there possible distortions hidden in the medical genetics of the late twentieth century? Without a systematic examination of the social context in which the programmes take place we cannot know.

The Nuffield report on the ethics of genetic screening and the HFEA sex selection consultation document both assure us that 'it [the Nazi eugenic horrors] could not happen here'. The Nuffield report says that

> genetic testing in medicine in the UK is used to help individuals and their families avoid the occurrence of serious inherited disorders or their associated complications . . . [as are] . . . those wider population-based genetic screening programmes that have so far been established.   *(Nuffield Council on Bioethics, 1993, para. 8.19)*

The Nuffield report (para. 8.18) reminds us that the UK has never had any eugenic legislation, although there was support for eugenic ideas (but see Chapter 10). The HFEA consultation document on sex selection also assures us we need not worry – in this case because we have legislation to control medical actions (HFEA, 1993, p. 24). But legislation can be changed. Governments of the past 15 years have, ironically, increased central government powers at the same time as they have encouraged free competition in the economy: our civil liberties have been reduced.

16.17 Causes for concern

Some things make one anxious. I have demonstrated above the failure to pay attention to the gender order when considering the ethics of genetics and also to the way in which women are depersonalised in parts of the genetic programme – not thought of as persons but as vessels for embryos and thus 'naturally' the subjects of medical experiment. Is not eugenicism lurking here?

At a meeting to discuss genetic screening, a young woman affected by Down syndrome heard a geneticist proposing screening with the effective aim of reducing or eliminating the disease. The young woman was appropriately angry and said so. After the meeting more than one doctor member of the audience said to me that it had been inappropriate that she had been present. Why? Because it made them unhappy to be reminded of the essential humanity of a category of people they hoped would in future not be born? Because her intervention made it difficult to stick to merely rational argument? Because, as one doctor made plain to me, he could not see her as a real person, but as a typical bundle of Down syndrome symptoms, mentioning among other things her fidgetiness? Because she was a non-person and only persons should have been admitted? Many of the impaired are people who can speak for themselves and answer back: they have a long experience of oppression.

This has also been the case with women. While initially asking for obstetric advances to be made available, many women discovered disadvantages and their organisations turned to protesting against many aspects of the active management of labour (the Association for the Improvement of the Maternity Services, for example). Fear of unacceptable third-party interference with and control over reproduction later led other women to form a radical oppositional organisation (Feminist International Network Against Reproductive and Genetic Engineering – FINNRAGE). Hearing such unlikely companions as radical feminists and devout Catholic women making similar protests against aspects of the new genetics and the NRTs suggest that some fundamental issues are involved.

In terms of genetic population screening, more is needed than reassurance as to the good intentions of practitioners. What arrangements could be put in place that would convincingly prevent any slide from helping individuals to 'improving the quality of the population'? Germany in the 1920s, 1930s and 1940s was undoubtedly a special case. Were the problem to arise here it would be in a different form. Complacency is dangerous in these matters: gross distortions could always happen here. Resting on past laurels, not speaking the names of the unspeakable, such as 'ethnic cleansing', 'improving the population', failing to discuss in public difficult issues – like possible connections between genetics and racism or sexism – is not enough to ensure that we prevent distortions. That is the ostrich act. Such acts are already happening: not only in the public documents that I have discussed, but also in consultations when practitioners attempt to be neutral for fear of appearing eugenist. Neutrality in the unequal relationship of a doctor–patient consultation is not possible.

16.18 Concluding comments

Research evidence makes plain that the scientific way of thinking, although it has become dominant over the last 200 years, does not answer all the problems that social life presents. Nor is it simply a question of people not understanding about science, although that may be part of it. Life has aspects that are experienced in other ways, including by scientists and medical practitioners – the artistic, the spiritual, for example. The dilemmas of everyday life that people have to include in reproductive decisions have more dimensions than can be expressed in the logic of genetic chances. Perhaps this is why scientists themselves do not always behave in an expected way. Only 20% of the staff in a hospital department of molecular and medical genetics in London volunteered for cystic fibrosis carrier testing when it was offered (Flinter *et al.*, 1992).

Genetics shares the triumphs and hazards of the biological vision (Stacey, 1993). Many of the problems I have raised cannot be solved within genetics alone. Genetics is part of the whole society, indeed is an active agent in its persistence and change. Looking into the future, Reid (1990, p. 319), after reviewing a variety of issues related to genetic testing and screening says, 'What we regarded as miracles of science today will no doubt become the everyday choices of tomorrow.' No doubt they may, but how will this have changed our world, our – especially us women's – lifestyles, the meanings we give to our lives and to that of others? Peter Harper, a medical geneticist with great experience of Huntingdon's disease, speaking of possible public health applications of genetics as in mass screening programmes, concludes that as well as the help that medical genetics can offer (of which he is proud), its 'potential for harm is correspondingly great if we do not think carefully about the effects of the new developments' (Harper, 1992, p. 721).

There is no doubt that with the extensive third-party involvement in highly personal matters that it involves, the new genetics could unintentionally create further problems: further increasing the oppression of women could be one outcome and of the disabled another. Genetics may also find it difficult to maintain an appropriate scientific distance from profit-making industry; or it could take on an important role in constraining commercial developments that exploit people's fears and desires. High moral statements will not send these problems away. To maximise all of our chances of living reasonable lives, recognition is needed that genetics is impinging on an untidy world – and one where the non-rational is as important as the rational. Tidying away some hereditable diseases will not make the society tidy, nor will it eliminate suffering.

Has the time perhaps come when it is necessary to revise the scientific goal to one which would work with nature rather than attempting to beat it? Inevitably we shall be born and die. That is how the human organism is. Inevitably we shall become ill and some of us will be disabled. Medicine would have much to offer in such a cooperative enterprise. At a minimum it is crucial that genetics itself and clinical practice arising from it are placed in a social context.

Lessor (1993), in discussing problems arising in sister-to-sister egg donation in California, comments: 'In medicine, a problem is something to be overcome or eliminated, not usually negotiated.' In my view as a woman and a sociologist all problems are for negotiation: this chapter is offered in that spirit. That the new genetics is already having important social consequences and might have serious unfortunate social consequences does not imply that the new genetics should necessarily be 'overcome or eliminated'. It does, however, mean that there is a good deal of difficult negotiating and also a lot more good empirical work to be done.

16.19 References

Alderson, P. (1990) *Choosing for Children: Parents' Consent to Surgery*. Oxford: Oxford University Press.

Alderson, P. (1993) *Children's Consent to Surgery*. Buckingham: Open University Press.

Anon. (1994) *The Times Higher Education Supplement*, 30 September, pp. 17–19.

Arditti, R., Klein, R. D. and Minden, S. (eds.) (1984) *Test-tube Women: What Future for Motherhood*. London: Pandora.

Arney, W. R. (1982) *Power and the Profession of Obstetrics*. Chicago and London: University of Chicago Press.

Bhaskar, R. (1989) *Reclaiming Reality: A Critical Introduction to Contemporary Philosophy*. London: Verso.

Birke, L., Faulkner, W., Best, S., Janson-Smith, D. and Overfield, K. (eds.) (1980) (Brighton Women and Science Group) *Alice Through the Microscope: The Power of Science over Women's Lives*. London: Virago.

Bleier, R. (1984) *Science and Gender: A Critique of Biology and its Theories on Women*. New York: Pergamon Press.

Carr-Hill, R., Samphier, M. and Suave, B. (1982) Socio-demographic variations in the sex composition and preferences of Aberdeen families. *Journal of Biosocial Sciences*. 14, 429–43.

Clothier Report (1992) *Report of the Committee on the Ethics of Gene Therapy* [Clothier Report], Cm 1788. London: HMSO.

Corea, G., Klein, R. D., Hanmer, J. *et al.* (1985) *Man-made women: How New Reproductive Technologies affect Women*. London: Hutchinson.

Crombie, A. (1993) *Styles of Scientific Thinking in the European Tradition: The History of Argument and Explanation Especially in the Mathematical and Biomedical Sciences*. London: Butterworth.

Edwards, J., Hirsch, E., Franklin, S., Price, F., and Strathern, M. (1993) *Technologies of Procreation: Kinship in the Age of Assisted Conception*. Manchester: Manchester University Press.'

Farrant, W. (1985) Who's for amniocentesis? The politics of prenatal screening. In H. Homans (ed.), *The Sexual Politics of Reproduction*. Aldershot: Gower.

Smith, D. (1987) *The Everyday World as Problematic*. Milton Keynes: Open University Press.

Spallone, P. (1989) *Beyond conception: The New Politics of Reproduction*. Basingstoke: Macmillan.

Spallone, P. (1992) *Generation Games: Genetic Engineering and the Future for our Lives*. London: Women's Press.

Stacey, M. (1960) *Tradition and Change: A Study of Banbury*. London: Oxford University Press.

Stacey, M. (1981) The division of labour revisited or overcoming the two Adams. In P. Abrams, R. Deem, J. Finch and P. Rock (eds.), *Development and Diversity: British Sociology 1950–1980*. London: George Allen & Unwin.

Stacey, M. (1988) *Sociology of Health and Healing: a Textbook*. London: Unwin Hyman.

Stacey, M. (ed.) (1992) *Changing Human Reproduction: Social Science Perspectives*. London: Sage.

Stacey, M. (1993) Editorial: the biological vision: triumphs and hazards. *Social Science and Medicine*, **37**(7), v–ix.

Stacey, M. (1994) About genetics: aspects of social culture worth considering. Paper presented at the conference Culture, Kinship and Genes, Abergavenny, April 1994 (to be published by Macmillan).

Stacey, M. and Davies, C. (1983) *Division of Labour in Child Health Care: Final Report to the SSRC*. Coventry: Department of Sociology, University of Warwick.

Statham, H., Green, J., Snowdon, C. and France-Dawson, M. (1993) *Lancet*, **341**, 564–5.

Steinbacher, R. and Holmes, H. B. (1985) Sex choice: survival and sisterhood. In G. Corea, R. Duelli Klein, J. Hanmer *et al.* (eds.), *Man-made Woman: How New Reproductive Technologies Affect Women*. London: Hutchinson.

Strathern, M. (1992*a*) The meaning of assisted kinship. In M. Stacey (ed.), *Changing Human Reproduction: Social Science Perspectives*, pp. 148–169. London: Sage.

Strathern, M. (1992*b*) *After Nature: English Kinship in the Late Twentieth Century*. Cambridge: Cambridge University Press.

Strathern, M. (1992*c*) *Reproducing the Future: Anthropology, Kinship and the New Reproductive Technologies*. Manchester: Manchester University Press.

Tuckett, D. A., Boulton, M. and Olson, C. (1985) A new approach to the measurement of patients' understanding of what they are told in medical consultations. *Journal of Health and Social Behaviour*, **26**, 27–38.

Weindling, P. (1993) Professional degradation: the German medical profession 1900–1945. Paper presented at the Human Values in Health Care Forum 'Milieu and ethics in health care', 17 November 1993, London.

West, C. (1993) Reconceptualizing gender in physician–patient relationships. *Social Science and Medicine*, **36**, 57–66.

Wright, P. and Treacher, A. (eds.) (1982) *The Problem of Medical Knowledge: Examining the Social Construction of Medicine*. Edinburgh: Edinburgh University Press.

Young, M. and Willmott, P. (1957) *Family and Kinship in East London*. London: Routledge & Kegan Paul.

17

Afterword

MARTIN RICHARDS and THERESA MARTEAU

The rate of change in the field of the new human genetics appears hectic. Each week brings reports of the localisation and cloning of new genes. The accelerating rate of change is well illustrated by the history of research on hereditary breast and ovarian cancer. Although it has been long accepted that a family history of breast cancer indicates some increase in risk for all women in the family, it was generally assumed that this was the result of many genetic and environmental factors working together. Few people believed until very recently that it was likely that single-gene effects were significant. In fact, as long ago as 1866, Broca, in his *Traité des Tumeurs*, published his wife's family tree showing ten women in four generations who had died of the disease. While today this pedigree would be interpreted as evidence of an autosomal dominantly inherited susceptibility, at the time its significance was not understood. It was not until 1990 that Mary-Claire King and her colleagues used genetic linkage analysis to identify a location on chromosome 17q for a gene that was named *BRCA1*. This observation was soon confirmed and extended to include ovarian cancer. By the time work had begun on this book, linkage testing for *BRCA1* was just starting to be offered to members of suitable families. Indeed, one of the personal accounts in the book describes what may well have been the first occasion on which the results of linkage testing were used in making a decision about treatment. As we now write in late 1994, *BRCA1* has been cloned and some of its apparently numerous mutations have been described. A second gene (*BRCA2*) associated with these cancers has been located on another chromosome, and there is evidence for others. Direct tests for *BRCA1* mutations are expected in the near future.

Although change has been fast from a scientific perspective, from the perspective of many families there has been little more than promises of future developments. Certainly linkage testing for *BRCA1* has been used by a handful of women worldwide in making decisions about the mode of treatment for tumours and about prophylactic surgery as well as other risk-reducing strategies. Very many more have now experienced prenatal diagnosis for a wide range of genetic conditions. The first experimental steps have been taken on what now promises to be a long road to gene therapy. But there are no new cures for genetic disease, and such treatments that are currently available owe almost nothing to the new genetics. Most developments have been in testing techniques, providing information about the gene mutations that people may or may not carry rather than medical treatment.

For testing there are currently two key issues: how services should be delivered, and for which conditions testing should be available. The first of these is more sharply focused at the moment. To take one example, whereas some clinicians now are planning population-based screening for *BRCA1* gene mutations, there are others who argue that there are many questions that need to be resolved before this should begin (Ponder, 1994). Clincially, for example, it is not known what is the most effective way to manage those at increased risk, in terms of screening, treatment or surgery. From a psychological perspective, preliminary reports of DNA testing for this and other late onset cancers show that reactions to testing are not simple. For instance, those receiving negative results are not always reassured, and may wish to continue with medical surveillance (Lynch, 1993; Michie *et al.*, 1995).

Many research programmes are already under way aimed at finding the most appropriate and effective ways of offering tests to both general populations, and those from high-risk families. There are key questions about genetic counselling: who should provide it; who needs to see a clinical genetic specialist rather than, for instance, a nurse specialist; should this work be concentrated in genetic departments or dispersed to places like breast or well women clinics and obstetric departments. Genetic counselling is rather unusual among medical encounters in that the patients generally get a rather long appointment in which to discuss their problems. They are also given much more detailed and scientific explanations than is common elsewhere in medicine. If we take the example of predictive testing for Huntington's disease, as the two personal accounts in this book show, many hours of professional and patient time are involved before and after testing under current protocols. But as further genetic tests are developed, current resources will not be sufficient. This is leading to proposals for new and briefer ways of preparing people for testing and supporting them afterwards – some have even suggested that the whole counselling enterprise should be abandoned and tests should be available for anyone to purchase across the counter.

The questions about who should be offered which tests are less focused at present. Harold Klawans, an American neurologist, tells the story (Klawans, 1988) of a woman who came to see him to ask whether or not she would develop Huntington's disease. At the time no tests were available. As she had already completed her family he was intrigued to know why she wanted the information.

'If I am not to get Huntington's chorea, then I will get a divorce as soon as my son graduates
    from high school.'
'And if not?'
'Then I will stay married to the bastard to whom I am married and let him support me for the
    rest of my life.'

We wonder how the clinicians who now offer predictive testing would have responded to such a request.

Perhaps even more pressing than this are questions about which conditions should be tested for. As discussed in this book, one area of controversy concerns the testing of children for conditions for which there is no treatment. The identification of genes for late onset conditions for which there may be treatment but no cure, such as for some familial cancers, raises the issue of how serious a condition needs to be to warrant the offer of a termination of pregnancy, and who should be making such decisions. Some countries are currently looking for national regulations, whereas others are leaving the decision to doctors, or to specially convened hospital ethics committees. One of the major challenges is how to involve the public in discussions about possible regulation. Although the public have largely been absent from such debates thus far, there seems to be growing interest in engaging them. There are also signs that some of those who carry genetic conditions are becoming more outspoken in questioning the value of new developments. The recent identification of the gene associated with achondroplasia (which restricts growth) was described by one person of small stature as a dangerous and unnecessary development – a view widely reported in the media. The new genetics is a double-edged sword. Although research to date has mapped out some of the areas of benefit and harm, there is still much work to be done to determine the relative balance between these.

At a broader societal level there are worries that new developments will lead to a society that is less tolerant of disability and diversity. There is a conflict between those who argue that some forms of genetic testing should be promoted to reduce the incidence of genetic disease in the population and others who insist that the use of such tests must always be a matter for individual choice. When testing and prenatal diagnosis are available for a condition, will health and other services always be provided for the children of those who chose not to use testing and abortion? Can we dismiss all the claims that we are creating a new eugenics (see, for example, Duster, 1990)?

There is an optimistic feel about the future from those who argue for an increase in education in genetics so that we may all come to share a 'genetic literacy', which may better enable us to take appropriate decisions about the use of new genetic techniques. It is an attractive argument, not least because it imagines a future world in which we are in a position to make our own decisions with much less need of professional help. But perhaps this argument misses the point. The difficulties in arriving at a decision to take, say, a carrier detection test for cystic fibrosis, or a predictive test for hereditary breast cancer or Huntington's disease, are not to do with a knowledge of Mendelian genetics, or at least only in a fairly trivial way. The decisions are much more based on our capacities to imagine possible futures, particularly future social relationships. Perhaps we delude ourselves if we believe that we will develop a better way of coping with the use of new tests by increasing the general knowledge of genetics. There may, of course, be many other arguments for education in scientific genetics. However, we should also note that current discussion of genetic literacy and education largely concerns Mendelian genetics. This is knowledge that was gained by the scientific community the best part of a century ago, yet is still not part of the common culture. Beyond a few brave efforts on the part of the popularisers of science, we have not begun to contemplate the creation of a literacy in molecular genetics or the broader field of molecular medicine. Indeed, we might say that what the public is currently being offered is a rather confusing mixture of nineteenth-century science with a few elements being described with the vocabulary of the new genetics.

There is a pessimistic feel about the future from those who argue that, once feasible, there is a pressure for people to make use of genetic screening, particularly in pregnancy, and this undermines those with disabilities and may reinforce existing discrimination. This must be seen in the context of the reduced public spending on health and welfare that is taking place in most of the industrial world. On a more positive note, though, there seems to be a growing public debate with a questioning of these pressures and the assumptions of these arguments, not least from those who have disabilities, as we have already noted. There is now an umbrella organisation in the UK for all the groups concerned with specific genetic conditions (GIG: Genetic Interest Group), and with similar developments elsewhere it seems likely that in the future we shall hear a lot from those most directly involved. Perhaps we shall see a situation akin to that with HIV–AIDS where, especially in the USA, those who are HIV-positive have successfully challenged research agendas and previous ideas about service provision. These people are now very much part of not only the debate, but of the decision-making process.

Few, however, argue against continued research and the monitoring of developments. The key question at the moment is whether there should be regulation and, if so, how this should be done. Whereas gene therapy research is subject to regulations in all countries where it is being conducted, the approach to genetic testing has not been so systematic. In Denmark there is a moratorium on all population-based genetic screening programmes until the Council of Ethics advises otherwise. There are now calls in Canada, the USA, the UK and elsewhere to regulate genetic testing. It remains to be seen how effective such regulation will be. Earlier legislation in the USA aimed at regulating sickle cell screening was ineffective largely because the mechanisms for policing practice did not work.

Although it may be easier to regulate clinical settings, it is more difficult to see how such regulations will apply to commercial outlets. As tests become simpler, for example by the use of desktop analysers, it will become easier for them to be conducted 'over the counter', as with, for instance, pregnancy tests.

But whatever the future holds, psychologists and social scientists have an important role to play in conducting research on developments in the new human genetics. We should not, and probably cannot anyway, stop further technical developments in the field. But we must not be passive and let the technological tail wag the societal dog. We must join the debate and help to negotiate the kind of future we all want.

References

Duster, T. (1990) *Backdoor to Eugenics*. New York: Routledge.
Klawans, H. L. (1988) *Toscanini's Fumble and Other Tales of Clinical Neurology*. London: Bodley Head.

Lynch, H. T. (1993) DNA Screening for breast/ovarian cancer: susceptibility based on linked markers – a family study. *Archives of Internal Medicine*, **153**, 1979–87.

Michie, S., McDonald, V. and Marteau, T. M. (1995) Understanding responses to predictive genetic testing: a grounded theory approach. *Psychology and Health* (in press).

Ponder, B. (1994) Breast cancer genes: searches begin and end. *Nature*, **271**, 279.

Index

Page references to figures and tables are printed in italic type.

SWINDON COLLEGE

LEARNING RESOURCE CENTRE

SWINDON COLLEGE

LEARNING RESOURCE CENTRE